Essentials of Probability for Statistics

A primer

Taeryon Choi

자유아카데미

Preface

This book is based on a set of lecture notes I have used to teach an introductory probability course, STAT 221, given at the department of statistics in Korea University. STAT 221 is a one-semester course in elementary probability where there is a strong emphasis on problem solving and applications. Except for statistics, economics, mathematics, computer science, biology, physics, engineering, political sciences, and social sciences, the students come from many departments of the University. No previous knowledge of probability is assumed. However, students should have a working knowledge of calculus.

This book is designed for a first course in probability at an undergraduate level, targeting students who have studied calculus at the university level and are seeking an introduction to probability with mathematical content. This book covers all of the standard material for such courses, but it also contains many topics that are usually not found in introductory probability books – such as simulation using R and additional material useful for mathematical statistics. Where possible, I tried to provide mathematical details, and it is expected that students are seeking to gain some mastery over these. For use in a standard one-semester course, in which both discrete and continuous probability is covered, students should have taken a prerequisite of calculus, including an introduction to multiple integrals, for example, STAT 201 provided by the department of statistics in Korea University. The main mathematical challenge lies not in performing technical calculus derivations, but in translating between abstract concepts and concrete examples. This book is also intended to prepare students for advanced work in probability, who are intending to go on to take mathematical statistics, stochastic processes, computational methods, or any other advanced probability course.

The emphasis throughout the book is on probability, but attention is also given to statistics. The book can be used in a variety of course lengths, levels and areas of emphasis, and is suitable not only for introductory courses, but also for self-study. It presents a comprehensive treatment of probability ideas and techniques essential for a firm understanding of probability and statistics. **Chapter 1** introduces the concept of probability and covers sample spaces and events, the axioms of probability, and counting. The basic properties of a probability measure are also developed. **Chapter 2** deals with conditional probability, law of total probability, Bayes' theorem, and independence of events. Discrete random variables and their distributions are the subject of **Chapter 3**. **Chapter 4** introduces continuous random variables and their distributions. Joint continuous distributions are the focus of **Chapter 5**, including marginal distributions, expectations of a function of random variables, and the effects of change of variables. **Chapter 6** mainly discusses conditional distributions, including conditional expectations and variances. The last two sections in Chapter 6 deal with bivariate normal distribution theory and multivariate distributions as advanced topics. All the chapters include material on Monte Carlo methods using a statistical software, **R**. Simulation is a key aspect of the application of probability theory, and it is my view that its teaching can be integrated with the theory from the start. This reveals the power of probability to solve real-world problems and helps convince students that it is far more than just an interesting mathematical result. This material can be skipped, if a student or an instructor believes otherwise or feels there is not enough time to cover it effectively.

The book contains more than the usual number of examples worked out in detail, ranging from straightforward to reasonably challenging. It is not possible to go through all these examples in class. Rather, a student is expected to deal quickly with the main points of theory, then spend working time on examples and exercises from the book. A student cannot hope to really learn the material simply by sitting passively in the classroom and listening to the instructor. The student must get actively involved in working problems.

I am indebted to my students without whom I could not have written this book. They helped me with writing and revising earlier versions of this text. In particular, Eunji Lee spent many, many hours editing LaTeX code for the book, which is a modified version of the Legrand Orange Book LaTeX Template Version 2.1.1.The best aspect of the book's layout are due to her hard work; any stylistic deficiencies are due to my lack of expertise. Sunsik Kim also worked hard to edit and produce many figures together with Eunji Lee. Dongu Han and Hyunsue Jung read through the entire book very carefully and gave me detailed comments for improvements and corrections. Also, I would like to thank all the students in STAT 221 for valuable comments, suggestions, and inspiration.

Finally, you should note that the book contains a random number of errors that may follow Poisson distributions. Errors unavoidably arise in any project like this (meaning a project in which I am involved). For this reason, I will try to post information about the book, including a list of corrections, on the available link of the publisher's website. Readers are encouraged to inform me of any errors that they discover.

February 2020 TAERYON CHOI

Contents

I Basic Probability

I Basic Probability

1. What is probability?

"Les questions les plus importantes de la vie ne sont en effet, pour la plupart, que des problèmes de probabilité." : "Life's most important questions are, for the most part, nothing but probability problems."

"On voit, que la théorie des probabilités n'est, au fond, que le bon sens réduit au calcul" : "One sees, that the theory of probabilities is basically just common sense reduced to calculus"

- Pierre-Simon Laplace

Uncertainty is a basic fact of life. Probability is a fundamental tool for quantifying uncertainty. The concept of probability agrees with our intuition. In everyday life, probability of an event is understood as a chance that this event will happen. Probability provides precise mathematical rules for understanding of uncertainty and analyzing our own ignorance. It also gives us a framework for working with our limited knowledge and for making sensible decisions based on what we do and do not know. Learning probability gives you a new way of reasoning, and we can learn how to reason conditionally and unconditionally.

When we think about probability, the first thing that comes to mind would be about tossing a coin. If a fair coin is tossed, we say that it has a 50-50 (equal) chance of getting heads or tails. Hence, the probability of each side equals $1/2$. How do we interpret this *probability* $1/2$? There are many distinct interpretations of probability. Here, we will only discuss three interpretations.

- **Frequency interpretation of probability** : A probability of an event is the proportion of times that we expect the event to occur, if the experiment were repeated a large number of times. The probability $1/2$ can be viewed as a proportion of times getting heads, or its relative frequency in a long run experiments, say tossing a coin 1,000 times.

- **Classical interpretation of probability** : A probability of an event is the ratio of the number of favorable outcomes and possible outcomes in a (symmetric) experiment. The classical interpretation of probability is based on the concept of *equally likely outcomes*. Since there are two equally likely outcomes in tossing a coin, a head and a tail, we can assign the equal probability to two events, $1/2$.

- **Subjective interpretation of probability** : A subjective (personal) **probability** is an individual's degree of knowledge or belief in the occurrence of an event. For example, the weather forecaster who predicts rain tomorrow with a probability of $1/2$ is based on particular atmospheric conditions that will never be repeated. Instead, the forecaster is expressing his confidence of rain, based on all the available information, in terms of a value between 0 and 1 that represents from absolute certainty of no rain to rain respectively.

Each of these interpretations can be very useful in applying probability theory to statistical inferences and practical problems. However, the theory of probability does not depend on interpretation.

Remark The mathematical theory of probability can be traced back to the mid-seventeenth century. Early probabilists focused on games of chance.

- In the history of probability, the physician and mathematician Gerolamo Cardano (1501– 1575) was among the first to attempt a systematic study of the calculus of probabilities. His book *Liber de Ludo Aleae* (Book on Games of Chance) was written perhaps in the 1560s but published posthumously in 1663.

- It is generally believed that the mathematical theory of probability was started by the French mathematicians, Blaise Pascal(1623-1662) and Pierre de Fermat(1601-1665). The calculus of probability is conventionally dated from July 1654, when Blaise Pascal wrote Pierre de Fermat with a question raised by a friend, the Chevalier de Méré, concerning a dice game, called the *problem of points*.

- Jacob Bernoulli (1654-1705) attempted to develop the theory of games of chance in his work *Ars Conjectandi* (1713), the first systematic and analytic discussion of the probability calculus.

- Abraham De Moivre (1667-1754) extended the mathematics in 1718, incorporating improved combinatorial relations, new techniques of infinite series expansions and approximate distributions, and an early form of limit theory into his work, the *Doctrine of Chances* (1718).

Remark To formally quantify probabilities, we consider a *probability space*, which mathematically consists of three components $(\Omega, \mathscr{F}, P)$: (1) A *sample space* Ω, the set of all possible outcomes of a certain experiment (2) A collection of *events*, $\mathscr{F}$, each event being a subset of the sample space (3) A *probability measure* or *probability function* P, which indicates the chance of each possible event occurring.

1.1 Sample spaces and events

All of probability theory begins with the concept of an experiment. An *experiment* is any (well-defined) procedure for which you do not necessarily know the outcome ahead of time. Running a random experiment involves setting up a specific situation and watching what happens. Examples can be

- Toss a coin.
- Roll a dice.
- Run a poll to assess the popularity of several candidates.
- Gamble on the result of a race.
- Follow the price per share of a specific stock.
- Send a bit over a communications channel and see what is received.
- Run a computer program and want to know how long it takes.

Everything that you measure (keep track of) in an experiment is the *outcome*. Before an experiment is run, the collection of all possible outcomes is called the *sample space*.

Definition 1.1.1 The *sample space* is the set of all possible outcomes. This is sometimes called the outcome space or state space. We generally use Ω (the capital Greek letter) or S to denote the sample space. Elements of Ω are called *sample points* and typically denoted by ω (the lower case Greek letter) or s.

▪ **Example 1.1** Consider the random experiment in which a coin is tossed once. In this case, we have $\Omega = \{H, T\}$, where H denotes heads and T denotes tails. If you consider tossing a coin twice, then we have $\Omega = \{HH, HT, TH, TT\}$ by recording the two tosses, in the order in which they occurred. ▪

▪ **Example 1.2** The sample space for the random experiment in which a fair, six sided die is rolled is the 6-element set $\Omega = \{1,2,3,4,5,6\}$. Also, consider the random experiment where two, fair, six sided dice are rolled. Then, the sample space is represented as $\Omega = \{1,\ldots,6\} \times \{1,\ldots,6\}$, namely, the *Cartesian product* of the set of single roll outcomes with itself. That is, elements of the sample space are *tuples* of the form (i, j) with $i, j \in \{1,\ldots,6\}$. ▪

▪ **Example 1.3** Suppose the experiment consists of tossing a coin infinitely many times. We denote 1 by a head(H) and 0 for a tail(T). Then, an element (a possible outcome) or a sample point ω is an infinite sequence $\omega = (x_1, x_2, x_3, \ldots)$, where x_i is either 0 or 1. Thus, the sample space Ω is the space of all infinite sequences of 0's and 1's.

$$\Omega = \{\omega : \omega = (x_1, x_2, \ldots)\} = \{0,1\} \times \{0,1\} \times \{0,1\} \times \cdots \times \{0,1\} \times \cdots.$$ ■

■ **Example 1.4** Suppose we observe the number of customers that arrive at a help desk during a fixed time period. Then, the sample space is the set of non-negative integers, $\Omega = \{0,1,2,3,\ldots\}$. ■

■ **Example 1.5** The experiment is to measure the time until a machine breaks down, the life time of a machine. The sample space is the set of the positive real numbers, $\Omega = \{t : 0 < t < \infty\}$, where t indicates that it takes a time t until it is out of order. ■

Remark Examples 1.1 and 1.2 have finite sample spaces, whereas Examples 1.3, 1.4 and 1.5 are infinite sample spaces. The sample spaces in Examples 1.3 and 1.4 are countably infinite, and Example 1.5 has a continuous(uncountable) sample space. Finite and countably infinite spaces are called *discrete*.

In general, an **event** is any subset of the outcome space Ω, a collection of possible elementary outcomes that represent some particular result. We say that A is a *subset* of B, denoted $A \subset B$ if every outcome in A is also in B. An event represents something that might occur. Which elementary outcome occurs determines whether or not it does occur. Events are typically denoted by upper case letters A, B, C and others.

Definition 1.1.2 A subset of the sample space (a collection of possible outcomes) is known as an *event*. We write $A \subset \Omega$ if A is a subset of Ω.

■ **Example 1.6** Suppose that we flip a fair coin. The sample space is $\Omega = \{H, T\}$ (H for heads and T for tails). Then, $\emptyset$, $\{H\}$, $\{T\}$, Ω are all subsets of Ω and they are events. We take $\mathscr{F} = \{\emptyset, \{H\}, \{T\}, \{H,T\}\}$, the collection of all subsets of Ω. The term "fair coin" means that the two outcomes are equally likely. If the coin is to be tossed twice, then $\Omega = \{HH, TT, HT, TH\}$ where HT denotes a head followed by a tail. For example, we can consider the event A of getting at least one tails, $A = \{TT, HT, TH\} \subset \Omega$. ■

Definition 1.1.3 An event A occurs if the outcome of the experiment is an element of the set A.

- The empty subset $\emptyset$ is the event with no outcomes in it, often called the null event(set).
- The sample space Ω itself is also an event. It is the event that contains all possible outcomes, that is, the certain event.
- The complement of an event, A, denoted A^c, is the event that A does not occur, and hence consists of all those elements of the sample space that are not in A, $A^c = \Omega - A$. Obviously, $\emptyset = \Omega^c$.

Definition 1.1.4 Given any events (sets), we can combine them together with the following set operations :

- The union of two events A and B (written $A \cup B$) is the event that at least one of A or B occurs. The union of the events can be obtained by the union of the sets.
- The union of a sequence $A_1, A_2, \dots$ of events is denoted $\cup_{i=1}^{\infty} A_i$ and it contains all outcomes that are in at least one of the events,

$$\cup_{i=1}^{\infty} A_i = \{\omega \in \Omega \ : \ \omega \in A_i \text{ for at least one } i\}.$$

- The intersection of two events A and B (written $A \cap B$) is the event that both A and B occur. The intersection of two events can be obtained by forming the intersection of the sets.
- The intersection of a sequence $A_1, A_2, \dots$ of events is denoted $\cap_{i=1}^{\infty} A_i$, and it contains all outcomes that are in every event,

$$\cap_{i=1}^{\infty} A_i = \{\omega \in \Omega \ : \ \omega \in A_i \text{ for every } i\}.$$

- Given two events A and B, the difference $A - B$ is defined to be the event of outcomes of A that are not in B, $A - B = \{\omega \in \Omega \ : \ \omega \in A \text{ and } \omega \notin B\}$.
- Two events A and B are *disjoint* if $A \cap B = \emptyset$. Several events are *mutually disjoint* or *mutually exclusive* if no two of them have any outcomes in common, i.e., each pair is disjoint. That is, $A_1, A_2, \dots$ are mutually disjoint or mutually exclusive if $A_i \cap A_j = \emptyset$ whenever $i \neq j$.
- For any event A, the events A and A^c are disjoint, and their union is the whole of the sample space: $A \cap A^c = \emptyset$ and $A \cup A^c = \Omega$.
- If A is a subset of B, then occurrence of A necessarily implies occurrence of the event B. That is, $A \subset B$ if and only if $A \cap B = A$. Also, $A \subset B$ if and only if $B^c \subset A^c$. If $A \subset B$ and $B \subset A$, it follows that $A = B$.

Theorem 1.1.1 Some basic rules of set theory are :

- **Commutative** : $A \cup B = B \cup A$, $A \cap B = B \cap A$.
- **Associative** : $(A \cup B) \cup C = A \cup (B \cup C)$, $(A \cap B) \cap C = A \cap (B \cap C)$.
- **Distributive** : $(A \cup B) \cap C = (A \cap C) \cup (B \cap C)$, $(A \cap B) \cup C = (A \cup C) \cap (B \cup C)$. They are also valid for sequences of sets, $A_1, A_2, \dots$,

$$(\cup_{i=1}^{\infty} A_i) \cap B = \cup_{i=1}^{\infty} (A_i \cap B), \quad (\cap_{i=1}^{\infty} A_i) \cup B = \cap_{i=1}^{\infty} (A_i \cup B).$$

- **De Morgan's law** : $(A \cup B)^c = A^c \cap B^c$, $(A \cap B)^c = A^c \cup B^c$. They are also valid for sequences of sets, $A_1, A_2, \ldots$,

$$(\cup_{i=1}^{\infty} A_i)^c = \cap_{i=1}^{\infty} A_i^c \text{ and } (\cap_{i=1}^{\infty} A_i)^c = \cup_{i=1}^{\infty} A_i^c.$$

- **Disjoint union** : $A \cap B$ and $A \cap B^c$ are disjoint, with $(A \cap B) \cup (A \cap B^c) = A$. In addition, $A \cup B = (A \cap B^c) \cup (A^c \cap B) \cup (A \cap B)$, where $A \cap B^c$, $A^c \cap B$, and $A \cap B$ are disjoint.

Remark We can think of union as addition and intersection as multiplication.

The sample space Ω may be too big to consider all the subsets, and some smaller collection of subsets is useful. Thus, we focus on the following collection of subsets of Ω called a *σ-field* or an *event space* to define an event.

Definition 1.1.5 A *σ-field* (or *σ-algebra*) of subsets, $\mathscr{F}$, is a collection of some subsets $A_i \subseteq \Omega$ (where $i = 1, 2, \ldots, \infty$) that satisfies:

- $\emptyset \in \mathscr{F}$;
- if $A \in \mathscr{F}$, then $A^c \in \mathscr{F}$;
- if $A_1, A_2, \cdots \in \mathscr{F}$, then $\cup_{i=1}^{\infty} A_i \in \mathscr{F}$.

A subset A of the sample space Ω is said to be an event if it belongs to $\mathscr{F}$.

■ **Example 1.7** For the simplest case of tossing a coin once with $\Omega = \{H, T\}$, the set of all subsets of Ω, called the *power set* and denoted by 2^{Ω}, is a σ-field given by $\mathscr{F} = \{\emptyset, \{H\}, \{T\}, \Omega\}$. ■

Remark In discrete sample spaces the class $\mathscr{F}$ contains *all subsets of the sample space*. But in more complicated sample spaces this cannot be the case.

1.2 Probability measure

A measure is a mathematical object that tells us the size of a set. For example, one natural measure of a set A is its cardinality, #(A), a non-negative, possibly infinite, number. Another possible measure of a set A is the length of a set, which is known as *Lebesgue* measure in one dimension. In two dimensions, Lebesgue measure is the area of a region. Cardinality, length and area can be zero for empty set, non-negative and additive.

Probability measures the abstract concept of the chance that the event occurs. Thus it shares basic properties that all measures have. We assign a probability measure $P(A)$ to an event A. It is a real-valued function defined on a collection of events $\mathscr{F}$.

Definition 1.2.1 Let Ω be the sample space of a random experiment. A probability measure $P : \mathscr{F} \to [0,1]$ is a set function which assigns real numbers to the various events of Ω satisfying the following three axioms, **Axiom 1–Axiom 3**,

Axiom 1 : $P(A) \geq 0$ for $A \in \mathscr{F}$,

Axiom 2 : $P(\Omega) = 1$,

Axiom 3 : $P(\cup_{k=1}^{\infty} A_k) = \sum_{k=1}^{\infty} P(A_k)$ if $A_1, A_2, \dots$ are mutually disjoint events, that is, $A_i \cap A_j = \emptyset$, for all $i \neq j$.

Remark These axioms seem to be reasonable with our intuitive understanding of probability.

Axiom 1 simply states that we work only with positive(nonnegative) probabilities, because in some sense, probability measures the length, size, or volume of the set (event).

Axiom 2 states that one of the possible outcomes in the sample space must occur. However, there are often events of probability 1, which are not the sample space.

Axiom 3 states that probabilities are additive in a natural way. This axiom is known as the *countable additivity* assumption. However, this additivity property will not hold in general for an uncountable collection of events.

Any set function with the above three properties is a probability measure for Ω.

Theorem 1.2.1 There are commonly used properties in probabilities calculations :

- (Empty set) $P(\emptyset) = 0$.
- (Finite additivity) If $A_1, \dots, A_n$ are mutually disjoint events then

$$P\left(\bigcup_{i=1}^{n} A_i\right) = \sum_{i=1}^{n} P(A_i).$$

- (Complements) $P(A^c) = 1 - P(A)$.
- (Monotonicity) If $A \subseteq B$, then $P(A) \leq P(B)$. Thus, for every A, $0 \leq P(A) \leq 1$.

- (Addition Law) $P(A \cup B) = P(A) + P(B) - P(A \cap B)$.
- (Finite subadditivity) $P\left(\cup_{i=1}^{n} A_i\right) \leq \sum_{i=1}^{n} P(A_i)$, also known as *Boole's inequality*.

Proof. We can derive these properties easily from **Axiom1–Axiom 3**.

- (Empty set) Let $A_1 = \Omega$ and $A_i = \emptyset$ for $i = 2, 3, ..., \infty$. Then $\Omega = \bigcup_{i=1}^{\infty} A_i$ where $A_i \cap A_j = \emptyset$ for $i \neq j$. By **Axiom 2** and **Axiom 3**, we get

$$1 = P(\Omega) = P\left(\bigcup_{i=1}^{\infty} A_i\right) = \sum_{i=1}^{\infty} P(A_i) = P(A_1) + \sum_{i=2}^{\infty} P(A_i) = 1 + \sum_{i=2}^{\infty} P(\emptyset).$$

  Therefore $\sum_{i=2}^{\infty} P(\emptyset) = 0$. Since $P(\emptyset) \geq 0$ by **Axiom 1**, we have $P(\emptyset) = 0$. Alternatively, we can let $A_i = \emptyset$ for all i. Then the A_i are disjoint and their union also equals $\emptyset$. Thus, we get $P(\emptyset) = \sum_{i=1}^{\infty} P(\emptyset)$, which implies that $P(\emptyset) = 0$.

- (Finite additivity) Consider the collection $\{B_i\}_{i=1}^{\infty}$ of the subsets of the sample space Ω such that $B_1 = A_1, B_2 = A_2, \ldots, B_n = A_n$ and $B_{n+1} = B_{n+2} = B_{n+3} = \cdots = \emptyset$. Hence

$$\begin{aligned} P\left(\bigcup_{i=1}^{n} A_i\right) &= P\left(\bigcup_{i=1}^{\infty} B_i\right) = \sum_{i=1}^{\infty} P(B_i) = \sum_{i=1}^{n} P(B_i) + \sum_{i=n+1}^{\infty} P(B_i) \\ &= \sum_{i=1}^{n} P(A_i) + \sum_{i=n+1}^{\infty} P(\emptyset) = \sum_{i=1}^{n} P(A_i) + 0 = \sum_{i=1}^{n} P(A_i). \end{aligned}$$

- (Complements) Since $A \cap A^c = \emptyset$, and $\Omega = A \cup A^c$, **Axiom 3** tells us that $1 = P(\Omega) = P(A \cup A^c) = P(A) + P(A^c)$. Re-arranging gives $P(A^c) = 1 - P(A)$.

- (Monotonicity) Since $B = A \cup (B - A)$, and $A \cap (B - A) = \emptyset$, **Axiom 3** tells us that $P(B) = P(A \cup (B - A)) = P(A) + P(B - A)$. By **Axiom 1**, we know that $P(B - A) \geq 0$. Thus, we have $P(B) \geq P(A)$. In addition, note that $\emptyset \subset A \subset \Omega$, which implies that $0 \leq P(A) \leq 1$.

- (Addition Law) Since $A \cup B = (A \cap B) \cup (A \cap B^c) \cup (B \cap A^c)$, which are mutually disjoint, $A = (A \cap B) \cup (A \cap B^c)$, and $(A \cap B) \cap (A \cap B^c) = \emptyset$, **Axiom 3** tells us that $P(A) = P(A \cap B) + P(A \cap B^c)$ and that

$$\begin{aligned} &P(A \cup B) = P[(A \cap B) \cup (A \cap B^c) \cup (B \cap A^c)] = P(A \cap B) + P(A \cap B^c) + P(B \cap A^c) \\ &= P(A \cap B) + P(A) - P(A \cap B) + P(B) - P(B \cap A). \end{aligned}$$

- (Finite subadditivity) Let $C_n = \cup_{i=1}^{n} A_i$. Using the fact that $P(A_1 \cup A_2) \leq P(A_1) + P(A_2)$ from *Additive law* and mathematical induction, we obtain that $P\left(\cup_{i=1}^{n+1} A_i\right) = P(C_{n+1}) = P(C_n \cup A_{n+1}) \leq P(C_n) + P(A_{n+1}) \leq \sum_{i=1}^{n+1} P(A_i)$.

■

Remark **Principle of Mathematical Induction**

Let $P(n)$ denote a statement involving a positive integer n. If the following two conditions are satisfied:

① the statement $P(1)$ is true,
② whenever the statement $P(k)$ is true for a positive integer k, then the statement $P(k+1)$ is also true,

then the statement $P(n)$ is true for all positive integers n.

Example 1.8 Let A and B be two events with probabilities $P(A)=\frac{3}{4}$ and $P(B)=\frac{3}{8}$. Then, $P(A\cup B)\geq\frac{3}{4}$ and $\frac{1}{8}\leq P(A\cap B)\leq\frac{3}{8}$.

- Since $A\cap B\subset A$ and $A\cap B\subset B$, $P(A\cap B)\leq\min\{P(A),P(B)\}$, which implies that $P(A\cap B)\leq\frac{3}{8}$. Also, since $P(A\cup B)=P(A)+P(B)-P(A\cap B)$, $P(A\cup B)\geq P(A)+P(B)-\min\{P(A),P(B)\}=P(A)=\frac{3}{4}$.

- Since $A\cup B\subset\Omega$, we get $P(A\cup B)\leq P(\Omega)$. That is, $P(A)+P(B)-P(A\cap B)\leq P(\Omega)$, which implies that $\frac{3}{4}+\frac{3}{8}-P(A\cap B)\leq 1$. Hence, we obtain $\frac{1}{8}\leq P(A\cap B)$.

■

Example 1.9 Let A and B be two events with probabilities $P(A)>0$ and $P(B)=1$. Then, $P(A)=P(A\cap B)$.

- Note that $A\cap B$ and $A\cap B^c$ are disjoint and $A=(A\cap B)\cup(A\cap B^c)$. Thus $P(A)=P(A\cap B)+P(A\cap B^c)$.

- Since $A\cap B^c\subset B^c$, $P(A\cap B^c)\leq P(B^c)=1-P(B)=0$. Hence, $P(A\cap B^c)=0$ and $P(A)=P(A\cap B)$.

■

Theorem 1.2.2 We generalize the addition law and finite subadditivity to the case of an infinite sequence of events, called *countable subadditivity*. If $A_1,A_2,\ldots$ are any events (not necessarily disjoint), then

$$P(\cup_{i=1}^{\infty}A_i)\leq\sum_{i=1}^{\infty}P(A_i).$$

Proof. We construct a disjoint sequence of events $\{B_i\}_{i=1}^{\infty}$ such that $\cup_{i=1}^{\infty}B_i=\cup_{i=1}^{\infty}A_i$. Define B_i by $B_1=A_1$, $B_2=A_2\cap A_1^c=A_2-A_1$, $B_i=A_i-\left(\cup_{j=1}^{i-1}A_j\right)$, $i=2,3,\ldots$ Note that B_i's are mutually disjoint, $B_i\cap B_j=\emptyset$, $\forall\, i\neq j$, $B_i\subset A_i$, and that $\cup_{i=1}^{\infty}A_i=\cup_{i=1}^{\infty}B_i$. Thus, we have $P\left(\cup_{i=1}^{\infty}A_i\right)=P\left(\cup_{i=1}^{\infty}B_i\right)=\sum_{i=1}^{\infty}P(B_i)\leq\sum_{i=1}^{\infty}P(A_i)$. ■

■ **Example 1.10** Suppose that $P(A_i) = 1$ for each i. Then, $P(\cap_{i=1}^{\infty} A_i) = 1$.

- Since $P(A_i) = 1$ for each i, we have $P(A_i^c) = 1 - P(A_i) = 0$, for all $i \geq 1$.
- Thus, by DeMorgan's law and countable subadditivity, we have
  $P(\cap_{i=1}^{\infty} A_i) = 1 - P(\cup_{i=1}^{\infty} A_i^c) \geq 1 - \sum_{i=1}^{\infty} P(A_i^c) = 1$. Hence, $P(\cap_{i=1}^{\infty} A_i) = 1$.

■

Theorem 1.2.3 We have two results for monotone sequences of events, known as the *continuity of probability* theorem.

- (Continuity from below) If $A_1, A_2, \ldots, A_n, \ldots$ is an increasing sequence of events such that $A_1 \subset A_2 \subset \cdots \subset A_n \cdots$, then $P\left(\cup_{n=1}^{\infty} A_n\right) = \lim_{n\to\infty} P(A_n)$.
- (Continuity from above) If $B_1, B_2, \ldots, B_n, \ldots$ is a decreasing sequence of events such that $B_1 \supset B_2 \supset \cdots \supset B_n \cdots$, then $P\left(\cap_{n=1}^{\infty} B_n\right) = \lim_{n\to\infty} P(B_n)$.

Proof. We define $\lim_{n\to\infty} A_n = \cup_{n=1}^{\infty} A_n$ when $A_1 \subset A_2 \subset \cdots \subset A_n \cdots$ and $\lim_{n\to\infty} B_n = \cap_{n=1}^{\infty} B_n$ when $B_1 \supset B_2 \supset \cdots \supset B_n \cdots$.

- (Continuity from below) Given an increasing sequence of events $A_1 \subset A_2 \subset \cdots \subset A_n \subset \cdots$, we define a disjoint collection of events as in the proof of Theorem 1.2.2

  $$E_1 = A_1,\ E_n = A_n - \left(\cup_{i=1}^{n-1} A_i\right) = A_n - A_{n-1},\ n = 2, 3, \ldots,$$

  Then $\{E_n\}_{n=1}^{\infty}$ is a disjoint collection of events such that $\bigcup_{n=1}^{\infty} A_n = \bigcup_{n=1}^{\infty} E_n$. Further,

  $$\begin{aligned} P\left(\bigcup_{n=1}^{\infty} A_n\right) &= P\left(\bigcup_{n=1}^{\infty} E_n\right) = \sum_{n=1}^{\infty} P(E_n) = \lim_{n\to\infty} \sum_{k=1}^{n} P(E_k) \\ &= \lim_{n\to\infty} \left[P(A_1) + \sum_{k=2}^{n} [P(A_k) - P(A_{k-1})]\right] = \lim_{n\to\infty} P(A_n). \end{aligned}$$

- (Continuity from above) Taking B_n as a complementary event of A_n, $B_n = A_n^c$,

  $$\begin{aligned} P\left(\bigcap_{n=1}^{\infty} B_n\right) &= 1 - P\left(\bigcup_{n=1}^{\infty} A_n\right) = 1 - \lim_{n\to\infty} P(A_n) \\ &= \lim_{n\to\infty} [1 - P(A_n)] = \lim_{n\to\infty} P(B_n). \end{aligned}$$

■

Remark If a real sequence $\{a_n\}$ is monotone and bounded, then $\{a_n\}$ is convergent.

- If a sequence $\{a_n\}$ is increasing and bounded above, then it converges to its least upper bound, $\lim_{n\to\infty} a_n = \sup\{a_n : n = 1, 2, \ldots\}$.

- If a sequence $\{a_n\}$ is decreasing and bounded below, then it converges to its greatest lower bound, $\lim_{n\to\infty} a_n = \inf\{a_n : n = 1, 2, ...\}$.

A real-valued function f is continuous if and only if, for every convergent sequence $\{a_n\}_{n=1}^{\infty}$, $\lim_{n\to\infty} f(a_n) = f\left(\lim_{n\to\infty} a_n\right)$.

Theorem 1.2.4 — Inclusion-Exclusion principle. Let $A_1, A_2, \ldots, A_n$ be events. Then

$$\begin{aligned} P(\cup_{i=1}^{n} A_i) &= \sum_{i=1}^{n} P(A_i) - \sum_{i<j} P(A_i \cap A_j) + \ldots \\ &+ (-1)^{m+1} \sum_{i<j<\ldots<m} P(A_i \cap A_j \cap \ldots \cap A_m) \\ &+ \ldots + (-1)^{n+1} P(\cap_{i=1}^{n} A_i). \end{aligned}$$

Proof. The most basic case for $n = 2$ is given as *Addition law* in Theorem 1.2.1, $P(A \cup B) = P(A) + P(B) - P(A \cap B)$. This is called the inclusion–exclusion formula since we must exclude from $P(A) + P(B)$ the points that have been counted twice, i.e. where both A and B occur.

- The next case is for the union of three events.

$$P(A \cup B \cup C) = P(A) + P(B) + P(C) - P(A \cap B) - P(B \cap C) - P(A \cap C) + P(A \cap B \cap C).$$

  We verify graphically that the statement is correct for $n = 3$ as in the following Venn diagram.

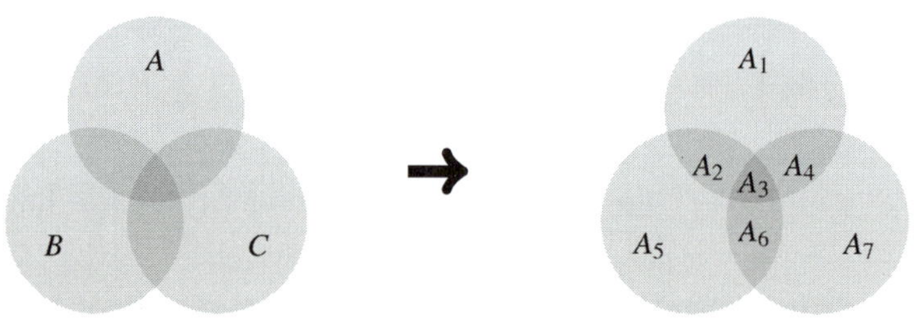

Figure 1.1: Venn diagram of three sets and their partition

- We see that $P(A) = P(A_1) + P(A_2) + P(A_3) + P(A_4)$ and analogously for $P(B)$ and $P(C)$, their sum generating A_2, A_4, A_6 twice and A_3 three times. Also, $P(A \cap B) = P(A_2) + P(A_3)$ and analogously for $P(A \cap C)$ and $P(B \cap C)$, their sum generating A_3 three times and no other repetitions. Finally, $P(A \cap B \cap C) = P(A_3)$. Hence, the result is true for $n = 3$.

- The diagram also indicates that the various intersections of A, B, and C and their complements will partition the sample space as well as A, B, and C in disjoint subsets. For example, $A_1 = A \cap B^c \cap C^c$, $A_2 = A \cap B \cap C^c$, $A_3 = A \cap B \cap C$, $A_4 = A \cap B^c \cap C$, and $A = A_1 \cup A_2 \cup A_3 \cup A_4$, which are mutually disjoint.

- We now prove the result by mathematical induction. We know that $P(A_1 \cup A_2) = P(A_1) + P(A_2) - P(A_1 \cap A_2)$. Assume that the relation holds for $n = m$. We will show that the relation then also holds for $n = m+1$. We have

$$P\left(\cup_{i=1}^{m+1} A_i\right) = P\left(\cup_{i=1}^{m} A_i \cup A_{m+1}\right) = P\left(\cup_{i=1}^{m} A_i\right) + P(A_{m+1}) - P\left(\left(\cup_{i=1}^{m} A_i\right) \cap A_{m+1}\right).$$

We now use the induction hypothesis twice. First,

$$P\left(\cup_{i=1}^{m} A_i\right) = \sum_{i=1}^{m} P(A_i) - \sum_{i<j\le m} P(A_i \cap A_j) + \sum_{i<j<k\le m} P(A_i \cap A_j \cap A_k) - \cdots.$$

Second,

$$\begin{aligned} P\left(\left(\cup_{i=1}^{m} A_i\right) \cap A_{m+1}\right) &= P\left(\cup_{i=1}^{m} (A_i \cap A_{m+1})\right) \\ &= \sum_{i=1}^{m} P(A_i \cap A_{m+1}) - \sum_{i<j\le m} P(A_i \cap A_j \cap A_{m+1}) + \cdots \\ &= \sum_{i<j=m+1} P(A_i \cap A_j) - \sum_{i<j<k=m+1} P(A_i \cap A_j \cap A_k) + \cdots. \end{aligned}$$

Inserting the last two equations into the first gives the desired result, which completes the proof by induction.

■

■ Example 1.11 Let A, B, and C be three events such that $P(A \cup B \cup C) = 1$, $P(A \cap B^c \cap C^c) = P(A^c \cap B \cap C^c) = P(A^c \cap B^c \cap C) = 0$ and $P(A \cap B \cap C) = 0$. If $P(A \cap B) = 0.10$ and $P(A \cap C) = 0.20$, find $P(B)$.

- As shown in the Venn diagram for the proof of Theorem 1.2.4, we have

$$\begin{aligned} P(A \cup B \cup C) &= P(A \cap B^c \cap C^c) + P(A^c \cap B \cap C^c) + P(A^c \cap B^c \cap C) \\ &\quad + P(A \cap B) + P(B \cap C) + P(A \cap C) - 2P(A \cap B \cap C). \end{aligned}$$

Thus, $P(B \cap C) + P(A \cap B) + P(A \cap C) = 1$.

- Hence, we have

$$\begin{aligned} P(B) &= P(B \cap A) + P(B \cap A^c \cap C) + P(B \cap A^c \cap C^c) \\ &= P(B \cap A) + P(B \cap C) - P(B \cap C \cap A) = 0.8. \end{aligned}$$

■

■ Example 1.12 Given three events A, B and C, the probability that exactly one of these three events occurs equals

$$P(A) + P(B) + P(C) - 2P(A \cap B) - 2P(A \cap C) - 2P(B \cap C) + 3P(A \cap B \cap C).$$

- We can also use the Venn diagram of Figure 1.1 given in the proof of Theorem 1.2.4 and see that the probability that exactly one of these three events occurs is $P(A_1)+P(A_5)+P(A_7)$.

- Note that $P(A_1)=P(A\cap(B\cup C)^c)$ and that

$$\begin{aligned} P(A\cap(B\cup C)^c) &= P(A)-P(A\cap(B\cup C))=P(A)-P((A\cap B)\cup(A\cap C)) \\ &= P(A)-P(A\cap B)-P(A\cap C)+P(A\cap B\cap C). \end{aligned}$$

- Thus, by symmetry, we get $P(A_5)=P(B\cap(A\cup C)^c)=P(B)-P(A\cap B)-P(B\cap C)+P(A\cap B\cap C)$ and $P(A_7)=P(C)-P(A\cap C)-P(B\cap C)+P(A\cap B\cap C)$.

- Combining these, the required probability is given by

$$P(A)+P(B)+P(C)-2P(A\cap B)-2P(A\cap C)-2P(B\cap C)+3P(A\cap B\cap C).$$

■

1.3 Counting methods

In experiments with finite sample spaces, it is often the case that many outcomes have the same probability so that we can find the probability of an event by multiplying the common probability by the number of outcomes with that probability. A *simple sample space* is a finite sample space in which each outcome has the same probability. If there are n distinct outcomes in a simple sample space, each must have probability $1/n$. Such outcomes are called *equally likely* or *equally probable*. When all outcomes in Ω have the same probability, then the probability of an event A is given by

$$P(A)=\frac{\#(A)}{\#(\Omega)}.$$

Theorem 1.3.1 The basic principle of counting (multiplication rule) : Suppose that two experiments are to be performed. Then if experiment 1 can result in any one of n_1 possible outcomes and if for each outcome of experiment 1 there are n_2 possible outcomes of experiment 2, then together there are $n_1\cdot n_2$ possible outcomes.

■ **Example 1.13 — The Birthday problem.** Suppose that there are k students in the classroom and that each has a birthday that is one of the 365 possible days of the year other than February 29. An experiment consists of recording each student's birthday. The sample space Ω could be the set of all k-tuples of integers from 1–365, where each integer stands for one day of the year and each coordinate is the birthday of one specific person in the group. There are 365^k possible outcomes, which are all equally likely. ■

■ **Example 1.14** KU metro made up of n cars is boarded by k passengers ($k<n$), each entering a car completely at random. That is, every car has the same probability of being

entered by a given passenger. Let i_1 be the number of the car entered by the first passenger, i_2 the number of the car entered by the second passenger, and so on. Then, the sample space Ω is the set of all k-tuples of positive integers, the arrangement of the passengers in the cars $(i_1, i_2, \ldots, i_k)$, where each of the numbers i_1, i_2, ..., i_k can range from 1 to n. Hence, there are n^k possible arrangements of the passengers in the cars. ■

When an experiment consists of sampling k items out of n distinct items and recording the k selected items in the order sampled, a special case of the multiplication rule is considered. In this case, $n_1 = n$, $n_2 = n-1$, ..., $n_k = n-k+1$. The number of outcomes is $n(n-1)\cdots(n-k+1)$. This number is denoted $P_{n,k}$ and is called the number of *permutations of n elements taken k at a time*.

Definition 1.3.1 Let A be any finite set. A *permutation of A* is a one-to-one mapping of A onto itself. The total number of permutations of a set A of n elements is given by $P_{n,n} = n\cdot(n-1)\cdot(n-2)\cdot\ldots\cdot 1 = n!$ called *n factorial* with the special convention $0! = 1$.

- Let A be an n-element set, and let k be an integer between 0 and n. Then a k-permutation of A is an ordered listing of a subset of A of size k. The total number of k-permutations of a set A of n elements is given by

$$P_{n,k} = n\cdot(n-1)\cdot(n-2)\cdot\ldots\cdot(n-k+1) = \frac{n!}{(n-k)!}.$$

■ **Example 1.15** We use permutations and multiplication rule to count the elements of an event and compute the probability of the event.

- How many different batting orders are possible for a baseball team with 9 players? 9!
- Joanne has 10 books that she is going to put on her bookshelf. Of these, 4 are mathematics books, 3 are statistics books, 2 are history books, and 1 is a language book. Joanne wants to arrange her books so that all the books dealing with the same subject are together on the shelf. What is the probability of having the books arranged per subject on the book shelf? $\frac{4!3!2!1!4!}{10!}$
- When n balls are randomly dropped into n urns, the probability that each urn contains a single ball is given by $\frac{n!}{n^n}$.
- If n is the number of urns and k the number of balls $(n > k)$, the probability that no urn contains more than a single ball is $\frac{P_{n,k}}{n^k}$.
- Seven passengers are in an elevator that stops at each of ten floors. The probability that no two passengers leave at the same floor is $\frac{P_{10,7}}{10^7}$.
- (Sampling with replacement) Consider n objects and making k choices from them, one at a time *with replacement*. The term "with replacement" refers to the fact that if a certain object is selected, it may be selected again at any future time. That is, we

choose, at random, an object from a population, check its properties, put the object back into the population. Then, there are n^k possible outcomes.

- (Sampling without replacement) Consider n objects and making k choices from them, one at a time *without replacement* (i.e. choosing a certain object precludes it from being chosen again. That is, repetition is not allowed). Then there are $P_{n,k} = n(n-1)\cdots(n-k+1)$ possible outcomes, for $k \leq n$.

■

■ **Example 1.16 — Birthday Problem.** Suppose that we are interested in the event A that at least two students have the same birthday among k students enrolled in STAT 221.

- If $k > 365$, then $P(A) = 1$. If $k < 2$, then $P(A) = 0$.
- If $2 \leq k \leq 365$, we can still compute $P(A) = \dfrac{\#(A)}{\#(\Omega)} = \dfrac{\#(A)}{365^k}$.
- We need to count how many of the 365^k outcomes are in A. It turns out to be slightly simpler to count the number of outcomes in A^c in which every student has a different birthday.
- Each outcome in A^c is an ordered list of k distinct birthdays out of 365. That is, a permutation of 365 elements taken k at a time. We used permutations to count the elements of A^c and the multiplication rule to count the elements of Ω. There are $P_{365,k}$ outcomes in A^c and 365^k outcomes in Ω.
- Thus, the required probability is given by

$$P(A^c) = \frac{P_{365,k}}{365^k} = \frac{365!}{(365-k)! \cdot 365^k} \text{ and } P(A) = 1 - P(A^c) = 1 - \frac{365!}{(365-k)! \cdot 365^k}.$$

# of People	Probability of at least two having the same birthday
4	0.0164
10	0.116
20	0.411
23	0.507
40	0.891
41	0.903
45	0.941
50	0.970

Table 1.1: Birthday Probability

■

R We may compute the probability of birthday problem when $k = 23$, $P(A) \approx 0.507$ using R :

```
k = 23; 1-exp(lfactorial(365)-lfactorial(365-k)-k*log(365))
[1] 0.5072972

pbirthday(k)  #an R built-in function for birthday problem
[1] 0.5072972

sapply(4:50,pbirthday) # birthday probabilities for k=4,...50.
```

Remark As in Table 1.1, if there are 23 students in the class, the probability that two of them will have the same birthday is over a 50% chance, and with 41 or more students, the probability is over 90%.

- The reason is with the *combination* of two students, that is, there are more pairs of two students than the number of students in the class.
- For example, suppose there are 23 students in the class. Then, there are $(23 \times 22)/2 = 253$ pairs of students, and any one of those pairs has probability $1/365$ of having the same birthday, so the average number of pairs with the same birthday is $253/365 \approx 0.69$.
- However, it is possible that several different pairs may have the same birthday, so the resulting probability would be 0.507 taking them into account.
- Note that the probability that at least one of the $k = 23$ students have *the same birthday as you* is $1 - \left(\frac{364}{365}\right)^k$.

Example 1.17 Suppose that KU-metro made up of n cars is boarded by k passengers ($k < n$), each entering a car completely at random. What is the probability of the passengers all ending up in different cars?

- There are n^k possible arrangements of the passengers in the cars. This is equivalent to sampling with replacement, making k choices from n objects.
- Let A be the event that no more than one passenger enters any car. Then, A occurs if the first passenger can enter one of n cars, but the second passenger can only enter one of $n-1$ cars, the third passenger one of $n-2$, and so on. This is equivalent to sampling without replacement, making k choices from n objects. Hence $\#(A) = n(n-1)\cdots(n-k+1) = P_{n,k}$. Therefore, $P(A) = \dfrac{n(n-1)\cdots(n-k+1)}{n^k}$.

■

■ Example 1.18 — Beethoven's symphonies. Suppose that your professor's playlist contains 100 classical orchestral symphonies in the *music app* and plays them in random order.

- Suppose that there are 9 symphonies by Ludwig van Beethoven in the playlist. We can compute the probability that the first Beethoven's symphony played is the fourth symphony played. Let A be the event that the first Beethoven's symphony played is the fourth symphony played. Then $\#(A) = 91 \cdot 90 \cdot 89 \cdot 9$. Thus, the required probability is $P(A) = \frac{91 \cdot 90 \cdot 89 \cdot 9}{100 \cdot 99 \cdot 98 \cdot 97} = \frac{P_{91,3} \cdot 9}{P_{100,4}}$.

- We can also compute the probability that the same symphony is played at least twice in the 20 symphonies played, which is $1 - \frac{100 \times 99 \times \cdots \times 81}{100^{20}}$. This is the birthday problem with 100 equally likely birthdays and 20 people.

■

Remark In a class of k students, each student writes down a number at random from $1, 2, \ldots, n$. What is the probability that two or more students have chosen the same number? This is the birthday problem with n equally likely birthdays and k people. Suppose that k is fixed and n is sufficiently large (that is, $n \to \infty$). We can approximate the probability as follows :

- Consider the probability that all the students (k) pick different numbers (n), denoted by $p(n,k)$. Then, $p(n,k)$ can be approximated by

$$\begin{aligned} p(n,k) &= \frac{n(n-1)\cdots(n-k+1)}{n^k} = 1 \cdot \left(1 - \frac{1}{n}\right)\left(1 - \frac{2}{n}\right)\cdots\left(1 - \frac{k-1}{n}\right) \\ &\approx \exp\left(-\frac{1}{n}[1+2+\ldots+(k-1)]\right) = \exp\left(-\frac{1}{2n}k(k-1)\right). \end{aligned}$$

  Here, we use the definition of the (Euler) number e, $e = \lim_{n\to\infty}\left(1 + \frac{1}{n}\right)^n$. In fact, for any real value of x, we define $e^x = \lim_{n\to\infty}\left(1 + \frac{x}{n}\right)^n$.

- Thus, the required probability can be approximated by $1 - \exp\left(-\frac{1}{2n}k(k-1)\right)$.

For the birthday problem with $n = 365$ in Example 1.16, we have

$$\begin{aligned} P(A^c) &= p(365,k) = \frac{P_{365,k}}{365^k} = \frac{365!}{(365-k)!365^k} = \prod_{i=0}^{k-1}\frac{365-i}{365} = \prod_{i=0}^{k-1}\left(1 - \frac{i}{365}\right) \\ &= \exp\left(\sum_{i=0}^{k-1}\log\left(1 - \frac{i}{365}\right)\right) \approx \exp\left(-\frac{1}{365}\sum_{i=0}^{k-1} i\right) = \exp\left(-\frac{k(k-1)}{2 \cdot 365}\right), \end{aligned}$$

- Here, we use the first order Taylor series approximation $\log(1-x) \approx -x$ for small x to get the last quantity.

1.3.1 Combination

Sometimes it is inconvenient or not useful to distinguish the same subset in different orders. For example, when we deal cards from a deck, we might be concerned only with the cards we draw and not the order in which we drew them.

Suppose we have a set with n distinct elements and we want to choose k items from the set. In problems of this type, we do not arrange, (ordering does not matter) and repetition is not allowed. Such a choice of items is called a *combination*. The number of ways of choosing k items from n distinct items without replacement(repetition) is $C_{n,k}$.

Definition 1.3.2 Let $C_{n,k}$ stand for the number of distinct subsets (combinations) of n elements taken k at a time. Each combination of n elements taken k at a time can be permuted $k!$ different ways to make different permutations of n elements taken k at a time. Similarly, every permutation is a particular permutation of some combination. It follows that $P_{n,k} = k!C_{n,k}$. Hence

$$C_{n,k} = \frac{n!}{k!(n-k)!},$$

Remark The symbol $\binom{n}{k}$ is also common notation for $C_{n,k}$. That is,

$$\binom{n}{k} = \frac{n!}{k!(n-k)!} = C_{n,k}.$$

Note that $\binom{n}{n-k} = \binom{n}{k}$, which is the number of distinct subsets of size k that can be chosen from a set of n items without replacement.

■ **Example 1.19** For every sequence of n items, there are $\binom{n}{k}$ distinct ways to pick k out of the n items for special treatment.

- If you flip a coin n times, there are $\binom{n}{k}$ sequences of flips that have exactly k heads.
- Successive patients in a clinical trial respond either positively or negatively to a new treatment. Suppose that there are n patients and k of them respond positively. Then, there are $\binom{n}{k}$ sets of k patients that can be chosen out of n.
- If 3 balls are randomly drawn from a box containing 7 white and 5 black balls, what is the probability that one of the balls is white and the other two black? Here, we consider the balls are all distinguishable. Thus, we can compute the probability using either $\binom{n}{k}$ (unordered set of selected balls) or $P_{n,k}$ (ordered set of selected balls).

① If we do not consider the ordering, we can use $\binom{n}{k}$. Thus, the required probability is $\frac{\binom{7}{1}\binom{5}{2}}{\binom{12}{3}}$.

② If we consider the ordering, we can use $P_{n,k}$. That is, the sample space Ω consists of $12\times 11\times 10$ possible outcomes, and there are three ways of choosing one white and two black balls. Thus, the required probability is

$$3\times\frac{7\times 5\times 4}{12\times 11\times 10}=\frac{\binom{7}{1}\binom{5}{2}}{\binom{12}{3}}.$$

■

■ **Example 1.20 — Beethoven's symphonies (revisited).** Suppose that your professor's playlist contains 100 classical orchestral symphonies in the *music app* and plays them in random order. Suppose that there are 9 symphonies by Ludwig van Beethoven in the playlist.

- Let A be the event that the first Beethoven's symphony played is the fourth symphony played. Then $\#(A)=91\cdot 90\cdot 89\cdot 9$. Then, $P(A)=\frac{91\cdot 90\cdot 89\cdot 9}{100\cdot 99\cdot 98\cdot 97}=\frac{P_{91,3}\cdot 9}{P_{100,4}}$ using permutation as computed in Example 1.18. Alternatively, we can consider the number of ways of choosing 9 positions for Beethoven's symphonies using combination, which is $\binom{100}{9}$. Then, if we choose 8 positions of the last 96 positions, there there are $\binom{96}{8}$ ways. Thus, $P(A)=\frac{\binom{96}{8}}{\binom{100}{9}}=\frac{\frac{96!}{8!\cdot 88!}}{\frac{100!}{9!\cdot 91!}}=\frac{91\cdot 90\cdot 89\cdot 9}{100\cdot 99\cdot 98\cdot 97}$.
- By similar reasoning, we can compute the probability that at least one of the first four symphonies played is Beethoven's. Let B be the event that at least one of the first four symphonies played is written by Beethoven and let A_i be the event that the first Beethoven's symphony is the ith symphony played. Then, $B=\cup_{i=1}^{4}A_i$ and A_i's are disjoint, $A_i\cap A_j=\emptyset$, $i\neq j$. Thus, $P(B)=P\left(\cup_{i=1}^{4}A_i\right)=\sum_{i=1}^{4}P(A_i)=\frac{\sum_{i=1}^{4}\binom{100-i}{8}}{\binom{100}{9}}$.

■

■ **Example 1.21** Suppose that you have a box with n smartphones, exactly one of which is your phone. You select one phone randomly at a time.

- If you fail to get your own phone, you replace it and try again (sampling with replacement). Let A_k be the event that you get your phone at the kth trial. Then, the probability that you get your phone at the kth trial is $P(A_k)=\dfrac{(n-1)^{k-1}\times 1}{n^k}$.
- If you fail to get your own phone, you do not replace it and try again (sampling without replacement). Then, the probability that you get your phone at the kth trial is

$$\begin{aligned}P(A_k) &= \frac{P_{n-1,k-1}\times 1}{P_{n,k}}=\frac{n-1}{n}\times\frac{n-2}{n-1}\times\cdots\times\frac{n-1-(k-1)+1}{n-k+2}\times\frac{1}{n-k+1}\\ &= \frac{1}{n}\left(=\frac{(n-1)!}{n!}\right).\end{aligned}$$

Here, you can think about your phone being at the kth position and the other $(n-1)$

phones being in any of $(n-1)!$ possible orders, giving $\frac{(n-1)!}{n!} = \frac{1}{n}$.

- Suppose that k of these phones are withdrawn at one time. What is the probability that your phone is chosen? Let A be the event that your phone is chosen out of k phones withdrawn.

① $P(A) = \dfrac{\binom{1}{1}\binom{n-1}{k-1}}{\binom{n}{k}} = \dfrac{k}{n}.$

② Since k phones are withdrawn at one time, the sampling is without replacement, and it is equivalent to selecting one phone randomly k times. Let A_i denote the event your phone is chosen at the ith selection, $i = 1, \ldots, k$. Then,

$$\begin{aligned} P(A_i) &= \frac{(n-1)(n-2)\cdots(n-i+1)(1)(n-i)\cdots(n-k+1)}{n(n-1)\cdots(n-k+1)} \\ &= \frac{(n-1)!}{n!} = \frac{1}{n},\ i = 1, \ldots, k. \end{aligned}$$

Note that $\{A_i\}_{i=1}^k$ are mutually disjoint and that $A = \cup_{i=1}^k A_i$. Hence, the required probability $P(A) = P\left(\cup_{i=1}^k A_i\right) = \sum_{i=1}^k P(A_i) = \sum_{i=1}^k \frac{1}{n} = \frac{k}{n}$.

■

■ **Example 1.22** A box contains 100 balls, of which r are red. Suppose that the balls are drawn from the box one at a time, at random, without replacement. Determine the probability that the 50th ball drawn will be red.

- We can imagine that the 100 balls are randomly ordered in a list, and then drawn in that order. Thus, the required probability is simply the probability that the 50th in the list is red, given by $\dfrac{99!\binom{r}{1}}{100!} = \dfrac{r}{100}$.
- In fact, the probability that the kth ball drawn will be red is the same $\dfrac{r}{100}$, $k = 1, \ldots, 100$, because of the random order of the list. Note that the number of possible outcomes in the event that the kth ball drawn will be red is the number of ways choosing $r-1$ position for red balls out of 99 positions with the kth position chosen for a red ball. Thus, it can be calculated in either of two ways.

$$\frac{99!\binom{r}{1}}{100!} = \frac{r}{100} = \frac{\binom{99}{r-1}}{\binom{100}{r}}.$$

■

Theorem 1.3.2 — Binomial theorem. Consider the polynomial $(x+y)^n$. This product expands into a sum of 2^n terms, each of which has n factors, 0 to n of which are x's and the rest are y's. Each term falls into one of $n+1$ classes where class k contains those terms with exactly k x's for $k = 0, \ldots, n$ and $n-k$ y's. How many terms are in class k?

In other words, how many ways are there to choose k of the n factors to equal the x's? The answer is $\binom{n}{k}$. It follows that

$$(x+y)^n = \sum_{k=0}^{n} \binom{n}{k} x^k y^{n-k}.$$

Proof. We know that

$$(x+y)^2 = x^2+2xy+y^2 = \binom{2}{0}x^2+\binom{2}{1}xy+\binom{2}{2}y^2 = \sum_{k=0}^{2}\binom{2}{k}x^{2-k}y^k.$$

Similarly, we have $(x+y)^3 = x^3+3x^2y+3xy^2+y^3 = \sum_{k=0}^{3}\binom{3}{k}x^{3-k}y^k$.

Assume the statement holds for $n=m$. Then for $n=m+1$,

$$\begin{aligned}(x+y)^{m+1} &= (x+y)(x+y)^m = (x+y)\sum_{k=0}^{m}\binom{m}{k}x^k y^{m-k}\\ &= \sum_{k=0}^{m}\binom{m}{k}x^{k+1}y^{m-k}+\sum_{k=0}^{m}\binom{m}{k}x^k y^{m+1-k} = \sum_{i=1}^{m+1}\binom{m}{i-1}x^i y^{m+1-i}+\sum_{i=0}^{m}\binom{m}{i}x^i y^{m+1-i}\\ &= x^{m+1}+\sum_{i=1}^{m}\left[\binom{m}{i-1}+\binom{m}{i}\right]x^i y^{m+1-i}+y^{m+1} = \sum_{i=0}^{m+1}\binom{m+1}{i}x^i y^{m+1-i},\end{aligned}$$

where the last inequality follows from the direct calculation (cf. Theorem 1.3.3)

$$\binom{m}{i-1}+\binom{m}{i} = \frac{m!}{(i-1)!(m-i+1)!}+\frac{m!}{i!(m-i)!} = \binom{m+1}{i}.$$ ■

Theorem 1.3.3 For any positive integer n and $r = 1,2,3,...,n$, we have

$$\binom{n}{r} = \binom{n-1}{r}+\binom{n-1}{r-1}.$$

Proof. First, note that $(1+y)^n = (1+y)(1+y)^{n-1} = (1+y)^{n-1}+y(1+y)^{n-1}$. Then, using the Binomial theorem, we have

$$\sum_{r=0}^{n}\binom{n}{r}y^r = \sum_{r=0}^{n-1}\binom{n-1}{r}y^r+y\sum_{r=0}^{n-1}\binom{n-1}{r}y^r = \sum_{r=0}^{n-1}\binom{n-1}{r}y^r+\sum_{r=0}^{n-1}\binom{n-1}{r}y^{r+1}.$$

Equating the coefficients of y^r from both sides of the above expression, we obtain

$$\binom{n}{r} = \binom{n-1}{r}+\binom{n-1}{r-1}$$

and the proof is now complete. Alternatively, we can prove this relation as follows. Suppose that in a box there are n balls, where one is white and $n-1$ are black. There are $\binom{n}{r}$ ways of choosing r balls. We can choose one white ball and $r-1$ black balls in $\binom{1}{1}\binom{n-1}{r-1}$ ways, or r black balls in $\binom{n-1}{r}$ ways, adding up to $\binom{n}{r}$. ■

Remark We can use this $\binom{n}{r}=\binom{n-1}{r}+\binom{n-1}{r-1}$ to determine the famous *Pascal's triangle*. The nth row of this triangle has the entries $\binom{n}{0}$, $\binom{n}{1}$, ..., $\binom{n}{n}$. We know that the first and last of these numbers are 1. The remaining numbers are determined by the recurrence relation $\binom{n}{r}=\binom{n-1}{r}+\binom{n-1}{r-1}$; that is, the entry $\binom{n}{r}$ for $0<r<n$ in the nth row of Pascal's triangle is the sum of the entry immediately to its right and the one immediately to its left in the $(n-1)$st row.

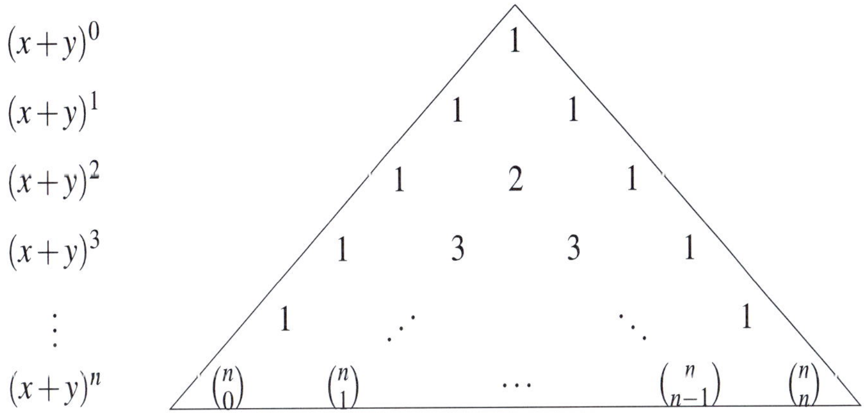

Figure 1.2: Pascal's triangle

Theorem 1.3.4 We have following combinatorial identities :

① There are 2^n subsets of a set consisting of n elements.

② $\binom{n}{1}-\binom{n}{2}+\binom{n}{3}-\cdots+(-1)^{n-1}\binom{n}{n}=1$.

③ $n\binom{n-1}{k-1}=k\binom{n}{k}$ for any positive integers n and k with $k\leq n$.

④ Vandermonde's identity : $\binom{m+n}{k}=\sum_{j=0}^{k}\binom{m}{j}\binom{n}{k-j}$.

Proof. We can prove them using the Binomial theorem or combinatorial argument.

① There are $\binom{n}{0}+\binom{n}{1}+\cdots+\binom{n}{n}=(1+1)^n=2^n$ subsets. Note that for each of the n elements in the set, there are two possibilities, inclusion or exclusion in the given subset. Thus, there are 2^n possibilities.

② From the Binomial theorem, we get $(1+x)^n = \sum_{r=0}^{n} \binom{n}{r} x^r$ for all real numbers x. Letting $x = -1$, we get $0 = \sum_{r=0}^{n} \binom{n}{r}(-1)^r$. Alternatively, let $a_n = \sum_{r=0}^{n} \binom{n}{r}(-1)^r$. Then, we have

$$\begin{aligned} a_n &= \sum_{r=0}^{n}\left[\binom{n-1}{r} + \binom{n-1}{r-1}\right](-1)^r \\ &= \sum_{r=0}^{n}\binom{n-1}{r}(-1)^r + \sum_{r=0}^{n}\binom{n-1}{r-1}(-1)^r = a_{n-1} - a_{n-1} = 0, \end{aligned}$$

where the last inequality follows from that $\binom{n-1}{-1} = \binom{n-1}{n} = 0$.

③ It is obvious to see that $n\binom{n-1}{k-1} = n\frac{(n-1)!}{(k-1)!(n-k)!} = k\frac{n!}{k!(n-k)!} = k\binom{n}{k}$. That is, among n people, we can construct a committee with k people by choosing one president out of n people and then choosing $k-1$ members out of the remaining $n-1$ people. Alternatively, we can choose k people for the committee out of n people and then choose one president among the chosen k people.

④ Since $(1+x)^{m+n} = (1+x)^m(1+x)^n$, we have

$$\sum_{r=0}^{m+n}\binom{m+n}{r}x^r = \left\{\sum_{r=0}^{m}\binom{m}{r}x^r\right\}\left\{\sum_{r=0}^{n}\binom{n}{r}x^r\right\}.$$

Equating the coefficients of x^k from the both sides of the above expression, we obtain

$$\binom{m+n}{k} = \binom{m}{0}\binom{n}{k} + \binom{m}{1}\binom{n}{k-1} + \cdots + \binom{m}{k}\binom{n}{k-k}$$

and the conclusion of the theorem follows. Alternatively, we can prove ④ as follows. Suppose that in a box there are $m+n$ balls, where m are white and n are black. There are $\binom{m+n}{k}$ ways of choosing k balls. We can choose j white balls and $k-j$ black balls in $\binom{m}{j}\binom{n}{k-j}$ ways. Summing over all the possible choices for j gives the required result.

■

Note that a given number $\binom{n}{k}$ in the Pascal's triangle equals the sum of the numbers in the diagonal string that starts with the number $\binom{n-1}{k-1}$ that is located above and to the left of the given number and then proceeds upward to the right $\binom{k-1}{k-1}$. For example, we have $6 = \binom{4}{2} = \binom{3}{1} + \binom{2}{1} + \binom{1}{1}$ as shown in Figure 1.3. In other words, each number in Pascal's triangle is the sum of the two numbers above it or just the "1" above it, if it occurs at the end of a line, which is given in the following theorem for the diagonal sum in the Pascal's triangle.

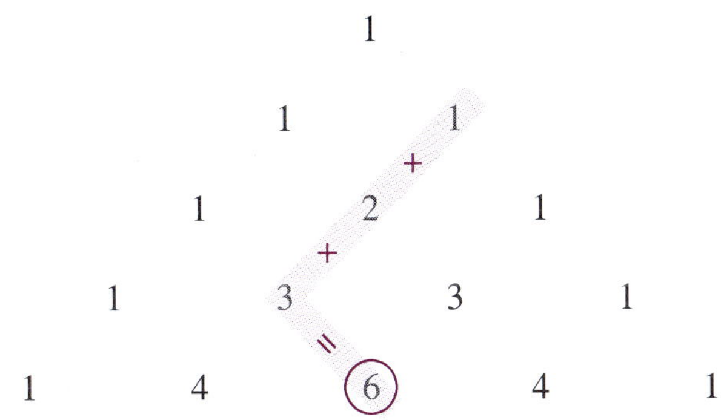

Figure 1.3: Pascal's triangle : Diagonal sum

Theorem 1.3.5 Let n be a positive integer and $k = 1,2,3,\ldots,n$. Then

$$\binom{n}{k} = \binom{n-1}{k-1} + \binom{n-2}{k-1} + \binom{n-3}{k-1} + \cdots + \binom{k}{k-1} + \binom{k-1}{k-1} = \sum_{\ell=k-1}^{n-1} \binom{\ell}{k-1}.$$

Proof. In order to establish the above identity, we can use two different ways.

- Note that $\binom{n}{k} = \binom{n-1}{k-1} + \binom{n-1}{k}$. Thus, if $n = 1$, the equation holds. Assume that it holds if $n = m$, $\binom{m}{k} = \sum_{\ell=k-1}^{m-1} \binom{\ell}{k-1}$. Then, we have
$$\binom{m+1}{k} = \binom{m}{k} + \binom{m}{k-1} = \sum_{\ell=k-1}^{m-1} \binom{\ell}{k-1} + \binom{m}{k-1} = \sum_{\ell=k-1}^{m} \binom{\ell}{k-1}.$$
Thus, it is proven by mathematical induction. In other words, we get the result by making repeated use of the equation, $\binom{n}{k} = \binom{n-1}{k-1} + \binom{n-1}{k}$.
$$\begin{aligned}\binom{n}{k} &= \binom{n-1}{k-1} + \binom{n-1}{k} = \binom{n-1}{k-1} + \binom{n-2}{k-1} + \binom{n-2}{k} \\ &= \binom{n-1}{k-1} + \binom{n-2}{k-1} + \binom{n-3}{k-1} + \cdots + \binom{k}{k-1} + \binom{k-1}{k-1}.\end{aligned}$$

- Alternatively, imagine picking a committee of k people from n people, and label the people as $1,2,3,\ldots,n$. When we list out the $\binom{n}{k}$ possible committees, we can arrange them in groups according to what the lowest number in the committee is. For example, if the lowest number is "1", then the other $k-1$ people on the committee must be chosen from the $n-1$ people who are 2 or higher, with $\binom{n-1}{k-1}$ ways. Similarly, if the lowest number is "2", then there are $\binom{n-2}{k-1}$ ways. In this way, this method of counting continues until we get the stage where the lowest number equals $n-(k-1)$, with only $\binom{k-1}{k-1} = 1$ way to choose the other $k-1$ people.

■

Remark We can also prove Theorem 1.3.5 using the general version of the Binomial theorem written as $(1-x)^{-n} = \sum_{k=0}^{\infty} \binom{n+k-1}{k} x^k$, for $n \geq 0$ and $|x| < 1$. Details can be found in Chapter 3 for the discussion of the negative binomial distribution.

Definition 1.3.3 If $n_1 + n_2 + \ldots + n_r = n$, we define $\binom{n}{n_1, n_2, \ldots, n_r}$ by

$$\binom{n}{n_1, n_2, \ldots, n_r} = \binom{n}{n_1} \cdot \binom{n-n_1}{n_2} \cdots \binom{n-n_1-\cdots-n_{r-2}}{n_{r-1}} = \frac{n!}{n_1! n_2! \ldots n_r!}.$$

- The number $\binom{n}{n_1, n_2, \ldots, n_r}$ is called a *multinomial* coefficient.
- The number $\binom{n}{n_1, n_2, \ldots, n_r}$ represents the number of possible divisions of n distinct objects into r distinct groups of respective sizes $n_1, n_2, \ldots, n_r$.

Remark Multinomial coefficients are a generalization of binomial coefficients, and the binomial coefficient $\binom{n}{k}$ can be written as $\binom{n}{k, n-k}$.

Suppose there are three summands, x, y, and z. Then, using the binomial theorem,

$$(x+y+z)^n = (x+(y+z))^n = \sum_{\substack{j,k=0 \\ j+k=n}}^{n} \binom{n}{j,k} x^j (y+z)^k = \sum_{\substack{j,k=0 \\ j+k=n}}^{n} \binom{n}{j,k} x^j \sum_{\substack{\ell,i=0 \\ \ell+i=k}} \binom{k}{\ell,i} y^\ell z^i.$$

Here, the inner sum is over all pairs (ℓ, i) of non-negative integers whose sum is k and that the outer sum is over all pairs of non-negative integers (j,k) whose sum is n. Note that this is the same as all triples (j, ℓ, i) whose sum is n and that the coefficients $\binom{n}{j,k}\binom{k}{\ell,i} = \frac{n!}{j!\ell!i!}$ is written as $\binom{n}{j,\ell,i}$. Thus, we have $(x+y+z)^n = \sum_{\substack{j,\ell,i=0 \\ i+j+\ell=n}}^{n} \binom{n}{j,k,i} x^j y^\ell z^i$, which extends the binomial theorem to three summands.

Theorem 1.3.6 The following result

$$(x_1 + x_2 + \cdots + x_r)^n = \sum_{n_1+n_2+\cdots+n_r=n} \binom{n}{n_1, n_2, \ldots, n_r} x_1^{n_1} x_2^{n_2} \cdots x_r^{n_r}$$

is known as the multinomial theorem and it generalizes the binomial theorem. The summation extends over all positive combinations of nonnegative integers $n_1, n_2, \ldots, n_r$ such that $n_1 + n_2 + \cdots + n_r = n$, and

$$\binom{n}{n_1, n_2, \ldots, n_r} = \frac{n!}{n_1! n_2!, \ldots, n_r!}.$$

This coefficient is known as the multinomial coefficient.

■ **Example 1.23** Suppose that a KU cheer-leading team contains 4 students from the freshman class, 6 students from the sophomore class, 6 students from the junior class, and 4 students from the senior class. They will be sitting at a long table with 20 seats. How many seating arrangements are possible, with regard to their classes? (Here, we do not care, for instance, about permuting the individual freshman within the seats assigned to freshmen.)

- There are $\binom{20}{4}$ ways to choose the freshman seats. Once those are chosen, there are $\binom{16}{6}$ ways to choose the sophomore seats. Once those are chosen, there are $\binom{10}{6}$ ways to choose the junior seats. The senior seats are then already determined, that is, there will be only one way at that point, $\binom{4}{4}$.

- Thus, the total number of seating arrangements is

$$\frac{20!}{4!16!}\cdot\frac{16!}{6!10!}\cdot\frac{10!}{6!4!}\cdot\frac{4!}{4!0!}=\frac{20!}{4!6!6!4!}=\binom{20}{4,6,6,4}.$$

■

1.4 Probability of a Union

For two events A and B, $P(A\cup B)=P(A)+P(B)-P(A\cap B)$. For three or more events, there is a generalization of this formula, called the inclusion-exclusion principle as in Theorem 1.2.4. We use the inclusion-exclusion principle to compute the probability of a union when the events are not mutually exclusive.

Theorem 1.4.1 For any finite sequence of events $A_1,\ldots,A_n$,

$$P(\cup_{i=1}^n A_i)=\sum_{i=1}^n P(A_i)-\sum_{i<j}P(A_i\cap A_j)+\sum_{i<j<k}P(A_i\cap A_j\cap A_k)\cdots+(-1)^{n+1}P(\cap_{i=1}^n A_i).$$

- First, add the probabilities of the individual events.
- Then subtract the probabilities of all the pairwise intersections since these have all been counted twice.
- Then add back in the probabilities of all intersections of three events since these got subtracted out three times instead of twice.
- Then subtract the probabilities of all intersections of four ...

Proof. Here, we provide an alternative way of explaining the inclusion-exclusion principle to the mathematical induction discussed in the proof of Theorem 1.2.4.

- Consider the case that the sample space is simple sample space. In this case, $P(A) = \#(E)/\#(\Omega)$, that is, $P(E) = \sum_{\omega \in E} P(\omega)$. Thus, it is enough to consider how to count the number of elements in the union of events. Let $N(A)$ be the number of elements in the set A. Then, the number of elements in $A \cup B$, $N(A \cup B)$ is computed $N(A \cup B) = N(A) + N(B) - N(A \cap B)$ because any element who is both A and B is counted once in all three terms on the right hand side, and so contributes just one to the total $N(A \cup B)$. More generally, if there are n sets $A_1, \ldots, A_n$, then the number of elements that are contained at least one of $A_1, \ldots, A_n$, i.e. $\cup_{i=1}^n A_i$ is

$$N\left(\cup_{i=1}^n A_i\right) = \sum_{i=1}^n N(A_i) - \sum_{i<j} N(A_i \cap A_j) + \cdots + (-1)^{n-1} N\left(\cap_{i=1}^n A_i\right).$$

- Fix $\omega \in A$. If $\omega \notin \cup_{i=1}^n A_i$, then ω does not belong to any of the sets A_i, so ω does not contribute to either the left-hand side or the right-hand side. Suppose that $\omega \in \cup_{i=1}^n A_i$. Then there is at least one set A_i to which ω belongs. Let k be the number of sets A_i to which ω belongs. Note that in $N\left(\cup_{i=1}^n A_i\right)$, ω is counted only once. In the first term of the right-hand side, ω is counted k times, in the second term $\binom{k}{2}$ times, in the third term $\binom{k}{3}$ times, and so on. Thus, the number of times that ω is counted on the right-hand side is $\binom{k}{1} - \binom{k}{2} + \binom{k}{3} - \cdots + (-1)^{k-1}\binom{k}{k} = 1$. as given in ② of Theorem 1.3.4. Alternatively, we can use mathematical induction as given in the proof of Theorem 1.2.4.

■

Example 1.24 In a certain city, three newspapers, A,B, and C are published. Suppose that 60 percent of the families in the city subscribe to newspaper A, 40 percent of the families subscribe to newspaper B, and 30 percent subscribe to newspaper C. Suppose also that 20 percent of the families subscribe to both A and B, 10 percent subscribe to both A and C, 20 percent subscribe to both B and C, and 5 percent subscribe to all three newspapers A, B, and C. What percentage of the families in the city subscribe to at least one of the three newspapers? What percentage of the families in the city subscribe to exactly one of the three newspapers?

- Let A, B, and C stand for the events that a randomly selected family subscribes to the newspaer with the same name. Then $P(A \cup B \cup C)$ is the proportion of families that subscribe to at least one newspaper. Thus, from the inclusion-exclusion principle with three sets, the probability that at least one of the three newspapers is subscribed by the families in the city is

$$\begin{aligned} & P(A \cup B \cup C) \\ = \; & P(A) + P(B) + P(C) - P(A \cap B) - P(A \cap C) - P(B \cap C) + P(A \cap B \cap C) \\ = \; & 0.6 + 0.4 + 0.3 - 0.2 - 0.1 - 0.2 + 0.05 = 0.85. \end{aligned}$$

- As shown in Example 1.12, given three events A, B, C, the probability that exactly one of the events occurs equals

$$P(A) + P(B) + P(C) - 2P(A \cap B) - 2P(A \cap C) - 2P(B \cap C) + 3P(A \cap B \cap C)$$

Thus, the probability that exactly one of the three newspapers is subscribed by the families in the city is $0.6+0.4+0.3-2\cdot(0.2+0.1+0.2)+3\cdot 0.05$.

■

■ **Example 1.25** Suppose that a KU cheer-leading team contains 4 students from the freshman class, 6 students from the sophomore class, 6 students from the junior class, and 4 students from the senior class. If 5 students are selected at random from the team, what is the probability that at least one student will be selected from each of the four classes? We can first determine the probability that at least one of the four classes will not be represented in the selection.

- Let A be the event that at least one student will be selected from each of the four classes. Then, $A = \cap_{i=1}^4 A_i$, where A_1 denotes the event at least one student from the freshman class is selected, and A_2, A_3 and A_4 denote the corresponding events for the sophomore, junior and senior classes.

- Let B denote the event that at least one of the four classes will not be represented in the selection, $B = A^c = \cup_{i=1}^4 A_i^c = \cup_{i=1}^4 B_i$. That is, B_1 denotes the event that no student from the freshman class is selected, and let B_2, B_3, and B_4 denote the corresponding events for the sophomore, junior, and senior classes, respectively.

- First, we determine $P(\cup_{i=1}^4 B_i)$ by applying inclusion-exclusion principle. The event B_i will occur if and only if no student from the ith class is selected. For example, $P(B_1) = \binom{16}{5}/\binom{20}{5}$ and $P(B_3) = \binom{14}{5}/\binom{20}{5}$. Next, the event $B_1 \cap B_2$ will occur if and only if the 5 selected students are juniors or seniors, and thus $P(B_1 \cap B_2) = \binom{10}{5}/\binom{20}{5}$. Next the event $B_1 \cap B_2 \cap B_4$ will occur if and only if 5 students are selected from juniors, and thus $P(B_1 \cap B_2 \cap B_4) = \binom{6}{5}/\binom{20}{5}$. Of course, $P(B_1 \cap B_2 \cap B_3 \cap B_4) = 0$. In this way, $P(\cup_{i=1}^4 B_i)$ can be computed.

- Hence, the required probability is

$$
\begin{aligned}
P(A) &= 1 - P(B) = 1 - P\left(\cup_{i=1}^4 B_i\right) \\
&= 1 - \left[\sum_{i=1}^{4} P(B_i) - \sum_{i<j} P(B_i \cap B_j) + \sum_{i<j<k} P(B_i \cap B_j \cap B_k) \cdots + (-1)^5 P(\cap_{i=1}^4 B_i)\right]. \\
&= 1 - \left[\frac{\binom{16}{5}}{\binom{20}{5}} + \frac{\binom{14}{5}}{\binom{20}{5}} + \frac{\binom{14}{5}}{\binom{20}{5}} + \frac{\binom{16}{5}}{\binom{20}{5}}\right] + \left[\frac{\binom{10}{5}}{\binom{20}{5}} + \frac{\binom{10}{5}}{\binom{20}{5}} + \cdots\right] - \left[\frac{\binom{6}{5}}{\binom{20}{5}} + \frac{\binom{6}{5}}{\binom{20}{5}}\right].
\end{aligned}
$$

■

■ **Example 1.26** A box contains 30 red balls, 30 white balls, and 30 blue balls. If 10 balls are selected at random, without replacement, what is the probability that at least one color will be missing from the selection?

- Let A_1 denote the event that no red balls are selected, let A_2 denote the event that no white balls are selected, and let A_3 denote the event that no blue balls are selected.

The desired probability is $P(A_1 \cup A_2 \cup A_3)$.

- The event A_1 will occur if and only if the ten selected balls are either white or blue. Since there are 60 white and blue balls, out of a total of 90 balls, we have $P(A_1) = \binom{60}{10}/\binom{90}{10}$. Similarly, $P(A_2)$ and $P(A_3)$ have the same value. The event $A_1 \cap A_2$ will occur if and only if all ten selected balls are blue. Therefore, $P(A_1 \cap A_2) = \binom{30}{10}/\binom{90}{10}$. Similarly, $P(A_2 \cap A_3)$ and $P(A_1 \cap A_3)$ have the same value. Finally, the event $A_1 \cap A_2 \cap A_3$ will occur if and only if all three colors are missing, which is impossible.

- Therefore, the required probability is $P(A_1 \cup A_2 \cup A_3) = 3 \cdot \frac{\binom{60}{10}}{\binom{90}{10}} - 3 \cdot \frac{\binom{30}{10}}{\binom{90}{10}}$. ■

■ **Example 1.27** A student buys a probability textbook that has 300 pages, each numbered with 3 digits, from 000 to 299 and randomly opens the book to a page and starts to read! Assume that any of the 300 pages are equally likely to be chosen. Find the probability that the page number the student chooses contains at least one "2" as a digit.

- Let A_1 be the event that the ones digit is a "2", let A_2 be the event that the tens digit is a "2", and let A_3 be the event that the hundreds digit is a "2". The desired probability is $P(A_1 \cup A_2 \cup A_3)$.

- By inclusion-exclusion principle, we have

$$
\begin{aligned}
&P(A_1 \cup A_2 \cup A_3) = P(A_1) + P(A_2) + P(A_3) \\
&-P(A_1 \cap A_2) - P(A_1 \cap A_3) - P(A_2 \cap A_3) + P(A_1 \cap A_2 \cap A_3) \\
&= \frac{30}{300} + \frac{30}{300} + \frac{100}{300} - \frac{3}{300} - \frac{10}{300} - \frac{10}{300} + \frac{1}{300}.
\end{aligned}
$$

■

■ **Example 1.28** Consider rolling a fair die 10 times. Let A be the event that at least one "6" is obtained. Compute the probability $P(A)$.

- Define a sequence of events $A_1, \ldots, A_n$, where A_i is the event that the ith trial is a success, that is, 6 is obtained in the ith trial. Then $A = \cup_{i=1}^{10} A_i$ and $P(A) = P\left(\cup_{i=1}^{10} A_i\right)$. We can apply the inclusion-exclusion principle since A_i's are not mutually exclusive.

- Alternatively, define a sequence of events $B_1, \ldots, B_n$, where B_i is the event that the *first* success (6 is obtained) on the ith trial. Then $A = \cup_{i=1}^{10} B_i$ and B_i's are mutually exclusive. Thus, $P(A) = \sum_{i=1}^{10} P(B_i)$.

- To find $P(B_i)$, note that if the first success occurs on the ith trial, then the first $i-1$ trials are necessarily not successes, the ith trial is a success, and the results in the remaining $n-i$ trials do not matter. Thus, by the multiplication principle, we have

$$P(B_i) = \frac{5^{i-1}\times 1\times 6^{10-i}}{6^{10}} = \frac{1}{6}\left(1-\frac{1}{6}\right)^{i-1}, \; i=1,\ldots,n.$$

Hence, the desired probability is

$$P(A) = \sum_{i=1}^{10} P(B_i) = \frac{1}{6}\left(\frac{1-(1-1/6)^{10}}{1-(1-1/6)}\right) = 1-\left(1-\frac{1}{6}\right)^{10}.$$

- On the other hand, we can use the probability of A^c, the complementary event of A that 6 is not obtained in the 10 trials. Therefore, we have

$$P(A) = 1-P(A^c) = 1-\left(1-\frac{1}{6}\right)^{10}.$$

■

■ **Example 1.29** In the birthday problem of Example 1.16, we have computed the probability that at least two students have birthdays on the same day in a class of k students, which is given $1-\frac{P_{365,k}}{365^k}$. Here, we compute the probability that in a class of k students exactly two students have birthdays on the same day in terms of the probability of union.

- Let $A_{i,j}$, $i\neq j$, be the event that two students i and j have the same birthday and all other students have different birthdays. Then, there are $\binom{k}{2}$ possible pairs of two students that have the same birthday. Note that all $A_{i,j}$ are disjoint and have the same probability equal to $P(A_{1,2}) = 365\times 1\times 364\times 363\times\cdots\times(365-k+1)/365^k$.

- Hence, the required probability is

$$P\left(\cup_{i\neq j}A_{i,j}\right) = \sum_{i\neq j} P\left(A_{i,j}\right) = \binom{k}{2}\frac{365\times 1\times 364\times 363\times\cdots\times(365-k+1)}{365^k}.$$

■

■ **Example 1.30 — Matching Problem.** Suppose that n students leave their smartphones with the course TA before taking the midterm in STAT 221. When the exam is over and students return to get their phones, the TA randomly chooses phones to hand back. What is the probability that at least one student gets the correct smartphone?

- Assume that the students in STAT 221 are numbered 1 to n in the order they request their phones. Let the sample space consist of all possible orders in which the phones are handed to the students. For example, with $n=3$, $(3,2,1)$ means that student 1 got phone 3, student 2 got the correct phone and student 3 got phone 1. There are $n!$ possible orderings.

- Let A be the event that at least one student gets the correct phone. Let A_i be the event that the ith student gets the correct phone, for $i=1,\ldots,n$. Then $A=\bigcup_{i=1}^{n} A_i$.

- We are asked to find $P\left(\bigcup_{i=1}^{n} A_i\right)$.

- We have assumed that each outcome is equally likely, so the probability of each set is $1/n!$ times the number of outcomes in the set.

- Each A_i has those outcomes in which i appears in the ith position. There are $(n-1)!$ ways to arrange the other $n-1$ phones in the other $n-1$ positions, so $P(A_i) = (n-1)!/n! = 1/n$ for each i. Alternatively, imagine a experiment taking one ball from a urn one by one without replacement, where there are one white ball and $n-1$ black balls, and consider an event A_1 that you get a white ball in the first place. Then, $P(A_1) = 1/n$.

- Each $A_i \cap A_j$ has those outcomes with i and j in the correct positions. There are $(n-2)!$ such outcomes, hence $P(A_i \cap A_j) = (n-2)!/n!$ for each i and j. Continuing in this way, we see that $P(A_{i_1} \cap \cdots \cap A_{i_k}) = (n-k)!/n!$ for all k and all distinct $i_1, \ldots, i_k$.

- Now apply the result to inclusion-exclusion principle. First, add the probabilities of all the A_i's. There are n of them with probability $1/n$ each. The sum is 1.

- Next, subtract the sum of all probabilities of pairwise intersections. There are $\binom{n}{2}$ pairwise intersections, each with probability $(n-2)!/n!$. The sum is $1/2!$. Next, add the sum of all probabilities of intersections of three. There are $\binom{n}{3}$ of these each with probability $(n-3)!/n!$. The sum is $1/3!$.

- Following this pattern, we see that the probability of at least one student getting the correct phone is $\frac{1}{1!} - \frac{1}{2!} + \frac{1}{3!} - \frac{1}{4!} + - \cdots + (-1)^{n+1}\frac{1}{n!}$. This is just one minus the sum of the first $n+1$ terms in a Taylor series expansion of e^{-1}. Indeed, the limit of the probability as $n \to \infty$ is $1 - e^{-1}$.

- Notice that the answers oscillate around the limit with the odd values of n having larger probabilities and the even values having lower probabilities.

■

R We may compute the probability of matching problem when $n = 50$ using R

$$P(A) = \frac{1}{1!} - \frac{1}{2!} + \frac{1}{3!} - \frac{1}{4!} + - \cdots + (-1)^{n+1}\frac{1}{n!} \approx 1 - e^{-1}$$

```
# direct computation
> n<-50;  a<-0
> for (k in 1:n) a<-a+(-1)^(k+1)/factorial(k)
> a
[1] 0.6321206
> 1-exp(-1)
[1] 0.6321206

# Numerical approximation by simulation
> n <- 50; r <- replicate(10^4,sum(sample(n)==(1:n)))
> sum(r>=1)/10^4
[1] 0.6342
```

Remark Let $f(x)$ be $(n+1)$-times differentiable on an open interval containing the points a and x. Then,

- $f(x) = f(a) + f'(a)(x-a) + \frac{f''(a)}{2!}(x-a)^2 + \cdots + \frac{f^{(n)}(a)}{n!}(x-a)^n + R_n(x)$, where $R_n(x) = \frac{f^{(n+1)}(c)}{(n+1)!}(x-a)^{n+1}$, and c is some point between a and x.
- Let f have derivatives of all orders on an open interval containing the points a and x. If $R_n(x) \to 0$ as $n \to \infty$, then $f(x) = \sum_{n=0}^{\infty} \frac{f^{(n)}(a)}{n!}(x-a)^n$.
- $e^x = 1 + x + \frac{x^2}{2!} + \frac{x^3}{3!} + \cdots = \sum_{n=0}^{\infty} \frac{x^n}{n!}$ for $x \in \mathbb{R}$.

Example 1.31 — Matching Problem (Continued). In the matching problem, compute the probability that exactly one student gets the correct phone.

- Let E_i be the event that the ith student gets the correct phone and nobody else does, for $i = 1, \ldots, n$. Let E be the event that exactly one student gets the correct phone. Then $E = \bigcup_{i=1}^{n} E_i$, and the E_i's are disjoint. Hence the desired probability is $P(E) = \sum_{i=1}^{n} P(E_i) = nP(E_1)$ by the symmetry of the problem.
- We know $P(E_1) = \#(E_1)/n!$, and $\#(E_1)$ is the number of ways of arranging the returning $n-1$ phones so that *nobody* gets the correct phone. This, in turn, is $(n-1)!$ times the probability that nobody gets the correct phone when there are only $n-1$ phones instead of n. Finally, this equals $(n-1)!$ times one minus the probability that at least one student gets the correct phone when there are $n-1$ phones. That is,

$$\#(E_1) = (n-1)!\left[1 - \left(\frac{1}{1!} - \frac{1}{2!} + \frac{1}{3!} - \frac{1}{4!} + - \cdots + (-1)^n \frac{1}{(n-1)!}\right)\right].$$

- Recall that $P(E) = nP(E_1) = \frac{n\#(E_1)}{n!} = \frac{\#(E_1)}{(n-1)!}$. Hence, the required probability is

$$P(E) = \frac{1}{2!} - \frac{1}{3!} + \frac{1}{4!} - + \cdots + (-1)^{n-1}\frac{1}{(n-1)!}.$$

The same reasoning would allow us to compute the probability that exactly k students get the correct phones.

- For example, we compute the probability that exactly two students get the correct phones. Let B be the event that exactly two students get the correct phones and let $E_{i,j}$ be the event that the ith and jth students get his/her own phones and nobody else does. Then, by the symmetry of the problem, we have $P(B) = \sum_{i<j} P(E_{i,j}) = \binom{n}{2} \times P(E_{1,2})$.

$$\begin{aligned} P(E_{1,2}) &= \frac{\#(E_{1,2})}{n!} = \frac{(n-2)!P(\text{no one gets the correct phone among } n-2 \text{ people})}{n!} \\ &= \frac{(n-2)! \times \left(\frac{1}{2!} - \cdots + (-1)^{n-2}\frac{1}{(n-2)!}\right)}{n!}. \end{aligned}$$

Hence the desired probability

$$P(B) = \binom{n}{2} \times P(E_{1,2}) = \frac{1}{2!} \times \left(\frac{1}{2!} - \cdots + (-1)^{n-2}\frac{1}{(n-2)!}\right) = \frac{1}{2!}\sum_{i=0}^{n-2}\frac{(-1)^i}{i!}.$$

- In general, the probability that exactly k students will get their own phones is

$$\frac{1}{k!}\sum_{i=0}^{n-k}\frac{(-1)^i}{i!}, \; k = 1, 2, \ldots, n,$$

for which we will discuss another way using the conditional probability and the law of total probability in Chapter 2.

■

Remark The matching problem is the best-known contribution to the theory of probability of the French mathematician Pierre Rémond de Montmort.

- It appears in the form of a game called *Treize* in his treatise *Essay d'Analyse sur des Jeux de Hazard* (Montmart, 1708), also known as the *Probléme des Rencontres*. Montmort solved this problem by the use of a recursion relation.

- In a simplified version of this game considered by de Montmort, one turns over cards numbered 1 to 13, calling out 1, 2, …, 13 as the cards are examined. The problem asks for the probability that, once at least, the number on a card coincides with the number of the order in which it was drawn.

1.5 Exercises

Problem 1.1 Suppose that the numbers $1,2,\ldots,n$ are arranged in random order. Find the probability that the digits 3,4, and 5 appear as neighbors in the order "3 4 5".

Problem 1.2 Jaden and Joanne are taking STAT 221. The course has only three grades: A, B, and C. The probability that Jaden gets a B is 0.5. The probability that Joanne gets a B is 0.5. The probability that neither gets an A but at least one gets a B is 0.4. What is the probability that at least one gets a B but neither gets a C?

Problem 1.3 At a parking lot there are ten places arranged in a row. A person observed that there were seven cars parked, and that the three empty places were adjacent to each other. Compute the probability that when there are three empty places out of ten parking places, they are adjacent to each other.

Problem 1.4 Using mathematical induction, show that

$$P(\cup_{i=1}^{n} A_i) \geq \sum_{i=1}^{n} P(A_i) - \sum_{i<j\leq n} P(A_i \cap A_j).$$

Problem 1.5 Suppose that you have one final exam in each of 4 courses. The 4 final exams are all scheduled during one week, from Monday to Friday, and each exam is allocated randomly to one of the 5 possible days. What is the probability that you will end up with at least 2 exams on one of the 5 days?

Problem 1.6 Prove or disprove that

$$n - 2\binom{n}{2} + 3\binom{n}{3} - \cdots + (-1)^{n+1} n \binom{n}{n} = 0.$$

Problem 1.7 A fair die is rolled six times. What is the probability that each of the numbers 5 and 6 appears at least once?

Problem 1.8 What is the probability that in a class of 20 students exactly two students have birthdays on the same day? What is the probability that in a class of 20 students at least two students have birthdays on the same day?

Problem 1.9 Suppose that in a group of 10 Korean people, 8 speak English, 3 speak German, 3 speak French, 2 speak English and French, 1 speaks French and German, and 2 speak English and German. No one is trilingual. How many of them speak no foreign language?

Problem 1.10 Suppose that you belong to a class of $n(\geq 2)$ students. Your TA hands n graded assignments back to the students in random order. What is the probability that you are the only student who receives his or her own assignment? What is the probability that exactly two students out of n students get their own assignments back? In addition, find the limit of the probability that exactly two students out of n students get their own assignments back as $n \to \infty$.

Problem 1.11 If a set has $2n$ elements, show that it has more subsets with n elements than with any other number of elements. In addition, prove the following binomial identity

$$\binom{2n}{n} = \sum_{j=0}^{n} \binom{n}{j}^2.$$

Problem 1.12 Suppose that three dice are thrown independently. What is the probability that exactly one shows a "4"?

Problem 1.13 Find integers n and r such that the following equation is true. Here, we follow the convention that $\binom{k-1}{-1} = \binom{k-1}{k} = 0$.

$$\sum_{j=0}^{11} \left[\binom{10}{j-1} + \binom{10}{j} \right] \binom{11}{j} = \binom{n}{r}.$$

Problem 1.14 Suppose a fair die is rolled 15 times. We would like to calculate the probability that at least one of the six faces of the die never shows up in these 15 rolls.

Problem 1.15 Suppose that $P(A) = 0.6$, $P(B) = 0.3$, $P(A \cap B^c) = 0.4$, and $B \subset C$, calculate $P(A \cup B^c \cup C^c)$.

Problem 1.16 Using mathematical induction, show that

$$\sum_{k=0}^{n} (-1)^k \binom{a}{k} = (-1)^n \binom{a-1}{n},$$

where n is a natural number and a is a natural number $a \geq 1$.

Problem 1.17 KU movie cinema has n numbered seats, and n tickets are distributed among n students. What is the probability that at least two students will be seated at seats corresponding to their ticket numbers if all the seats are occupied at random?

Problem 1.18 Suppose that Seoul-metro consists of n cars, and $k(k \geq n)$ passengers get on it by selecting one of their cars randomly. Find the probability that there will be at least one passenger in each car.

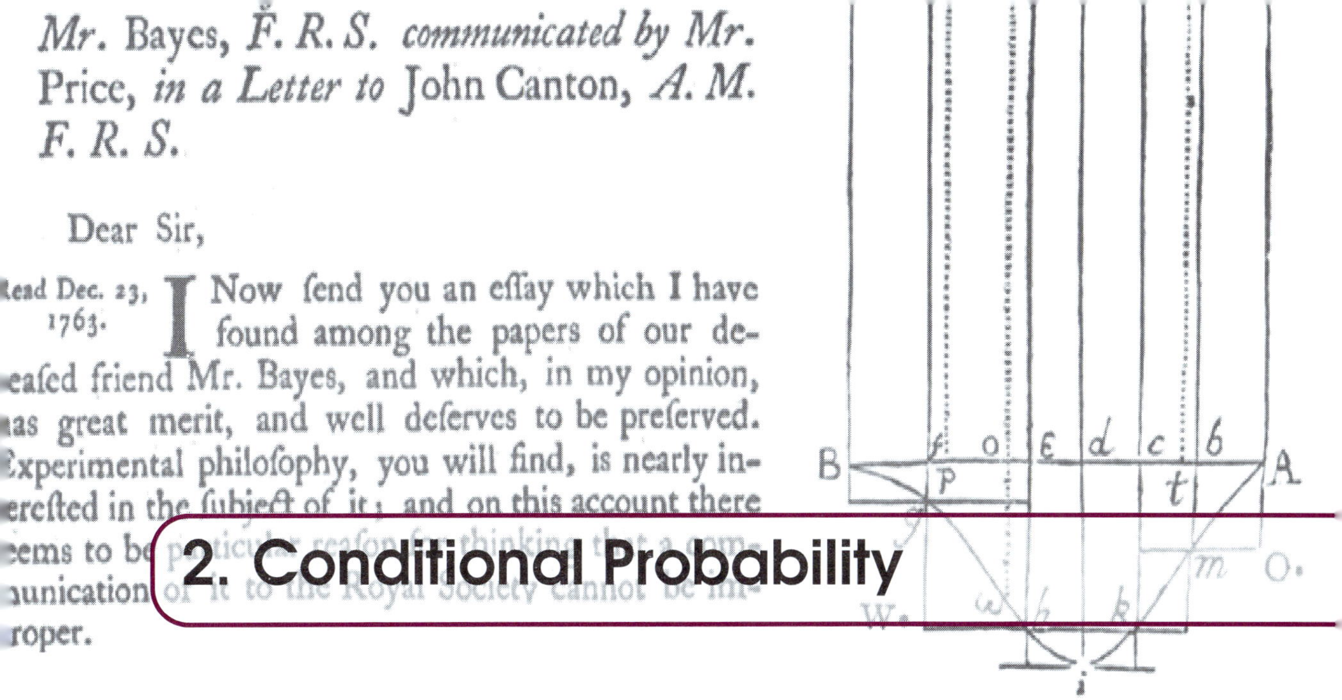

2. Conditional Probability

2.1 Conditional probability

Conditional probability provides us with a way to reason about the outcome of an experiment and update the probability of one event, after we learn another event has occurred. That is, suppose that we learn that an event B has occurred and that we wish to revise the probability of another event A taking into account that the event B has occurred. The *conditional probability of A given B* is the updated probability that we would (will) assign to A if we learn that B occurs. We denote this $P(A|B)$, the *probability of A conditional on B*.

Definition 2.1.1 The conditional probability of A given B, written $P(A|B)$ is defined by

$$P(A|B) = \frac{P(A \cap B)}{P(B)}, \quad \text{provided } P(B) > 0.$$

Remark Under the classical interpretation of probability, we can see that if we are told that B has occurred, then all outcomes in B are equally likely, and all outcomes not in B have zero probability — so B is the new (smaller) sample space.

- The number of ways that A can occur is now just the number of ways $A \cap B$ can occur, and these are all equally likely. Consequently we have

$$P(A|B) = \frac{\#(A \cap B)}{\#B} = \frac{\#(A \cap B)/\#\Omega}{\#B/\#\Omega} = \frac{P(A \cap B)}{P(B)}.$$

- Two events A and B are said to be *dependent* if they affect each other, that is, if the occurrence of event B affect the probability that the event A occurs.

- If we know that the event B has occurred, the probability that the event A will occur will be evaluated for the set of those outcomes in B that also result in the occurrence of A, precisely the set $A \cap B$ as shown in Figure 2.1. Thus, $A \cap B$ is evaluated under the new sample space B.

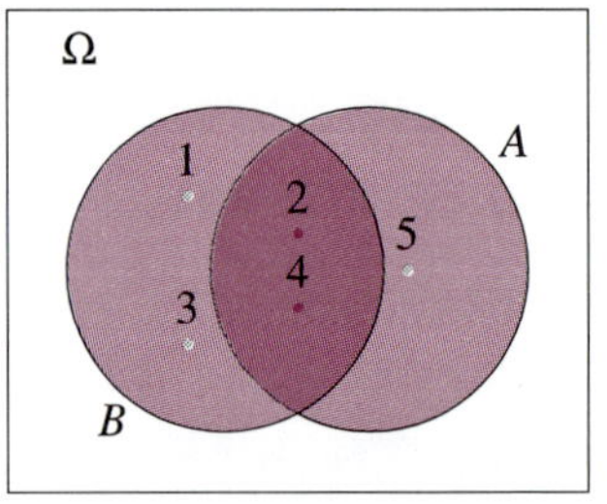

Figure 2.1: The outcomes in the event B that also belong to the event A

Example 2.1 A fair coin is flipped twice. What is the conditional probability that both flips land on heads, given that ① the first flip lands on heads ② at least one flip lands on heads? In a family with 2 children (b/g), if (a) the first child is a boy (b) at least one is a boy, what is the probability that both are boys?

- Four possible outcomes(either {HH,HT,TH,TT} of coin flips or {bb,bg,gb,gg} of two children) are all equally likely.

- Let A be the event that the both flips land on heads. Then, $P(A) = 1/4$.
- For ①, $P(B) = 2/4$, where B denotes the event that the first flip lands on heads. Thus, $P(A|B) = 1/2$.

- For ②, $P(B) = 3/4$, where B denotes the event that at least one flip lands on heads. Thus, $P(A|B) = 1/3$.

■

Remark $P(A|B)$ is the fraction of the probability of B that is also in A.

- $P(A) = P(A|\Omega) \neq P(A|B)$ in general.

- We can only condition on events with strictly positive probability.

- It is clear that $P(B^c|B) = 0$, and $P(B|B) = 1$.

- Sometimes, it is easy to determine a conditional probability from the problem description without the definition. Then we can write $P(A \cap B) = P(B)P(A|B)$.

- When $P(B) = 0$, $0 \leq P(A \cap B) \leq P(B) = 0$, and we may still write $P(A \cap B) = P(A|B)P(B) = 0$.

■ **Example 2.2** According to the genetics of color blindness, it is believed to be more common in men than in women. It is usually said that 5 men in 100 and 25 women in 10,000 suffer from the condition. Thus, the probability of being color blind is conditional on whether it is for men or women.

- Let A denote the event of being color blind and B denote the event of being male. Then, $P(A|B) = \frac{5}{100}$, $P(A|B^c) = \frac{25}{10000}$.
- Assuming $P(B) = P(B^c) = \frac{1}{2}$, we have

$$\begin{aligned} P(A) &= P(A \cap B) + P(A \cap B^c) = P(A|B)P(B) + P(A|B^c)P(B^c) \\ &= \frac{5}{100} \cdot \frac{1}{2} + \frac{25}{10000} \cdot \frac{1}{2} \neq P(A|B). \end{aligned}$$

■

■ **Example 2.3** Suppose that we have 11 boxes, each with 10 balls. Box i contains i red and $10-i$ blue balls for $i = 0, \ldots, 10$. We will select a box at random and select one ball at random from the selected box. Find the probability that we get a red ball and it comes from box 3.

- Let B be the event that we select box 3, and let A be the event that we get a red ball.
- We want $P(A \cap B)$.
- We can easily compute $P(B) = 1/11$ since all 11 boxes are equally likely. Similarly, $P(A|B) = 3/10$ since box 3 has 3 red and 7 blue balls, each equally likely.
- It follows that $P(A \cap B) = \frac{1}{11}\frac{3}{10} = \frac{3}{110}$.

■

Theorem 2.1.1 — Multiplication rule of probability. From the definition of conditional probability, we can write $P(A \cap B) = P(B)P(A|B)$, known as a *multiplication rule of probability*. The multiplication rule extends to more than two events.

- $P(A \cap B \cap C) = P(A|B \cap C)P(B|C)P(C)$.
- More generally, let $A_1, \ldots, A_n$ be events such that $P(A_1 \cap A_2 \cap \cdots \cap A_{n-1}) > 0$. Then,

$$P(A_1 \cap \cdots \cap A_n) = P(A_1)P(A_2|A_1) \cdots P(A_n|A_1 \cap \cdots \cap A_{n-1}).$$

■ **Example 2.4** A total of n balls are sequentially and randomly chosen, without replacement, an urn containing r red and b blue balls ($n \leq r + b$). Given that k of the n balls are blue, what is the conditional probability that the first ball chosen is blue?

- Let B be the event that the first ball chosen is blue and B_k be the event that a total of k blue balls are chosen. Then the desired probability is $P(B|B_k)$,

$$P(B|B_k) = \frac{P(B\cap B_k)}{P(B_k)} = \frac{P(B)P(B_k|B)}{P(B_k)} = \frac{\frac{b}{r+b}\frac{\binom{b-1}{k-1}\binom{r}{n-k}}{\binom{r+b-1}{n-1}}}{\frac{\binom{b}{k}\binom{r}{n-k}}{\binom{r+b}{n}}} = \frac{k}{n}.$$

- $P(B_k|B)$ is the probability that $k-1$ blue balls are chosen out of $n-1$ balls chosen from an urn containing $r+b-1$ balls with r red and $b-1$ blue balls.

- Alternatively, we can work with the smaller sample space with k blue balls and $n-k$ red balls and conclude that the required probability is k/n.

■

■ **Example 2.5** Suppose that an urn contains 8 red balls(R) and 4 white balls(W). We draw 2 balls from the urn without replacement. If we assume that at each draw each ball in the urn is equally likely to be chosen, what is the probability that both balls drawn are red?

- The sample space is the set of consisting four elements $\{R_1R_2, R_1W_2, W_1W_2, W_1R_2\}$. The first component of each element indicates the color of the first ball chosen and the second component the color of the second ball chosen. For example, R_1 denotes that the first ball chosen is red, and W_2 denotes the second ball chosen is white, and so on.

- $P(\{R_1\}) = P(\text{the 1st ball chosen is red}) = \frac{8}{12}$ and $P(\{W_2\}|\{R_1\}) = \frac{4}{11}$, which is the probability that the second ball chosen is white given that the first ball chosen is red. Note that if the first ball chosen is red then there remain 11 balls with 7 red and 4 white in the urn. Using multiplication rule, we have $P\{R_1W_2\} = P(\{R_1\})\cdot P(\{W_2\}|\{R_1\}) = \frac{8}{12}\times\frac{4}{11}$.

- In the same way, we have

$$P(\{R_1R_2\}) = \frac{8}{12}\times\frac{7}{11} = \frac{\binom{8}{2}}{\binom{12}{2}}, \quad P(\{R_1W_2\}) = \frac{8}{12}\times\frac{4}{11} = \frac{\binom{8}{1}\binom{4}{1}/2!}{\binom{12}{2}},$$

$$P(\{W_1W_2\}) = \frac{4}{12}\times\frac{3}{11} = \frac{\binom{4}{2}}{\binom{12}{2}}, \quad P(\{W_1R_2\}) = \frac{4}{12}\times\frac{8}{11} = \frac{\binom{4}{1}\binom{8}{1}/2!}{\binom{12}{2}}.$$

■

■ **Example 2.6** A box contains $r(\geq 7)$ red and $g(\geq 7)$ green balls. We will draw 7 balls at random without replacement. What is the probability that we get 4 red and 3 green?

- Let the sample space consist of all sequences of 7 characters each of which is either R or G. These outcomes will be sequences of draws, where R stands for a red ball and G stands for a green ball.

- Let E be the event that we draw 4 red and 3 green balls. Let R_i be the event that a red ball is drawn on draw i and let $G_i = R_i^c$. Then E is the union of $\binom{7}{4}$ disjoint events each of which is the intersection of 4 different R_i's and 3 different G_i's. One such event is $R_1 \cap R_2 \cap R_3 \cap R_4 \cap G_5 \cap G_6 \cap G_7 = \{RRRRGGG\}$, whose probability is $\frac{r}{r+g}\frac{r-1}{r+g-1}\frac{r-2}{r+g-2}\frac{r-3}{r+g-3}\frac{g}{r+g-4}\frac{g-1}{r+g-5}\frac{g-2}{r+g-6}$.

- All of the other events that make up E have the same probability. For example

$$R_1 \cap G_2 \cap G_3 \cap R_4 \cap R_5 \cap G_6 \cap R_7 = \{RGGRRGR\}$$

has probability $\frac{r}{r+g}\frac{g}{r+g-1}\frac{g-1}{r+g-2}\frac{r-1}{r+g-3}\frac{r-2}{r+g-4}\frac{g-2}{r+g-5}\frac{r-3}{r+g-6}$, which is the same as the probability of $\{RRRRGGG\}$. So, the probability of E is $\binom{7}{4}$ times the value above.

- Alternatively, we can imagine drawing all seven balls at once. Each outcome will be a subset of seven balls drawn from the $r+g$ available balls. There are $\binom{r+g}{7}$ such outcomes that are all equally likely. There are $\binom{r}{4}$ subsets of 4 red balls, and no matter which of those we get, there are $\binom{g}{3}$ subsets of 3 green balls. According to the multiplication rule, there are $\binom{r}{4}\binom{g}{3}$ subsets with 4 red and 3 green. The probability we want is then

$$\begin{aligned} P(E) &= \frac{\binom{r}{4}\binom{g}{3}}{\binom{r+g}{7}} \\ &= \binom{7}{4}\frac{r}{r+g}\frac{r-1}{r+g-1}\frac{r-2}{r+g-2}\frac{r-3}{r+g-3}\frac{g}{r+g-4}\frac{g-1}{r+g-5}\frac{g-2}{r+g-6}. \end{aligned}$$

■

■ **Example 2.7 — Picking seats with heights in a line.** Four chairs are arranged in a line, and four students randomly take seats. What is the probability that they end up in order of increasing height, from left to right?

- There are $4! = 24$ possible arrangements of the four students in the seats. But there is only one arrangement where they end up in order of increasing height. Thus, the probability is $\frac{1}{24}$.

- If we randomly assign the smallest student to a seat, there is a $\frac{1}{4}$ chance that the student ends up in the leftmost seat. Assuming that the student ends up there, there is a $\frac{1}{3}$ chance that the second tallest student ends up in the second leftmost seat (because there are only three seats left). In this way, we get the desired probability $\frac{1}{4}\cdot\frac{1}{3}\cdot\frac{1}{2}\cdot 1 = \frac{1}{24}$.

- Indeed, this is an application of multiplication rule of probability. Let A_i denote the ith smallest student ends up in the ith seat from left to right.

$$P(A_1 \cap A_2 \cap A_3 \cap A_4) = P(A_1)P(A_2|A_1)P(A_3|A_2 \cap A_1)P(A_4|A_3 \cap A_2 \cap A_1)$$
$$= \frac{1}{4} \cdot \frac{1}{3} \cdot \frac{1}{2} \cdot 1 = \frac{1}{24}.$$

■

■ **Example 2.8 — Picking seats with heights in a circle.** Suppose that four chairs are arranged in a circle and that four students randomly take seats. What is the probability that they end up in order of increasing height, going clockwise? The increasing sequence of students can start anywhere in the circle. That is, it doesn't matter which seat has the smallest student.

- As in the previous example, there are $4! = 24$ possible arrangements of the four students in the seats. But now there are four arrangements where they end up in order of increasing height since the smallest student can take four possible seats, and once the seat is picked, the positions of the other students are uniquely determined. Hence, the probability is $\frac{4}{24} = \frac{1}{6}$.

- If we randomly assign the smallest student to a seat, it can be any of all four seats in the circle with probability $1 = \frac{4}{4}$. But given that the smallest student ends up in a certain seat, the second smallest student needs to end up in the seat next to the smallest student in the clockwise direction, with probability $\frac{1}{3}$. Likewise, the third smallest student has a $\frac{1}{2}$ chance of ending up in the next seat in the clockwise direction. And then $\frac{1}{1}$ for the tallest student. The probability is therefore $1 \cdot \frac{1}{3} \cdot \frac{1}{2} \cdot 1 = \frac{1}{6}$.

■

■ **Example 2.9 — Pólya's urn.** An urn contains b black balls and r red balls. One ball is selected at random, its color noted, and the selected ball and c balls of the same color are added to the urn; the process is then repeated. Here b, r and c are positive integers. Determine the probability that, in three draws, the first two balls are black and the third is red.

- For each $j \in \mathbb{N}$, let B_j denote the event that the jth ball selected is black and let R_j denote the event that the jth ball selected is red. We want to compute $P(B_1 \cap B_2 \cap R_3)$. Applying the general multiplication rule for the case of three events, we get

$$P(B_1 \cap B_2 \cap R_3) = P(B_1)P(B_2|B_1)P(R_3|B_1 \cap B_2).$$

- Recalling that the urn initially contains b black balls and r red balls, we have that $P(B_1) = b/(b+r)$. Given that the first ball selected is black, the urn then contains $b+c$ black balls and r red balls; hence $P(B_2|B_1) = (b+c)/(b+r+c)$. And, given that the first two balls selected are black, the urn then contains $b+2c$ black balls and

r red balls; hence $P(R_3|B_1 \cap B_2) = r/(b+r+2c)$. Consequently,

$$P(B_1 \cap B_2 \cap R_3) = \frac{b}{b+r} \cdot \frac{b+c}{b+r+c} \cdot \frac{r}{b+r+2c}.$$

■

■ Example 2.10 — Matching problem (revisited). Consider the probability of the event that the student 1 is the only one who got the correct phone in the matching problem, denoted as E_1. We can compute $P(E_1)$ using the conditional probability and the multiplication rule. In this way, we can compute the probability that exactly one student gets the correct phone. More generally, we can compute the probability that exactly k students get the correct phones.

- We can write $E_1 = A_1 \cap G_1$, where A_1 is the event that student 1 gets the correct phone and G_1 is the event that none of the other $n-1$ students get the correct phone. Then, $P(E_1) = P(A_1 \cap G_1) = P(A_1)P(G_1|A_1)$.

- We computed $P(A_1) = 1/n$ earlier. Also, the conditional probability $P(G_1|A_1)$ is interpreted as the unconditional probability that nobody gets the correct phone in the matching problem with only $n-1$ people, $P(G_1|A_1) = \sum_{i=0}^{n-1} \frac{(-1)^i}{i!}$.

- The product of these two is the same as the $P(E_1)$ by multiplication rule, which we have already computed in Example 1.31,

$$\begin{aligned} P(E_1) &= \frac{\#(E_1)}{n!} = \frac{(n-1)!\left[1-\left(\frac{1}{1!}-\frac{1}{2!}+\frac{1}{3!}-\frac{1}{4!}+-\cdots+(-1)^n\frac{1}{(n-1)!}\right)\right]}{n!} \\ &= P(A_1 \cap G_1) = P(A_1)P(G_1|A_1) = \frac{1}{n}\sum_{i=0}^{n-1}\frac{(-1)^i}{i!}. \end{aligned}$$

Hence the probability that exactly one student gets the correct phone is

$$P(E) = \sum_{i=1}^{n} P(E_i) = nP(E_1) = \sum_{i=0}^{n-1} \frac{(-1)^i}{i!}.$$

- More generally, let B_k denote the event that in a particular set of k students (for example from student 1 to student k), all of them get correct phones, and let G_{n-k} be the event that none of the other $n-k$ students have a match. Then, $P(B_k) = P\left(\cap_{i=1}^{k} A_i\right)$ and $P(G_{n-k}|B_k)$ is the event that no one gets a correct phone out of $n-k$ students.

- Thus, we have

$$P(B_k) = P(A_1)P(A_2|A_1)\cdots P(A_k|\cap_{i=1}^{k-1} A_i) = \frac{1}{n}\frac{1}{n-1}\cdots\frac{1}{n-k+1} = \frac{(n-k)!}{n!},$$

$$P(G_{n-k}|B_k) = \sum_{i=0}^{n-k} \frac{(-1)^i}{i!}.$$

- Since there are $\binom{n}{k}$ ways of choosing a set of k students, we get

$$P(\text{exactly } k \text{ matches}) = \binom{n}{k} P(B_k)P(G_{n-k}|B_k) = \frac{1}{k!}\left(\sum_{i=0}^{n-k}\frac{(-1)^i}{i!}\right).$$

■

■ **Example 2.11 — No coincident birthday.** Suppose you meet two arbitrarily chosen people. What is the probability that their birthdays are different?

- Let B_2 denote the event that this happens. Then, $P(B_2) = 1 - \frac{1}{365}$.

- When the same question is asked with three people, $P(B_3) = P(A_3 \cap B_2)$, where A_3 is the event that the third person has a birthday that does not coincide with that of one of the first two persons.

- Using the multiplication rule :

$$P(B_3) = P(A_3|B_2)P(B_2) = \left(1 - \frac{2}{365}\right)\cdot\left(1 - \frac{1}{365}\right)$$

$$P(B_k) = \left(1 - \frac{k-1}{365}\right)\cdots P(A_3|B_2)P(B_2) = \prod_{i=1}^{k-1}\left(1 - \frac{k-i}{365}\right) = \frac{P_{365,k}}{365^k}.$$

■

■ **Example 2.12** Suppose that a KU cheer-leading team contains 4 students from the freshman class, 6 students from the sophomore class, 6 students from the junior class, and 4 students from the senior class. Suppose that these 20 students are randomly divided into 4 groups of 5 students. What is the probability that each group includes a senior student?

- We denote the four senior students by 1,2,3,4 and consider the following three events

$$\begin{aligned} A_1 &= \{\text{senior students 1 and 2 are in different groups}\} \\ A_2 &= \{\text{senior students 1, 2, and 3 are in different groups}\} \\ A_3 &= \{\text{senior students 1, 2, 3, and 4 are in different groups}\} \end{aligned}$$

- Now, note that

$$P(A_1) = \frac{15}{19},\ P(A_2|A_1) = \frac{10}{18},\ P(A_3|A_1 \cap A_2) = \frac{5}{17}.$$

That is, there are 15 slots for the senior student 2 other than 5 slots in the group that the senior student 1 belongs to, and there are 19 slots overall when the senior student 1's slot is taken. Similarly, there are 10 slots for the senior student 3 other than 10 slots in two groups that the senior student 1 and the senior student 2 belong to respectively, and there are 18 slots overall, excluding those for the senior students 1 and 2. The same reasoning can be applied for $P(A_3|A_1 \cap A_2)$.

- Hence, the required probability is $P(A_3) = P(A_1 \cap A_2 \cap A_3)$, and by the multiplication rule,

$$P(A_1 \cap A_2 \cap A_3) = P(A_1)P(A_2|A_1)P(A_3|A_1 \cap A_2) = \frac{15}{19} \cdot \frac{10}{18} \cdot \frac{5}{17}.$$

■

Remark Because conditional probabilities really just correspond to a new probability measure defined on a smaller sample space, this conditional probability measure $P(A|B)$ satisfies all three axioms of a probability measure. That is,

- $P(A|B) \geq 0$ for all event A.
- $P(B|B) = 1$, $P(\emptyset|B) = 0$.
- If $A_1, A_2, \ldots, A_k, \ldots$ are mutually exclusive events, then $P\left(\bigcup_{k-1}^{\infty} A_k \middle| B\right) = \sum_{k=1}^{\infty} P(A_k|B)$.

Thus, it is a probability measure with respect to the new sample space B. Therefore, every result that we learn about probability applies to conditional probability given an event B as well.

2.2 Law of total probability

Definition 2.2.1 A collection $\mathscr{P} = \{B_1, \ldots, B_k\}$ of disjoint events is called a *partition* of Ω if $\Omega = B_1 \cup \cdots \cup B_k$. That is, a partition of a sample space is simply the decomposition of the sample space into a collection of disjoint(mutually exclusive) events with positive probability. That is,

$$\Omega = \cup_{i=1}^{k} B_i, \quad B_i \cap B_j = \emptyset, \quad \forall i \neq j, \text{ where } P(B_i) > 0, \forall i.$$

We can make use of partitions to help compute probabilities of events, known as the *law of total probability*. That is, as shown in Figure 2.2, we calculate the probability of A by partitioning the sample space into disjoint events $\{B_i\}_{i=1}^{k}$ and computing the probability of the intersections of A with events $B_1, \ldots, B_k$ of a partition.

Theorem 2.2.1 — Law of total probability. Let A be an event and let $B_1, \ldots, B_k$ be a partition. We can write $A = (A \cap B_1) \cup \cdots \cup (A \cap B_k)$, and these events are disjoint. Then, we have

$$P(A) = P(B_1)P(A|B_1) + \cdots + P(B_k)P(A|B_k).$$

This is called the *law of total probability*.

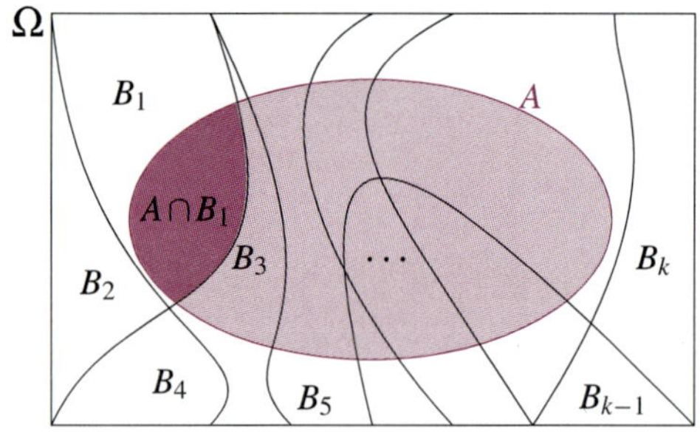

Figure 2.2: The intersections of event A with the partition $\{B_i\}_{i=1}^k$ of the sample space Ω.

Proof. It follows that $P(A) = P(A \cap B_1) + \cdots + P(A \cap B_k)$. Now, use the multiplication rule to write $P(A \cap B_i) = P(B_i)P(A|B_i)$. We now have $P(A) = P(B_1)P(A|B_1) + \cdots + P(B_k)P(A|B_k)$. ■

The law of total probability can also be applied when the sample space is decomposed by a countably infinite number of disjoint subsets. It is often the case that the unconditional probability $P(A)$ is computed easily by finding an appropriate partition $\{B_1, \ldots, B_k\}$ and conditioning on them.

■ **Example 2.13** You enter a squash tournament where your probability of winning a game is 0.3 against half the players (group 1), 0.4 against a quarter of the players (group 2), and 0.5 against the remaining quarter of the players (group 3). You play a squash against a randomly chosen opponent. What is the probability that you will win a game?

- Let B_i be the event of playing with an opponent from the group i, $P(B_1) = 0.5$, $P(B_2) = 0.25$, $P(B_3) = 0.25$.
- Let A be the event of winning. Then, we have $P(A|B_1) = 0.3$, $P(A|B_2) = 0.4$, and $P(A|B_3) = 0.5$.
- Thus, by the law of total probability, the probability that you will win a game is

$$P(A) = \sum_{i=1}^{3} P(A|B_i)P(B_i) = 0.5 \cdot 0.3 + 0.25 \cdot 0.4 + 0.25 \cdot 0.5 = 0.375.$$

■

■ **Example 2.14** Joanne is taking STAT 221, and at the end of each week she can be either up-to-date or she may have fallen behind with her classwork in STAT 221 in the week. If she is up-to-date in a given week, the probability that she will be up-to-date (or behind) in the next week is 0.8 (or 0.2, respectively). If she is behind in a given week, the probability that she will be up-to-date (or behind) in the next week is 0.6 (or 0.4, respectively). Joanne is (by default) up-to-date when she starts the class. What is the probability that she is up-to-date after two weeks?

- Let B_i and U_i be the events that Joanne is behind and up-to-date, respectively, after i weeks.

- The required probability is $P(U_2)$, and by the law of total probability, we have

$$P(U_2) = P(U_2|U_1)P(U_1) + P(U_2|B_1)P(B_1) = 0.8 \cdot 0.8 + 0.2 \cdot 0.6 = 0.76.$$

- In general, we have

$$\begin{aligned} P(U_{i+1}) &= P(U_{i+1}|U_i)P(U_i) + P(U_{i+1}|B_i)P(B_i) = 0.8P(U_i) + 0.6P(B_i) \\ P(B_{i+1}) &= P(B_{i+1}|U_i)P(U_i) + P(B_{i+1}|B_i)P(B_i) = 0.2P(U_i) + 0.4P(B_i). \end{aligned}$$

■

■ **Example 2.15 — Picking seats with heights in a line (revisited).** Three chairs are arranged in a line, and three students randomly take seats. What is the probability that the person with the middle height ends up in the middle seat?

- There $3! = 6$ possible orderings and $2!$ possible orderings that have the middle-height student in the middle seat. Thus, the required probability is $\frac{2!}{3!} = \frac{1}{3}$.

- If you want to assign the smallest student first, then there is a $\frac{1}{3}$ chance that the student ends up in the middle seat, in which case there is zero chance that the middle height student ends up there. There is a $\frac{2}{3}$ chance that the smallest student does not end up in the middle seat, in which case there is a $\frac{1}{2}$ chance that the middle-height student ends up there. Thus, by the law of total probability, we get

$$P(A) = P(A|B)P(B) + P(A|B^c)P(B^c) = 0 \cdot \frac{1}{3} + \frac{1}{2} \cdot \frac{2}{3} = \frac{1}{3},$$

  where A is the event that the middle-height student ends up in the middle seat and B is the event the smallest student ends up in the middle seat.

- Alternatively, imagine assigning the students randomly to the seats and asking the middle-height student to pick the seat first. It is obvious that there is a $\frac{1}{3}$ chance that the middle-height student ends up in the middle seat.

■

■ **Example 2.16 — Which comes first?.** The game of *craps* involves rolling dice until one or the other of two events occurs. This is a special case of running subexperiments repeatedly until one of two special events occurs. Consider rolling a pair of fair dice until either the sum 7 or the sum 8 is observed. Compute the probability that the sum 7 is observed first.

- We could try to construct a probability space with finite sequences of rolls that end in either 7 or 8. Or, we could argue that the probability is not affected by how many rolls it takes until we see the 7 or 8.

- Here are the possible outcomes from a single roll expressed as pairs of numbers giving the displayed faces of the two dice with those pairs that yield 7 or 8 underlined:

$$\begin{array}{cccccc}
(1,1) & (1,2) & (1,3) & (1,4) & (1,5) & \underline{(1,6)} \\
(2,1) & (2,2) & (2,3) & (2,4) & \underline{(2,5)} & \underline{(2,6)} \\
(3,1) & (3,2) & (3,3) & \underline{(3,4)} & \underline{(3,5)} & (3,6) \\
(4,1) & (4,2) & \underline{(4,3)} & \underline{(4,4)} & (4,5) & (4,6) \\
(5,1) & \underline{(5,2)} & \underline{(5,3)} & (5,4) & (5,5) & (5,6) \\
\underline{(6,1)} & \underline{(6,2)} & (6,3) & (6,4) & (6,5) & (6,6)
\end{array}$$

- Now, we can pretend as if the final roll is the only roll and compute the conditional probability that we roll 7 given that we roll either 7 or 8. Let F_i be the event that the final roll is i, and let $F_{i,j} = F_i \cup F_j$ be the event that the final roll is either i or j. Then the desired probability is $P(F_7|F_{7,8})$, $F_7 \cap F_{7,8} = F_7$, and there are 11 outcomes in $F_{7,8}$, and 6 outcomes in F_7. So

$$P(F_7|F_{7,8}) = \frac{P(F_7)}{P(F_7)+P(F_8)} = \frac{6/36}{11/36} = \frac{6}{11}.$$

- On the other hand, we can use the law of total probability by conditioning on the first roll. Let W be the event that the player will win, that is the event that the sum 7 is observed first. Then, by the law of total probability, we have $P(W) = \sum_{i=2}^{12} P(S_i)P(W|S_i)$, where S_i is the event that the first roll is i. Note that $P(W|S_8) = 0$ and $P(W|S_7) = 1$. For each sum of first roll $i \in \{2,3,4,5,6,9,10,11,12\}$, the game will start over and thus $P(W|S_i) = P(W)$. Therefore

$$\begin{aligned}
P(W) &= P(S_7) + P(W)(1 - P(S_7) - P(S_8)), \\
P(W) &= \frac{P(S_7)}{P(S_7)+P(S_8)}.
\end{aligned}$$

■

Remark In repeated independent trials, if A and B are mutually exclusive events, the probability that A occurs before B is $\frac{P(A)}{P(A)+P(B)}$.

■ **Example 2.17** Suppose that we roll a die and then flip that number of coins. What is the probability that we get exactly 4 tails?

- Let A be the event that we get exactly 4 tails and let B_i be the event that the die shows i, $i = 1, \ldots, 6$, which is a partition of Ω.

- Then $P(A) = \sum_{i=1}^{6} P(A \cap B_i) = \sum_{i=1}^{6} P(B_i)P(A|B_i)$, where $P(A|B_1) = 0$, $P(A|B_2) = 0$, $P(A|B_3) = 0$, $P(A|B_4) = 2^{-4}$, $P(A|B_5) = \binom{5}{4}2^{-5}$, and $P(A|B_6) = \binom{6}{4}2^{-6}$.

- Hence, $P(A) = \frac{1}{6}\left\{\frac{1}{16} + \frac{5}{32} + \frac{15}{64}\right\}$.

∎

∎ **Example 2.18 — Pólya's urn (continued).** An urn contains b black balls and r red balls. One ball is selected at random, its color noted, and the selected ball and c balls of the same color are added to the urn; the process is then repeated. Here b, r and c are positive integers. What is the probability that the first ball drawn is red given that the second ball drawn is red?

- Let R_2 denote the event that the 2nd drawn ball is red. Note that given R_1, the urn contains $r+c$ red balls and $P(R_2|R_1) = (r+c)/(b+r+c)$, whereas the urn contains r red balls given $R_1^c (= B_1)$, the event that the 1st drawn ball is black, so that $P(R_2|R_1^c) = r/(b+r+c)$. Thus,

$$\begin{aligned} P(R_2) &= P(R_2|R_1)P(R_1) + P(R_2|R_1^c)P(R_1^c) \\ &= \frac{r+c}{b+r+c}\frac{r}{b+r} + \frac{r}{r+b+c}\frac{b}{r+b} = \frac{r}{b+r} = P(R_1). \end{aligned}$$

- Hence, the probability that the first ball drawn is red given that the second ball drawn is red is $P(R_1|R_2)$,

$$P(R_1|R_2) = \frac{P(R_2|R_1)P(R_1)}{P(R_2)} = \frac{\frac{r+c}{b+r+c}\frac{r}{b+r}}{\frac{r}{b+r}} = \frac{r+c}{b+r+c}.$$

- In addition, the probability of selecting a red ball in the nth drawing is invariant with respect to n. That is, $P(R_1) = P(R_n) = \frac{r}{b+r}$, for all $n \geq 1$, which is established by mathematical induction.

- Note that $P(R_1) = \frac{r}{b+r}$ and $P(R_2) = P(R_2|R_1)P(R_1) + P(R_2|B_1)P(B_1) = \dfrac{r}{b+r}$ by the law of total probability, as before.

- Assume that $P(R_k) = \frac{r}{b+r}$ for $n = k$ and any choices of r, b and c. Then, once again by the law of total probability, it follows that

$$\begin{aligned} P(R_{k+1}) &= P(R_{k+1}|R_1)P(R_1) + P(R_{k+1}|R_1^c)P(R_1^c) \\ &= \frac{r+c}{r+b+c}\frac{r}{b+r} + \frac{r}{r+b+c}\frac{b}{r+b} = \frac{r}{r+b}, \end{aligned}$$

where the second equality holds since we can regard $P(R_{k+1}|R_1)$ as the probability that we get a red ball in the kth drawing when there are initially $r+c$ red balls and b black balls and $P(R_{k+1}|B_1)$ as the probability that we get a red ball in the kth drawing when there are initially r red balls and $b+c$ black balls.

- Alternatively, we can consider the sequences of balls in each drawing by denoting r_n as the number of red balls and b_n the number of black balls contained in the urn for the nth drawing, respectively. That is, for $n = 1$, $r_1 = r$ and $b_1 = b$. If the first ball selected is red, then $r_2 = r+c$ and $b_2 = b$ since c red balls are added to the urn for the 2nd drawing. Then, it is easy to see that $r_n + b_n = r + b + (n-1)c$ for $n \geq 1$.

- Clearly, $P(R_1) = \dfrac{r_1}{r_1 + b_1} = \dfrac{r}{r+b}$. Assume that $P(R_k) = \dfrac{r_k}{r_k + b_k} = \dfrac{r}{r+b}$ for $n = k$. Then, by the law of total probability, we have

$$
\begin{aligned}
P(R_{k+1}) &= P(R_{k+1}|R_k)P(R_k) + P(R_{k+1}|B_k)P(B_k) \\
&= \frac{r_k + c}{r_k + c + b_k} \times \frac{r_k}{r_k + b_k} + \frac{r_k}{r_k + c + b_k} \times \frac{b_k}{r_k + b_k} \\
&= \frac{r_k}{r_k + b_k} = \frac{r}{r+b}.
\end{aligned}
$$

Hence, by the mathematical induction, we establish that $P(R_n) = \frac{r}{r+b}$ for all $n \geq 1$.

■

■ **Example 2.19** Suppose that urn 1 contains ten blue marbles and eight yellow ones. In urn 2, there are six blue marbles and six yellow marbles. We draw a marble at random from urn 1, transfer it to urn 2, and draw a marble at random from urn 2. What is the probability that a blue marble is chosen from urn 2?

- Let A be the event that a blue marbel is chosen from urn 1 and B be the event that a blue marble is chosen from urn 2. Then,

$$
P(B) = P(B|A)P(A) + P(B|A^c)P(A^c) = \frac{7}{13}\frac{10}{18} + \frac{6}{13}\frac{8}{18} = 0.5042735
$$

■

R We may compute the probability of Example 2.19 by simulation using R

$$
P(B) = \frac{7}{13}\frac{10}{18} + \frac{6}{13}\frac{8}{18} = 0.5042735
$$

```
# User-defined function to compute P(B) by simulation

sim1 <- function(nreps){
 nb1 <- 10; n1 <-18; n2<-13; count <-0
 for (i in 1:nreps) {
  nb2 <- 6
 if (runif(1) < nb1/n1) nb2 <- nb2 +1
 if (runif(1) < nb2/n2) count <- count+1
}
return(count/nreps)
}

print(sim1(10^5))
[1] 0.50379
```

2.3 Bayes' theorem

■ **Example 2.20 — Medical test.** The concept of conditional probability is fundamental to the understanding of a number of routine medical practices. Suppose that a medical test is newly developed for a certain disease. The test is 99% reliable (accurate) in the following sense: If a person has the disease, there is a probability of 0.99 that the test will give a positive response; whereas, if a person does not have the disease, there is a probability of only 0.01 that the test will give a positive response. Suppose that your chances of having the disease are only 1 in 1000. What is the probability that a person who has a positive response to the test has the disease? This is a typical question for which Bayes' theorem was designed. ■

For two events A, B such that $P(B) > 0$ we have $P(A|B) = \dfrac{P(A \cap B)}{P(B)} = \dfrac{P(B|A)P(A)}{P(B)}$ by the definition of conditional probability and multiplication rule of probability. This is one of the most important and useful results with conditional probability as it tells us how to update conditional probabilities, that is, how to work out $P(A|B)$ from $P(B|A)$.

Theorem 2.3.1 — Bayes' Theorem. Suppose that we assign probabilities to events $B_1, \ldots, B_k$ in a partition and then we learn that some event A occurs. How do we revise the probabilities of the partition events given A?

$$P(B_i|A) = \frac{P(A \cap B_i)}{P(A)} = \frac{P(B_i)P(A|B_i)}{\sum_{j=1}^{k} P(B_j)P(A|B_j)},$$

which is called the Bayes' Theorem. Particularly, if the partition is simply $\{B, B^c\}$,then we have

$$P(B|A) = \frac{P(A|B)P(B)}{P(A|B)P(B) + P(A|B^c)P(B^c)}.$$

Proof. Using the definition of conditional probability and the law of total probability, we get $P(B_i|A) = \dfrac{P(A \cap B_i)}{P(A)} = \dfrac{P(A|B_i)P(B_i)}{\sum_{j=1}^{k} P(B_j)P(A|B_j)}$. The probability $P(B_k)$ is called *prior probability*. The probability $P(B_k|A)$ is called *posterior probability*. ■

Remark Thomas Bayes (1702-1761) was an English Presbyterian minister and mathematician, born in 1702 (not certain) and died April 17, 1761. In 1763, his paper titled "An Essay Toward Solving a Problem in the Doctrine of Chances" was published posthumously, which laid out the binomial probability estimation using celebrated *Bayes' theorem*, introducing the concept of inverse probability. Bayes' theorem is designed to compute the conditional probability, which was also considered in Laplace (1774), "Mémoire sur la probabilité des causes par les événements,", and Laplace (1821), "Théorie analytique des probabilités", independently of Bayes' work. Laplace used conditional probability to formulate an updated conditional probability, i.e. posterior probability from a prior probability, given evidence.

■ **Example 2.21** KU car company produces four different models: KU1, KU2, KU3, and KU4. The company has three different plants. Each plant produces the same number of cars per day, but they don't all produce the same kinds of models. For each plant, the following table tells how much of that plant's car production is of each kind.

	Model			
Plant	KU1	KU2	KU3	KU4
1	75%	25%	0%	0%
2	35%	5%	60%	0%
3	70%	0%	0%	30%

All models are gathered at a central facility each day and sorted by kind, regardless of where they were produced. Suppose that a car is selected at random from those produced on a single day before they are sorted. Let A_i be the event that the car is of model i, where $i = 1$ means the model KU1, $i = 2$ means KU2, $i = 3$ means KU3, and $i = 4$ means KU4. Let B_j be the event that the car was produced at plant j for $j = 1, 2, 3$.

- Since the car will be chosen at random from all cars produced, the proportions given in the table above can be taken as probabilities and conditional probabilities. For example, since each plant produces the same number of cars per day, $P(B_j) = 1/3$ for $j = 1, 2, 3$. The numbers in row j of the table translate to conditional probabilities of the A_i events given B_j. For example, $P(A_1|B_1) = 0.75$ and $P(A_4|B_1) = 0$.

- Since we have the conditional probabilities of the A_i's given the B_j's, and the B_j's form a partition, we can use the law of total probability to compute the probabilities of the A_i's, $P(A_i) = \sum_{j=1}^{3} P(A_i|B_j)P(B_j)$. Thus, we have

$$
\begin{aligned}
P(A_1) &= 0.75 \times \frac{1}{3} + 0.35 \times \frac{1}{3} + 0.7 \times \frac{1}{3} = 0.6, \\
P(A_2) &= 0.25 \times \frac{1}{3} + 0.05 \times \frac{1}{3} + 0 = 0.1, \\
P(A_3) &= 0 + 0.6 \times \frac{1}{3} + 0 = 0.2, \quad P(A_4) \;=\; 0 + 0 + 0.3 \times \frac{1}{3} = 0.1.
\end{aligned}
$$

- Suppose that the selected car is KU1. Then we can compute the probability that it came from plant 2, which is $P(B_2|A_1)$. By Bayes' theorem, it follows that

$$P(B_2|A_1) = \frac{P(B_2)P(A_1|B_2)}{P(A_1)} = \frac{\frac{1}{3} \times 0.35}{0.6} = \frac{7}{36}.$$

■

■ **Example 2.22** An urn contains two type A coins and one type B coin. When a type A coin is flipped, it comes up a head with probability $1/4$, whereas when a type B coin is flipped, it comes up a head with probability $3/4$. A coin is randomly chosen from the urn and flipped. Given that the flip landed on a head, what is the probability that it was a type A coin?

- Let H denote the event that the flip lands on a head and let A be the event that a type A coin is selected. Then, the desired probability is $P(A|H)$,

$$P(A|H) = \frac{P(H|A)P(A)}{P(H|A)P(A) + P(H|A^c)P(A^c)} = \frac{1/4 \times 2/3}{1/4 \times 2/3 + 3/4 \times 1/3}.$$

■

■ **Example 2.23** An urn contains three coins, c_1, c_2 and c_3. The coin c_1 has heads in both faces, c_2 has tails in both faces, and c_3 has a head in one face and a tail in the other. A coin is randomly chosen from the urn and flipped. Given that it comes heads, what is the probability that the opposite face is tails?

- Let A be the event that heads come up in the flip and C_i be the event that the coin c_i is chosen, $i = 1, 2, 3$. Then, $P(A|C_1) = 1$, $P(A|C_2) = 0$, and $P(A|C_3) = 0.5$.

- Thus, it follows from Bayes' theorem that

$$P(C_1|A) = \frac{P(A|C_1)P(C_1)}{P(A)} = \frac{P(A|C_1)P(C_1)}{\sum_{i=1}^{3} P(A|C_i)P(C_i)} = \frac{1 \cdot \frac{1}{3}}{1 \cdot \frac{1}{3} + 0 \cdot \frac{1}{3} + \frac{1}{2} \cdot \frac{1}{3}} = \frac{2}{3}.$$

- Note that $1 = P\left(\cup_{i=1}^{3} C_i|A\right) = \frac{2}{3} + P(C_3|A)$ since c_2 has tails on both faces and thus $P(C_2|A) = 0$. Also, given that it comes heads, the event that the opposite face is tails is equivalent to the event that C_3 is chosen. Hence, the required probability is $P(C_3|A) = \frac{1}{3}$.

- We might reason that the desired probability is $\frac{1}{2}$ since the opposite face is either heads or tails, which is wrong. Given that heads came up, it is more likely that the two-headed coin was chosen rather than a fair coin. The correct reasoning is to calculate the conditional probability.

■

■ **Example 2.24** Students who study for the midterm (which they do with probability 0.3) in STAT 221 are likely to get an A with probability 0.9. Students who do not study are likely to get an A with probability 0.05. Given that a student gets an A, what is the probability that the student studied?

- Let B be the event that they studied, and A be the event that a student get an A on the midterm. Then, the desired probability is $P(B|A)$,

$$P(B|A) = \frac{P(A|B)P(B)}{P(A|B)P(B) + P(A|B^c)P(B^c)} = \frac{0.9 \times 0.3}{0.9 \times 0.3 + 0.05 \times 0.7}.$$

- In answering a question on a multiple-choice problem given in the midterm of STAT 221, a student either would know the answer if the student studied for the midterm or would guess otherwise. Let p be the probability that the student knows the answer and $1 - p$ the probability that the student guesses. Assume that a student who guesses at the answer will be correct with probability $1/m$, where m is the number of multiple-choice alternatives. What is the conditional probability that a student knew the answer to a question, giving that he or she answered it correctly?

- Let B denote the event that the student answers the question correctly and let A denote the event that the student actually knows the answer. Then, the desired probability is $P(A|B)$,

$$P(A|B) = \frac{P(B|A)P(A)}{P(B|A)P(A)+P(B|A^c)P(A^c)} = \frac{1\times p}{1\times p+1/m\times(1-p)}.$$

■

■ **Example 2.25** Jaden searches for his term paper in his computer, which has n folders. Jaden knows that he stored his term paper in folder j with probability $p_j > 0$. Since there are lots of files stored in each folder, the probability that he finds the paper is only q_j even if he correctly guesses that the term paper is stored in folder j, $j = 1,\dots,n$. Suppose that Jaden searches in a particular folder, say folder k, but he could not find the term paper. Given this event, we find the probability that the term paper is stored in folder j.

- Let A_k be the event that Jaden finds his paper in folder k and let B_j be the event that Jaden's term paper is stored in folder j. Then, $P(A_k) = P(A_k|B_k)P(B_k) + P(A_k|B_k^c)P(B_k^c) = q_k p_k$ and $P(A_k^c) = 1 - q_k p_k$.

- Note that the desired probability is $P(B_j|A_k^c)$ and that $P(A_k^c|B_j) = 1$ if $j \neq k$ and $P(A_j^c|B_j) = 1 - q_j$. Hence, we have

$$P(B_j|A_k^c) = \frac{P(A_k^c|B_j)P(B_j)}{P(A_k^c)} = \begin{cases} \frac{p_j}{1-q_k p_k}, & \text{if } j \neq k, \\ \frac{(1-q_k)p_k}{1-q_k p_k}, & \text{if } j = k. \end{cases}$$

■

■ **Example 2.26 — Medical test (revisited).** For Example 2.20, we can compute the probability that a person who has a positive response to the test has the disease using Bayes' theorem.

- Let B_1 denote the event that a person has the disease, and let B_2 denote the event that a person does not have the disease, both of which forms a partition. Let A denote the event that the response to the test is positive. The required probability is $P(B_1|A)$ given by

$$\begin{aligned} P(B_1|A) &= \frac{P(A|B_1)P(B_1)}{P(A|B_1)P(B_1)+P(A|B_2)P(B_2)} \\ &= \frac{0.99\times 0.001}{0.99\times 0.001+0.01\times 0.999} \doteq 0.0901. \end{aligned}$$

■

■ **Example 2.27** Consider a communication channel involving a stream of transmitted bits (0's and 1's), where 70% of the bits are 1, and the rest 0. The channel is unstable due to physical disturbances and that bits received are sometimes distorted. Specifically, there is a chance 10% of interpreting a bit as 1 when it is actually 0, and similarly, there is a chance 15% of interpreting a bit as 0 when it is actually 1. Suppose that we received a bit and interpreted it as 1. What is the chance that it was actually transmitted as 1?

- Let B be the event a bit sent is interpreted as 1 and A be the event a bit 1 is actually transmitted. The desired probability is $P(A|B)$,

$$P(A|B) = \frac{P(B|A)P(A)}{P(B|A)P(A) + P(B|A^c)P(A^c)} = \frac{0.85 \times 0.7}{0.85 \times 0.7 + 0.1 \times 0.3}.$$

■

■ **Example 2.28** A box contains 10 green balls, 10 black balls, and 10 red balls. Two balls are selected at random without replacement from the box. Find the probability that the first ball is red given that the second ball is red.

- Let A_1 be the event that the first ball is red and A_2 be the event that the second ball is red. If the first ball is red, then 29 balls are left in the box of which 9 are red, which implies that $P(A_2|A_1) = 9/29$. Then, $P(A_1 \cap A_2) = \left(\frac{10}{30}\right)\left(\frac{9}{29}\right) = \frac{3}{29}$.

- On the other hand, if the first ball is not red, then 29 balls are left in the box of which 10 are still red. Thus, $P(A_1^c \cap A_2) = P(A_1^c)P(A_2|A_1^c) = \left(\frac{20}{30}\right)\left(\frac{10}{29}\right) = \frac{2\times 10}{3\times 29}$. Therefore, by the law of total probability, it follows that $P(A_2) = P(A_1 \cap A_2) + P(A_1^c \cap A_2) = \frac{1}{3}$, which is also obtained with the reasoning $\frac{29!\binom{10}{1}}{30!} = \frac{1}{3}$ given in Example 1.22.

- Hence, we have $P(A_1|A_2) = \dfrac{P(A_2|A_1)P(A_1)}{P(A_2)} = \dfrac{3/29}{1/3} = \dfrac{9}{29}$. Note that since $P(A_1) = P(A_2) = \cdots = \frac{1}{3}$, it follows that $P(A_1|A_2) = P(A_2|A_1)$.

■

■ **Example 2.29 — Monty Hall Problem.** Monty Hall, the host of a game show called "Let's make a deal", shows you three doors, behind one of which is a valuable prize (a luxurious car, for example). The other two have worthless stuff. You then choose one of the doors and Monty Hall opens one of the doors that you did not choose. The opened door reveals worthless stuff. Monty Hall then gives you the option to keep the door you chose or switch to the other unopened door. What should you do?

- We assume that the car is put behind one of doors equally likely.

- Monty knows where the car is, and always opens a door with a goat behind it.

- If Monty has a choice of doors (i.e., the contestant has picked the door with the car behind it), he chooses each door with probability 1/2.

- We will address this problem by computing the probabilities that your door and the "other" door have the valuable prize conditional on which door Monty Hall opens.

- Let B_1 be the event that the door you chose has the valuable prize. Let B_2 and B_3 be the events that each of the other two doors has the valuable prize. Suppose that the doors all seem the same to you so that $P(B_i) = 1/3$ before Monty Hall opens a door.

- Let C_2 and C_3 be the events that Monty Hall opens each of the other two doors that you did not choose. We know that $P(C_2|B_3) = P(C_3|B_2) = 1$. Let $P(C_j|B_1) = 1/2$ for $j = 2,3$. We want to compute $P(B_1|C_j)$ for each of $j = 2,3$. The law of total probability says that

$$P(C_2) = \sum_{j=1}^{3} P(B_j)P(C_2|B_j) = \frac{1/2}{3} + 0 + \frac{1}{3}, \quad P(C_3) = 1 - P(C_2).$$

Then, Bayes' theorem says that $P(B_1|C_j) = \dfrac{1/6}{1/6 + 1/3} = \dfrac{1}{3}$. You ought to switch.

- If we let $P(C_j|B_1) = q_j$ for $j = 2,3$. (Clearly, $q_2 + q_3 = 1$), we have

$$P(B_1|C_j) = \frac{q_j/3}{q_j/3 + 1/3} = \frac{q_j}{q_j + 1} \leq \frac{1}{2}, \quad 0 < q_j < 1.$$

So, no matter how Monty Hall chooses which door to open, when he has a choice, the probability that the "other" unopened door has the valuable prize is at least as large as the probability that your door has the prize. You ought to switch.

■

Remark If Monty has a choice of doors (i.e., the contestant has picked the door with the car behind it), he chooses each door with probability 1/2. In other words, suppose that when Monty has a choice of which door to open, he flips a fair coin. Also suppose that even when he has no choice, he still flips the coin. Then, the possible outcomes would be

Your choice	Car Location	Flip	Monty Opens
Door 1	Door 1	H	Door 2
Door 1	Door 1	T	Door 3
Door 1	Door 2	H	Door 3
Door 1	Door 2	T	Door 3
Door 1	Door 3	H	Door 2
Door 1	Door 3	T	Door 2

Table 2.1: Possible Monty Hall Outcomes : your first choice is Door 1

- Using the "stay" strategy, you will win the car with probability 1/3, since 1/3 of the time the door he picks will have the car behind it. On the other hand, if you plays the "switch" strategy, then you will win whenever the door he originally picked does not have the car behind it, which happens 2/3 of the time.

- Alternatively, we can apply the law of total probability to Monty Hall problem. Let E be the event that you get a car by using the "stay" strategy and let C_i be the event that the car is behind door i, $i = 1, 2, 3$. Then,

$$P(E) = \sum_{i=1}^{3} P(E|C_i)P(C_i) = 1 \cdot \frac{1}{3} + 0 \cdot \frac{1}{3} + 0 \cdot \frac{1}{3} = \frac{1}{3}.$$

- Imagine the same situation with 100 doors containing one luxurious car and 99 goats. After you pick a door, Monty opens 98 other doors and shows you 98 goats. In this case, if you do not switch, your chance of winning is still $\frac{1}{100}$. But if you do switch, your chance of winning is $\frac{99}{100}$, which can be seen by listing out (or imagining listing out) the 100 cases. It is clear that the switch should be made, because the only case where you lose after you switch is the case where you had initially picked the prize, and this happens only $\frac{1}{100}$ of the time. Additional information is gained by seeing 98 doors open and revealing 98 goats. There is initially a $\frac{99}{100}$ chance that the car is somewhere behind the other 99 doors, and Monty is kindly giving you the information of exactly which door it is (Monty actually must open all the (98) doors that do not contain the car but goats).

R We can perform a simulation of the Monty Hall problem using R.

```
Nsim <- 10^4
car <- sample(1:3, Nsim, replace = TRUE)
your_choice <- sample(1:3, Nsim, replace = TRUE)

Monty_open <- rep(0, Nsim); switch <- rep(0, Nsim)

for(i in 1:Nsim) {
  x <- c(1:3)[-c(car[i], your_choice[i])]
  Monty_open[i] <- x[sample.int(length(x), size = 1)]
  switch[i] <- c(1:3)[-c(Monty_open[i], your_choice[i])]
}

switch_win <- ifelse(switch == car, 1, 0)
stay_win <- ifelse(your_choice == car, 1, 0)

mean(switch_win); mean(stay_win)
[1] 0.673
[1] 0.327
```

We consider some other paradoxical probability problems similar to Monty Hall problem, applying Bayes' theorem.

■ **Example 2.30 — Puzzling Probability (1).** Your professor is known to have two children. Suppose that he is encountered walking with one of his children on the campus. If this child is a girl, what is the probability that both children are girls?

- Let $G_1(B_1)$ be the event that the first child is a girl(boy), and let $G_2(B_2)$ be the second child is a girl(boy). Let G be the child seen with your professor is a girl. The desired probability is $P(G_1 \cap G_2|G)$. Thus,

$$\begin{aligned}
P(G_1 \cap G_2|G) &= \frac{P(G_1 \cap G_2 \cap G)}{P(G)} = \frac{P(G_1 \cap G_2)}{P(G)}, \\
P(G) &= P(G|G_1 \cap G_2)P(G_1 \cap G_2) + P(G|G_1 \cap B_2)P(G_1 \cap B_2) \\
&+ P(G|B_1 \cap G_2)P(B_1 \cap G_2) + P(G|B_1 \cap B_2)P(B_1 \cap B_2), \\
P(G_1 \cap G_2|G) &= \frac{1/4}{1/4 + P(G|G_1 \cap B_2)/4 + P(G|B_1 \cap G_2)/4}.
\end{aligned}$$

Note that $P(G|G_1 \cap G_2) = 1$ since you will see the female child for sure if both children are girls, and $P(G|B_1 \cap B_2) = 0$ in the other way. Thus, the final answer depends on the assumptions we make about $P(G|G_1 \cap B_2)$ and $P(G|B_1 \cap G_2)$. For example, if the professor always choose to walk with a daughter, then $P(G|G_1 \cap B_2) = P(G|B_1 \cap G_2) = 1$.

■

■ **Example 2.31 — Puzzling Probability (2).** The French mathematician, Joseph L.F. Bertrand posed the following problem in 1889. There are three boxes. Box 1 contains two gold coins; box 2 contains two silver coins; and box 3 contains one gold and one silver coin. A box is picked at random. A coin is picked from the box and it is gold. What is the probability that the other coin in the box is also gold?

- Let B_i be the event that box i is chosen, $i = 1, 2, 3$. Let G_1 and G_2 denote events that the first coin and second coin chosen are gold, respectively. The desired probability is $P(G_2|G_1)$.

- Then, by the law of total probability and Bayes' theorem, we have

$$\begin{aligned}
P(G_1) &= \sum_{i=1}^{3} P(G_1|B_i)P(B_i) = \frac{1}{3}\left(1 + 0 + \frac{1}{2}\right) = \frac{1}{2}, \\
P(G_1 \cap G_2) &= \sum_{i=1}^{3} P(G_1 \cap G_2|B_i)P(B_i) = \frac{1}{3}(1 + 0 + 0) = \frac{1}{3}, \\
P(G_2|G_1) &= \frac{P(G_1 \cap G_2)}{P(G_1)} = \frac{1/3}{1/2} = \frac{2}{3}.
\end{aligned}$$

■

■ **Example 2.32 — Simpson's Paradox.** When tabulating discrete data, we need to be careful about aggregating groups. *Simpson's paradox*, a counterintuitive situation, occurs when the presence of the third variable is unexpectedly responsible for a change, or even reversal, of the relationship between two categorical variables.

- The following data comes from one of the studies into potential racial inequalities of the death penalty (Radelet, M. (1981). Racial characteristics and imposition of the dealth penalty, *American sociological review*, 46, 981–927).

- The 326 subjects were defendants convicted of homicide in murder trials in the state of Florida in US. The variables are the defendant's race and whether the defendant received the death penalty. The following contingency table shows that about 12% of white defendants and about 10% black defendants received the death penalty.

	Death Penalty			Percentage
Defendant's Race	Yes	No	**Total**	**Yes**
White	19	141	160	11.9
Black	17	149	166	10.2

Table 2.2: Results of death penalty from Radelet, M. (1981)

- The difference between the percentages of white defendants and black defendants who received the death penalty is small, but the percentage was lower for black defendants. Is there some explanation for these results? It turns out that the victim's race is often an important factor as shown in Table 2.3, whose values in Table 2.2 are disaggregated by victim's race. That is, overall from Table 2.2, relatively more white defendants got the death penalty. Considering victim's race, however, relatively more black defendants got the death penalty.

Victim's	Defendant's	Death Penalty			Percentage
Race	Race	Yes	No	**Total**	**Yes**
White	White	19	132	**151**	12.6
	Black	11	52	**63**	17.5
Black	White	0	9	**9**	0
	Black	6	97	**103**	5.8

Table 2.3: Table disaggregated by victim's race

- We can express Simpson's paradox in probability terms : Let A denote the event that victim's race is black, and let A^c denote the event that victim's race is white. Also, let B denote the event that defendant's race is black, and let B^c denote the event that defendant's race is white. Finally, let C denote the event that defendant will have a death penalty.

- Simpson's paradox reflects the fact that it is possible for all three of the following inequalities to hold simultaneously

$$P(C|A\cap B) > P(C|A\cap B^c),\ P(C|A^c\cap B) > P(C|A^c\cap B^c),\ P(C|B) < P(C|B^c).$$

- The law of total probability tells us why this can happen

$$\begin{aligned} P(C|B) &= P(C|A\cap B)P(A|B) + P(C|A^c\cap B)P(A^c|B), \\ P(C|B^c) &= P(C|A\cap B^c)P(A|B^c) + P(C|A^c\cap B^c)P(A^c|B^c). \end{aligned}$$

It can also be understood from the fact that $\frac{A}{B} > \frac{a}{b}$ and $\frac{C}{D} > \frac{c}{d}$ does not imply

$$\frac{A}{B} \cdot \frac{B}{B+D} + \frac{C}{D} \cdot \left(1 - \frac{B}{B+D}\right) = \frac{A+C}{B+D} > \frac{a}{b} \cdot \frac{b}{b+d} + \frac{c}{d} \cdot \left(1 - \frac{b}{b+d}\right) = \frac{a+c}{b+d}$$

because the weights $\frac{B}{B+D}$ and $\frac{b}{b+d}$ can be different so that the inequality flips.

■

2.4 Independent events

Two events are independent if knowing that one has occurred doesn't tell us anything new about whether the other will occur. There are several equivalent ways to make this precise.

Definition 2.4.1 — Independence of two events. Two events A and B are *independent* if $P(A \cap B) = P(A)P(B)$. This holds whenever either $P(A) = 0$ or $P(B) = 0$. If neither is 0, then A and B are independent if and only if $P(A|B) = P(A)$ and $P(B|A) = P(B)$.

In other words, A and B are independent if learning that one of them occurs will not change my probability for the other.

Theorem 2.4.1 If A and B are independent then A and B^c are independent.

Proof. Since A and B are independent, we have

$$P(A \cap B^c) = P(A) - P(A \cap B) = P(A) - P(A)P(B) = P(A)[1 - P(B)] = P(A)P(B^c).$$

Thus, the events A and B^c are independent. ■

Definition 2.4.2 — Independence of more than two events. Events A, B and C are *mutually* independent if

① $P(A \cap B) = P(A)P(B)$, $P(A \cap C) = P(A)P(C)$, $P(B \cap C) = P(B)P(C)$,

② $P(A \cap B \cap C) = P(A)P(B)P(C)$.

More generally, let $A_1, \ldots, A_k$ be events. They are *mutually* independent if for every subset $\{i_1, \ldots, i_j\}$ of $\{1, \ldots, k\}$ we have $P(A_{i_1} \cap \cdots \cap A_{i_j}) = P(A_{i_1}) \cdots P(A_{i_j})$.

Remark In Definition 2.4.2, the condition ① is called the *pairwise independence* of three events A, B and C.

- Pairwise independence does *not* imply (mutual) independence. That is, ① does not imply ②, and ② does not imply ①, either.

- We can easily find a counter example as in Example 2.33.

If $A_1, \ldots, A_n$ are (mutually) independent, then so are $B_1, \ldots, B_n$ where each B_i is either A_i or A_i^c.

■ **Example 2.33** Consider the results of three consecutive tosses of a fair coin. Let A be the event that the first toss results in a head, B the event that the second toss results in a head, and C the event that the third toss results in a head.

- Writing the outcomes of the three tosses in succession, the sample space consists of the eight points $\Omega = \{TTT, TTH, THT, THH, HTT, HTH, HHT, HHH\}$.

- Note that $A = \{HTT, HTH, HHT, HHH\}$, $B = \{THT, THH, HHT, HHH\}$, and $C = \{TTH, THH, HTH, HHH\}$. Thus,

$$P(A) = P(B) = P(C) = \frac{1}{2}, \; P(A \cap B) = P(A \cap C) = P(B \cap C) = \frac{1}{4}, \; P(A \cap B \cap C) = \frac{1}{8}.$$

Hence, the three events A, B, and C are independent.

■

■ **Example 2.34** Suppose that two balanced dice are rolled, one black and one gray.

A : event the black die comes up even, B : event the gray die comes up even,
C : event the sum of the dice is even, D : event the black die comes up 1,2, or 3,
E : event the black die comes up 3,4, or 5, F : event the sum of the dice is 5

- A, B and C are pairwise independent but not mutually independent since

$$\frac{9}{36} = P(A \cap B \cap C) \neq P(A)P(B)P(C) = \frac{3}{6} \cdot \frac{3}{6} \cdot \frac{18}{36}.$$

- $P(D \cap E \cap F) = P(D)P(E)P(F)$, but D, E, and F are not pairwise independent since

$$\frac{1}{6} = P(D \cap E) \neq P(D)P(E) = \frac{3}{6} \cdot \frac{3}{6}.$$

■

■ **Example 2.35** A box contains r red and g green balls. We will draw 7 balls at random with replacement. What is the probability that we get 4 red and 3 green?

- We can use the same sample space, namely all sequences of 7 characters each of which is either R or G. These outcomes will be sequences of draws, where R stands for a red ball and G stands for a green ball.

- Let E be the event that we draw 4 red and 3 green balls. Let R_i be the event that a red ball is drawn on draw i and let $G_i = R_i^c$. Then E is the union of $\binom{7}{4}$ disjoint events each of which is the intersection of 4 different R_i's and 3 different G_i's. One such event is $R_1 \cap R_2 \cap R_3 \cap R_4 \cap G_5 \cap G_6 \cap G_7 = \{RRRRGGG\}$. Using the multiplication rule, the probability of this event is $\frac{r}{r+g}\frac{r}{r+g}\frac{r}{r+g}\frac{r}{r+g}\frac{g}{r+g}\frac{g}{r+g}\frac{g}{r+g}$.

- All of the other events that make up E have the same probability. For example $\{RGGRRGR\}$ has probability $\frac{r}{r+g}\frac{g}{r+g}\frac{g}{r+g}\frac{r}{r+g}\frac{r}{r+g}\frac{g}{r+g}\frac{r}{r+g}$, which is the same as the probability of $\{RRRRGGG\}$, $\frac{r}{r+g}\frac{r}{r+g}\frac{r}{r+g}\frac{r}{r+g}\frac{g}{r+g}\frac{g}{r+g}\frac{g}{r+g}$. So, the probability of E is $\binom{7}{4}$ times the value above.

- Note that each event R_i has probability $r/(r+g)$ and each G_j has probability $g/(r+g)$. The calculations we did in the example easily generalize to show that the events $R_1, \ldots, R_7$ are mutually independent when we sample with replacement. This result easily generalizes to sampling with replacement from arbitrary finite collections with any number of different types of elements.

- If we consider sampling without replacement, then the desired probability is $\frac{\binom{r}{4}\binom{g}{3}}{\binom{r+g}{7}}$, which is equivalent to $\binom{7}{4}\frac{r}{r+g}\frac{r-1}{r+g-1}\frac{r-2}{r+g-2}\cdots\frac{g-1}{r+g-5}\frac{g-2}{r+g-6}$.

∎

Example 2.36 An infinite sequence of independent trials is to be performed. Each trial results in a success with probability p and a failure with probability $1-p$. What is the probability that at least one success occurs in the first n trials? To compute the probability of at least one success in the first n trials, we consider the complementary event that no success is obtained in the first n trials. Thus, the required probability is $1-(1-p)^n$. ∎

Example 2.37 Suppose that you roll a die and your score is the number shown on the die. On the other hand, suppose that your friend rolls five dice and his score is the number of 6's shown out of five rollings. Compute the probability that the two scores are equal.

- Let A be the event that the two scores are equal and let S_i be the event your friend gets score i, $i = 0, 1, \ldots, 5$. Then, $P(A|S_i) = \frac{1}{6}$ for $i = 1, \ldots, 5$, and $P(A|S_0) = 0$.

- By the law of total probability, we have

$$P(A) = \sum_{i=0}^{5} P(A|S_i)P(S_i) = \frac{1}{6}\sum_{i=1}^{5} P(S_i) = \frac{1}{6}[1 - P(S_0)].$$

- Note that $P(S_0) = \left(\frac{5}{6}\right)^5$ is the probability that 6 does not appear out of five rollings.

- Hence, the required probability is $P(A) = \frac{1}{6} - \frac{5^5}{6^6}$.

∎

■ **Example 2.38 — Which comes first?(revisited).** A random experiment is performed repeatedly. Events A and B are mutually exclusive. What is the probability that A occurs before B?

- We assume that $B \neq A^c$. That is, $p = P(A)$ and $q = P(B)$ with $p+q < 1$. Otherwise, the solution is just $P(A)$.

- Let E be the event that A occurs before B, and let E_k denote the event that A occurs on the kth trial and neither A nor B occurs on the first $k-1$ trials. Then, E_k's are mutually exclusive and $E = \cup_{k=1}^{\infty} E_k$.

- Thus, by the third axiom of probability and independence, it follows that
$$\begin{aligned} P(E) &= P(\cup_{k=1}^{\infty} E_k) = \sum_{k=1}^{\infty} P(E_k) \\ &= \sum_{k=1}^{\infty} p(1-p-q)^{k-1} = \frac{p}{1-(1-p-q)} = \frac{p}{p+q}. \end{aligned}$$

- Consider rolling a pair of fair dice until either the sum 7 or the sum 8 is observed. Then, the probability that the sum 7 is observed first is
$$(6/36)/(6/36+5/36) = 6/11.$$

■

Remark In repeated independent trials, if A and B are mutually exclusive events, the probability that A occurs before B is $\dfrac{P(A)}{P(A)+P(B)}$.

■ **Example 2.39 — Craps.** The popular dice game of *CRAPS* involves rolling dice until one or the other of two events occurs. If the sum on the first roll is 7 or 11, the player wins the game immediately. If the sum on the first roll is 2,3, or 12, the player loses the game immediately. If the sum on the first roll is 4,5,6,8,9, or 10, then the two dice are rolled again and again until the sum is either 7 or the original value. If the original value is obtained a second time before 7 is obtained, then the player wins. If the sum 7 is obtained before the original value is obtained a second time, then the player loses. Is the probability of winning in the game of craps is greater than 1/2?

- We compute the probability $P(W)$, where W is the event that the player will win. Let E_i be the event that the first roll is i, $i = 2, \ldots, 12$. Then, by the law of total probability, we have $P(W) = \sum_{i=2}^{12} P(E_i)P(W|E_i)$.
- For $i = 2, 3, 12$, $P(E_i) = \frac{1}{36}$, $\frac{2}{36}$, and $\frac{1}{36}$ respectively, and $P(W|E_i) = 0$ since the player loses if the first roll is 2, 3, or 12.

- For $i = 7, 11$, $P(E_i) = \frac{6}{36}$ and $\frac{2}{36}$ respectively, and $P(W|E_i) = 1$.

- For $i = 4,5,6,8,9,10$, $P(W|E_i) = \dfrac{P(E_i)}{P(E_i)+P(E_7)}$ as in Examples 2.16 and 2.38,

$$\begin{aligned} P(W|E_4) &= P(W|E_{10}) = \frac{1}{3},\ P(W|E_5) = P(W|E_9) = \frac{2}{5}, \\ P(W|E_6) &= P(W|E_8) = \frac{5}{11}. \end{aligned}$$

Thus, $P(W) = \sum_{i=2}^{12} P(E_i)P(W|E_i) \doteq 0.493$. That is, the probability of winning in the game of craps is slightly less than 1/2.

■

■ **Example 2.40 — Coupon collector's problem I.** There are n types of coupons, and each new one collected is independently of type i with probability p_i, $\sum_{i=1}^{n} p_i = 1$. Suppose k coupons are to be collected.

- If A_i is the event that there is at least one type i coupon among those collected, then, for $i \neq j$, find $P(A_i)$, $P(A_i \cup A_j)$, and $P(A_i|A_j)$.

$$\begin{aligned} P(A_i) &= 1 - P(A_i^c) = 1 - P\{\text{no coupon is type } i\} = 1-(1-p_i)^k, \\ P(A_i \cup A_j) &= 1 - P\left((A_i \cup A_j)^c\right) = 1-(1-p_i-p_j)^k, \\ P(A_i|A_j) &= \frac{P(A_i \cap A_j)}{P(A_j)} = \frac{P(A_i)+P(A_j)-P(A_i \cup A_j)}{P(A_j)}. \end{aligned}$$

■

■ **Example 2.41 — Coupon collector's problem II.** Suppose that each package of Cheetos snack contains the picture of one of twelve zodiac animal characters. The picture of each animal character is equally likely to be placed in any given package of Cheetos snack, and that pictures are placed in different packages independently of each other. Suppose that Joanne buys $k = 20$ packages of Cheetos. ① Determine the probability that at least one "Pig" character is contained in the 20 packages she buys. ② Determine the probability that Joanne will obtain a complete set of $n = 12$ different animal pictures.

- Note that the required probability for ① is the probability of A_i in Example 2.40 with $p_i = 1/12$. Thus, we focus on computing the probability for ②.

- Let A_i be the event that the ith animal character is missing out of $k = 20$ packages of Cheetos snack. Then, we need to compute $1 - P(\cup_{i=1}^{n} A_i)$, $n = 12$. Note that

$$P(A_i) = \frac{(n-1)^k}{n^k},\ \forall i,\ P(A_i \cap A_j) = \frac{(n-2)^k}{n^k},\ \forall\, i<j,\ \cdots,\ P(\cap_{i=1}^{n} A_i) = 0.$$

- Hence, by the inclusion-exclusion principle, the required probability is

$$1 - P(\cup_{i=1}^{n} A_i) = 1 - \left[n\left(\frac{n-1}{n}\right)^k - \binom{n}{2}\left(\frac{n-2}{n}\right)^k + \cdots + (-1)^n \binom{n}{n-1}\left(\frac{1}{n}\right)^k \right]$$

$$= \sum_{i=0}^{n} \binom{n}{i} (-1)^i \left(1 - \frac{i}{n}\right)^k.$$

- Alternatively, let $p_{k,r}$ be the probability that exactly r (out of a total of n) animal characters are represented among k packages; we call the event A. Note that $p_{k,n}$ is what we computed by the inclusion-exclusion formula. Let B_1 be the event that there are r different animal characters represented among the $k-1$ packages and let B_2 be the event that there are $r-1$ different animal characters represented among the $k-1$ packages. Let B_3 be the event that any other number of different animal characters occurs with $k-1$ packages.

- Then, we have $p_{0,0} = 1$, $p_{k,r} = 0$ for $0 \leq k < r$ and $p_{k,0} = 0$ for $k > 0$. In addition,

$$P(A|B_1) = \frac{r}{n},\ P(B_1) = p_{k-1,r},\ P(A|B_2) = \frac{n-r+1}{n},\ P(B_2) = p_{k-1,r-1},\ P(A|B_3) = 0.$$

- Hence, by the law of total probability, we have

$$p_{k,r} = P(A) = \sum_{i=1}^{3} P(A|B_i)P(B_i) = \frac{r}{n} p_{k-1,r} + \frac{n-r+1}{n} p_{k-1,r-1},\quad k,r \geq 1.$$

- Note that $\frac{n!}{n^n}$ for $k = n$, and 0 when $k < n$. This problem is equivalent to the probability that all of 12 birthmonths are represented in 20 students in the classroom.

■

■ **Example 2.42 — Gambler's ruin problem.** Consider the simplest game in which the gambler either wins \$1 or loses \$1 on each play. Let p be the probability that the gambler wins on a single play. Suppose that the gambler starts with \$$i$ dollars and the "bank" starts with \$$(N-i)$. That is, a total of \$$N$ is available for gambling. As soon as either the gambler or the bank has all the money, no more play is allowed. A gambler in this game is ruined when his/her fortune reaches \$0. The gambler wins if his/her fortune reaches \$$N$. Assuming that play continues until one or the other of these occurs, what is the probability that the gambler wins?

- Let a_i be the probability that the gambler wins given that the initial fortune is \$$i$. Clearly, $a_0 = 0$ and $a_N = 1$. For a_1, notice that the gambler cannot win unless he/she wins the first game, in which case the fortune goes up to \$2 and it is now the same as if the fortune had started at \$2. That is, $a_1 = pa_2$.

- In general, let W be the event that the gambler eventually wins, and let A be the event that the gambler wins the first game. Then

$$P(W) \quad = \quad P(A)P(W|A) + P(A^c)P(W|A^c) = pP(W|A) + (1-p)P(W|A^c).$$

If the gambler starts with \$$i$, then $P(W) = a_i$, $P(W|A) = a_{i+1}$, and $P(W|A^c) = a_{i-1}$. Hence, for $1 < i \le N-1$, $a_i = pa_{i+1} + (1-p)a_{i-1}$. These form a system of *difference equations*. Summarize these equations as

$$\begin{aligned}
a_1 &= pa_2 + (1-p)0, \\
a_2 &= pa_3 + (1-p)a_1, \\
a_3 &= pa_4 + (1-p)a_2, \\
&\vdots \\
a_{N-2} &= pa_{N-1} + (1-p)a_{N-3}, \\
a_{N-1} &= pa_N + (1-p)a_{N-2}.
\end{aligned}$$

- Rewrite a_i on the left as $pa_i + (1-p)a_i$ and divide each equation by p.

$$\begin{aligned}
a_2 - a_1 &= \frac{1-p}{p} a_1, \\
a_3 - a_2 &= \frac{1-p}{p}(a_2 - a_1) = \left(\frac{1-p}{p}\right)^2 a_1, \\
&\vdots \\
a_{N-1} - a_{N-2} &= \frac{1-p}{p}(a_{N-2} - a_{N-3}) = \left(\frac{1-p}{p}\right)^{N-2} a_1, \\
1 - a_{N-1} &= \frac{1-p}{p}(a_{N-1} - a_{N-2}) = \left(\frac{1-p}{p}\right)^{N-1} a_1.
\end{aligned}$$

- Add the left sides and right sides of these equations. Then, we have

$$\begin{aligned}
1 - a_1 &= a_1 \sum_{i=1}^{N-1} \left(\frac{1-p}{p}\right)^i \\
a_1 &= \frac{1}{\sum_{i=0}^{N-1} \left(\frac{1-p}{p}\right)^i} = \begin{cases} \frac{1-\left(\frac{1-p}{p}\right)}{1-\left(\frac{1-p}{p}\right)^N} & \text{if } p \ne 1/2, \\ \frac{1}{N} & \text{if } p = 1/2. \end{cases}
\end{aligned}$$

Finally,

$$a_k = \begin{cases} \frac{1-\left(\frac{1-p}{p}\right)^k}{1-\left(\frac{1-p}{p}\right)^N} & \text{if } p \ne 1/2, \\ \frac{k}{N} & \text{if } p = 1/2. \end{cases}$$

- Suppose that $p = 1/2$. Then a_i is less than or greater than 1/2 as k is less than or greater than $N/2$. The probability of winning is exactly the proportion of the total money that belongs to the gambler.

- We now turn to the problem of finding the probability of eventual ruin if the gambler is playing with an infinitely large N. Let A denote the event that he is eventually bankrupted and let $P_k(A)$ denote the probability calculated relative to the starting point k. This probability can be obtained by letting N go to ∞ in the expression for a_k calculated above. If $1-p<p$, then the ruin probability $P_k(A)=1-a_k$ approaches $(\frac{1-p}{p})^k$, and if $1-p>p$, it approaches 1. In the case $p=1-p=1/2$, we recall that $a_k=k/N$. Thus, if $N\to\infty$, we see that the probability of eventual ruin tends to 1,

$$P_k(A)=1-k/N \overset{N\to\infty}{\longrightarrow} 1.$$

■

■ Example 2.43 — Matching problem (revisited with letters and envelopes). Suppose n different letters are typed with their corresponding envelopes. If the letters are placed at random in the envelopes, we find that the probability that exactly k letters match their envelopes $p_{n,k}=\frac{1}{k!}\sum_{i=0}^{n-k}\frac{(-1)^i}{i!}$ using the law of total probability.

- Let E denote the event that no matches occur. Then $P(E)=p_{n,0}$. Let M denote the event that the first letter is correctly placed. Then,

$$p_{n,0}=P(E)=P(E|M)P(M)+P(E|M^c)P(M^c)=0\cdot\frac{1}{n}+P(E|M^c)\left(1-\frac{1}{n}\right).$$

- Conditioned on the event M^c that the first letter is not correctly placed but placed in the $j(j\neq 1)$, the event E can occur in two distinct ways depending on the matching of letter j: either letter j is matched with the envelope 1 or letter j is matched with a different envelope.

- In the first case, letter j has $n-1$ possible matches of which the envelope 1 is the correct one and so the conditional match of letter j to envelope 1 has probability $1/(n-1)$. Once letter j has been matched to envelope 1, the remaining $n-2$ letters have their own envelopes available for matching as letters j and 1 and their corresponding envelopes have been removed from the pool. Thus, the probability that E occurs with letter j matched to envelope 1 is $\frac{1}{n-1}p_{n-2,0}$.

- In the second case, with letter 1 and envelope j removed from the pool, letter j may be matched to an envelope other than 1. The probability that letter j is not matched to its envelope "1", and none of the other letters are matched to their envelopes is, by definition, $p_{n-1,0}$. Thus, by conditioning additionally upon whether letter j is matched to envelope 1 or not, we obtain

$$\begin{aligned}
P(E|M^c) &= \frac{1}{n-1}p_{n-2,0}+p_{n-1,0},\\
p_{n,0} &= P(E|M^c)\left(1-\frac{1}{n}\right)=\frac{1}{n}p_{n-2,0}+\frac{n-1}{n}p_{n-1,0},\\
p_{n,0}-p_{n-1,0} &= -\frac{1}{n}\left(p_{n-1,0}-p_{n-2,0}\right).
\end{aligned}$$

- While the recurrence holds formally for $n \geq 2$, it will be convenient to extend its domain of validity to all $n \geq 1$ with the natural choices $p_{0,0} = 1$, $p_{1,0} = 1$, and $p_{2,0} = 1/2$. Then, the general solution can be written

$$p_{n,0} - p_{n-1,0} = \left(-\frac{1}{n}\right)\left(-\frac{1}{n-1}\right)\cdots\left(-\frac{1}{2}\right)\left(-\frac{1}{1}\right)(p_{1,0} - p_{0,0}) = \frac{(-1)^n}{n!}$$

for each $n \geq 1$. As $p_{n,0} = p_{n-1,0} + \dfrac{(-1)^n}{n!}$, an easy induction now shows that

$$p_{n,0} = 1 - 1 + \frac{1}{2!} - \frac{1}{3!} + \cdots + (-1)^n \frac{1}{n!} = \sum_{i=0}^{n} \frac{(-1)^i}{i!} \qquad (n \geq 0).$$

■

There are many probability calculations that are very easy to do if we assume that certain events are independent, but we may not be comfortable assuming that they are independent. The tool of *conditional independence* allows us to use the "independence" calculations in more general settings.

Definition 2.4.3 — Conditional Independence. Two events A and B are called *conditionally independent given* C if $P(A \cap B|C) = P(A|C)P(B|C)$. Alternatively, A and B are *conditionally independent* given C if $P(A|B \cap C) = P(A|C)$.

Remark Events $A_1, \ldots, A_n$ are *conditionally independent* given B if, for every $j \leq n$ and all distinct $i_1, \ldots, i_j \in \{1, \ldots, n\}$, $P(A_{i_1} \cap \cdots \cap A_{i_j}|B) = P(A_{i_1}|B) \cdots P(A_{i_j}|B)$.

- Alternatively, events $A_1, \ldots, A_n$ are conditionally independent given B if, for every j, m such that $j + m \leq n$ and all distinct $i_1, \ldots, i_j, k_1, \ldots, k_m \in \{1, \ldots, n\}$, if E is an event that depends only on whether or not $A_{i_1}, \ldots, A_{i_j}$ occur and G is an event that depends only on whether or not $A_{k_1}, \ldots, A_{k_m}$ occur. Then $P(G|B \cap E) = P(G|B)$.

- From the definition of the conditional independence and the multiplication rule, we have

$$\begin{aligned} P(A|C)P(B|C) = P(A \cap B|C) &= \frac{P(A \cap B \cap C)}{P(C)} \\ = \quad \frac{P(C)P(B|C)P(A|B \cap C)}{P(C)} &= P(B|C)P(A|B \cap C), \end{aligned}$$

which yields the alternative definition of $P(A|C) = P(A|B \cap C)$. This relation states that if the event C is known to have occurred, the additional information that the event B also occurred does not change the probability of the event A.

■ **Example 2.44** Suppose that we inspect items sequentially while they are produced and each item has a probability p of being defective. Suppose (for now) that we assume that the items are defective or not independently of each other. That is, let A_i be the event that the ith item is defective. Then $A_1, A_2, \ldots$ are mutually independent. Suppose that we sample six of these items and record which are defective and which are not. Find the probability that exactly two of them will be defective.

- Let E be the event that exactly two of the six items are defective. Then E is the union of $\binom{6}{2}$ events each of which corresponds to two specific items being defective and the other four not.

- These events are all disjoint so the probability of the union is the sum of the probabilities. One of the events is $A_1 \cap A_2^c \cap A_3^c \cap A_4 \cap A_5^c \cap A_6^c$, whose probability is $p(1-p)(1-p)p(1-p)(1-p) = p^2(1-p)^4$. Clearly all $\binom{6}{2}$ events have this same probability, so $P(E) = \binom{6}{2}p^2(1-p)^4$.

- Let B be the event that the proportion of defectives in the very large collection equals p. We will model the six observed items as being conditionally independent given B with probability p of being defective, $P(A_1 \cap A_2 \cap A_3^c \cap A_4^c \cap A_5^c \cap A_6^c | B) = p^2(1-p)^4$.

- Suppose that p can be either 0.01 or 0.4. That is, let B_1 be the event that $p = 0.01$ and let B_2 be the event that $p = 0.4$. Let E be the event that two out of six items are defective. Then, $P(E|B_1) = \binom{6}{2}0.01^2 0.99^4$, and $P(E|B_2) = \binom{6}{2}0.4^2 0.6^4$.

■

■ **Example 2.45** Suppose that there are two coins in a box, one of which has probability 1/6 for heads, while the other has probability 5/6 for heads. One of the two is picked at random and tossed twice. Let A_1 and A_2 be the events that the first toss and the second toss, respectively, are heads and let C_1 denote the event that the first coin is picked.

- Given the event C_1, we have $P(A_1|C_1) = P(A_2|C_1) = 1/6$ and $P(A_1 \cap A_2|C_1) = 1/36$, that is to say, the events A_1 and A_2 are conditionally independent given that C_1 has occurred.

- Note that we have that $P(A_1) = P(A_2) = \frac{1}{2} \cdot \frac{1}{6} + \frac{1}{2} \cdot \frac{5}{6} = \frac{1}{2}$ by the law of total probability. Thus, $P(A_1 \cap A_2) = \frac{1}{2} \cdot \left(\frac{1}{6}\right)^2 + \frac{1}{2} \cdot \left(\frac{5}{6}\right)^2 = \frac{13}{36} \neq \frac{1}{4} = P(A_1)P(A_2)$, which implies that A_1 and A_2 are *not (unconditionally) independent*. Hence, we see that conditional independence does not imply independence and that independence does not imply conditional independence, either.

■

■ **Example 2.46** One box contains two blue balls, and another box contains one red ball and one blue ball. You pick one of the boxes at random and draw a ball n times, with replacement after each draw. If a blue ball is drawn all n times, what is the probability that you picked the box with two blue balls?

- Let B_1 be the event that you choose a box with two blue balls and B_2 be the event that you choose a box with one red ball and one blue ball. Also let F_n be the event that a blue ball is drawn all n times.

- Then $P(B_1) = P(B_2) = \frac{1}{2}$, and we also know that $P(F_n|B_1) = 1$ and $P(F_n|B_2) = \frac{1}{2^n}$ from the conditional independence. Thus, the required probability $P(B_1|F_n)$ is obtained as $P(B_1|F_n) = \frac{1 \cdot (1/2)}{1 \cdot (1/2) + (1/2)^n \cdot (1/2)} = \frac{2^n}{2^n + 1}$.

■

■ **Example 2.47 — Laplace's rule of succession.** Suppose that there are $k+1$ coins in a box. The ith coin will, when flipped, turn up heads with probability i/k, $i = 0, 1, \ldots, k$. A coin is randomly selected from the box and is then repeatedly flipped. If the first n flips all result in heads, what is the conditional probability that the $(n+1)$st flip will do likewise?

- Let C_i denote the event that the ith coin is initially selected, $i = 0, 1, \ldots, k$, let F_n denote the event that the first n flips all result in heads; and let H be the event that the $(n+1)$st flip is a head. The desired probability is $P(H|F_n)$.

- By the definition of conditional probability and conditional independence, we have

$$P(H|F_n) \quad = \quad \frac{P(H \cap F_n)}{P(F_n)} = \frac{\sum_{i=0}^{k} P(H \cap F_n|C_i)P(C_i)}{\sum_{i=0}^{k} P(F_n|C_i)P(C_i)} = \frac{\sum_{i=0}^{k} (i/k)^{n+1} \frac{1}{k+1}}{\sum_{i=0}^{k} (i/k)^n \frac{1}{k+1}}.$$

- When k is large, we see that $P(H|F_n) \approx \frac{n+1}{n+2}$ by the Riemann approximations of the integral $\frac{1}{k}\sum_{i=0}^{k}(i/k)^{n+1} \approx \int_0^1 x^{n+1}dx = \frac{1}{n+2}$, and $\frac{1}{k}\sum_{i=0}^{k}(i/k)^n \approx \int_0^1 x^n dx = \frac{1}{n+1}$. Hence, it follows that $P(H|F_n) \to 1$ as $n \to \infty$.

■

Remark Pierre Simon Laplace used these calculations (so called the Laplace's rule of succession) in 1812 as a basis for estimating the probability that the sun will rise tomorrow. He wrote "Placing the most ancient epoch of history at five thousand years ago, or at 1,826,213 days, and the sun having risen constantly in the interval at each revolution of twenty-four hours, it is a bet of 1,826,214 to 1 that it will rise again tomorrow."

■ **Example 2.48 — The secretary (best candidate) problem.** You wish to hire one secretary out of n candidates, interviewed in a random order. We assume that the best candidate is equally likely to be each of the n candidates in the sequence before the interviews start. Candidates are interviewed one by one, and at the interview you evaluate the candidate to give the job to the candidate with the highest score. After each interview, it is required that either we hire the current candidate immediately and stop the interviews, or we must let the current candidate go and we never call them back.

- Your strategy is to reject the first $0 \le m < n$ candidates after seeing them, regardless of how good they are. Starting with the next candidate $m+1$, we continue interviewing until the current candidate is the best we have seen so far. We then stop and hire the current candidate. If none of the candidates from $m+1$ to n is the best, we just hire n. We would like to compute the probability that we hire the best candidate.

- We start with a simple situation where $n=3$ and suppose that the scores of the three candidates are 10, 20 and 30. There are $3! = 6$ possible orders in which the candidates could be seen, $(10,20,30), (10,30,20), \ldots (30,20,10)$. If you simply chooses the first candidate or the last or any other particular candidate, then the probability of choosing the best candidate is $\frac{1}{3}$. But, suppose you let the first candidate leave and chooses the next one with a higher score, then the probability of selecting the most desirable candidate is $\frac{1}{2}$ with three cases $(10,30,20)$, $(20,10,30)$, $(20,30,10)$. In addition, skipping the first two candidates leads to the correct choice only in two cases $(10,20,30)$ and $(20,10,30)$ with probability $\frac{1}{3}$. Thus, for three candidates $(n=3)$ the best strategy is to reject the first candidate $(m=1)$ to be interviewed in any case and choose the next one with a higher score if this is possible, or otherwise choose the final candidate interviewed.

- Let A be the event that we hire the best candidate and let B_i be the event that the best candidate is in position i in the sequence of interviews. Define C_i be the event that the ith candidate is hired, that is, the event we keep interviewing until we see candidate i.

- $B_i \cap C_i \subset A$ and $B_i \cap C_i^c \cap A = \emptyset$ since if C_i fails, then we will hire someone before we interview candidate i when the best candidate is in position i, and thus A will not occur. Then, note that $B_i \cap A = B_i \cap A \cap C_i = B_i \cap C_i$, so $P(A|B_i) = P(C_i|B_i)$.

- Note that $P(B_i) = \dfrac{(n-1)!}{n!} = \dfrac{1}{n}$ and $P(C_i|B_i) = \dfrac{\binom{m}{1}(i-2)!}{(i-1)!} = \dfrac{m}{i-1}$ since conditional on B_i, C_i occurs if and only if the best of the first $i-1 (\ge m)$ candidates appears in the first m discarded positions that could not be hired. That is, $P(A|B_i) = 0$ for $i \le m$ and $P(A|B_i) = P(C_i|B_i) = \dfrac{m}{i-1}$, $i > m$. Thus, if $m > 0$, $p_{m,n} = P(A)$ is computed as

$$p_{m,n} = P(A) = \sum_{i=1}^{n} P(B_i)P(A|B_i) = \sum_{i=1}^{n} P(B_i)P(C_i|B_i) = \frac{m}{n} \sum_{i=m+1}^{n} \frac{1}{i-1}.$$

■

R In the secretary problem of Example 2.48, we can find m to make this probability as large as possible. That is, to determine the value of m that maximize the probability $p_{m,n}$, we need to look into the probability further.

- Note that $p_{0,n} = \frac{1}{n}$. Using the formula for $p_{m,n}$, we have $q_{m,n} = p_{m,n} - p_{m-1,n} = \frac{1}{n}\left(\sum_{i=m+1}^{n} \frac{1}{i-1} - 1\right)$, which decreases as m increases. For example, when $n=10$, the first few values of $q_{m,n}$ are

m	1	2	3	4
$q_{m,n}$	0.1829	0.0829	0.0329	-0.0004

```
q_mn <- function(m,n) {
  j <- (m+1):n
  return((1/n)*(sum(1/(j-1))-1)) }
n <- 10; q_mnvec <- numeric(n)
for (i in 1:n)  q_mnvec[i] <- q_mn(i,n)

> q_mnvec[1:4]
[1]  0.1828968254  0.0828968254  0.0328968254 -0.0004365079
```

- Since $p_{m,n} = q_{m,n} + p_{m-1,n}$ for $m \geq 1$, we have that $p_{m,n} = p_{0,n} + q_{1,n} + \cdots + q_{m,n}$. If there exists m such that $q_{m,n} \leq 0$, then $q_j < 0$ for all $j > m$ and $p_{j,n} \leq p_{m-1,n}$ for all $j \geq m$. On the other hand, for each m such that $q_{m,n} > 0$, $p_{m,n} > p_{m-1,n}$. Hence, we should choose m to be the last value such that $q_{m,n} > 0$. Thus, we need to reject the first $m = 3$ candidates to maximize p_m when $n = 10$. We can then compute $p_{m,n} = 0.3987$.

```
p_mn <- function(m,n) {
  j <- (m+1):n
  return((m/n)*sum(1/(j-1))) }
n <- 10; p_mnvec <- numeric(n); p_0n=1/n
for (i in 1:n)  p_mnvec[i] <- p_mn(i,n)

p_mnvec[1:4] # the same with p_0n+cumsum(q_mnvec)
[1] 0.2828968 0.3657937 0.3986905 0.3982540
which.max(p_mnvec)
[1] 3
```

- Assume that n is sufficiently large and $m/n \to t$, $0 < t < 1$, as $n \to \infty$. Then, we can approximate $p_{m,n} \to t \int_t^1 \frac{1}{x} dx = -t \log t$, $0 < t < 1$ as $n \to \infty$. Note that the function $f(t) = -t \log t$ is maximized by $e^{-1} \approx 0.368$ and that the corresponding probability is $p_{m,n} = f\left(e^{-1}\right) = e^{-1}$.

- Thus, we need to reject the first $m = ne^{-1}$ after seeing them, regardless of how good they are. In other words, the optimal strategy is for you to skip approximately the first 37% of candidates automatically. For example, if $n = 100$, the probability is maximized for $m = 37$, then you would interview the first 37 candidates and reject them, and then choose the next candidate whose score exceeds that of the preceding candidates.

```
n <- 100; p_mnvec <- numeric(n); p_0n=1/n
for (i in 1:n)  p_mnvec[i] <- p_mn(i,n)

which.max(p_mnvec)
[1] 37
```

- In addition to the exact form of the optimal proportion $1/e$ with a Riemann approximation to the probability, we can consider simulations and obtain the approximate probability, the optimal proportion $1/e \doteq 0.37$.

```
Nsim <- 1000; counts <- numeric(Nsim);
for (i in 1:Nsim) {
n <-100; score <- runif(n,10,100)
best.score <- max(score); m <- n/exp(1)
cutoff.score <- max(score[1:m])
candidate.index <- m + which(score[m+1:n] > cutoff.score)[1]
if (is.na(candidate.index)) { candidate.index <- n }
candidate.score <- score[candidate.index]
which(score==candidate.score)
counts[i] <- (candidate.score == best.score)}
mean(counts)
[1] 0.37
```

- We can find the optimal number m with n candidates by simulations.

```
simulate.candidates <- function(n,m,Nsim) {
for (i in 1:Nsim) {
score <- runif(n,10,100)
best.score <- max(score)
cutoff.score <- max(score[1:m])
candidate.index <- m + which(score[m+1:n] > cutoff.score)[1]
if (is.na(candidate.index)) {
candidate.index <- n
}
candidate.score <- score[candidate.index]
which(score==candidate.score)
counts[i] <- (candidate.score == best.score)
}
return(mean(counts))
}
simulate.candidates(100,37,1000)
#optimal number is m=37 when n=100
[1] 0.387
> simulate.candidates(100,7,1000)
[1] 0.207
> simulate.candidates(100,35,1000)
[1] 0.366
> simulate.candidates(100,36,1000)
[1] 0.381
> simulate.candidates(100,38,1000)
[1] 0.368
```

2.5 Exercises

Problem 2.1 Suppose we roll a fair die. If the result is 1 or 2, we roll once more but otherwise, we stop. What is the probability that the sum total of our rolls is at least 4?

Problem 2.2 If A, B, and C are mutually independent events, show that $A \cup B$ and C are independent.

Problem 2.3 A box contains 10 green balls, 6 black balls, and 14 red balls. Two balls are selected at random without replacement from the box. What is the probability that both balls have the same color?

Problem 2.4 Suppose A, B, and C are events of strictly positive probability in some probability space. If $P(A|C) > P(B|C)$ and $P(A|C^c) > P(B|C^c)$, is it true that $P(A) > P(B)$? If $P(A|C) > P(A|C^c)$ and $P(B|C) > P(B|C^c)$, is it true that $P(A \cap B|C) > P(A \cap B|C^c)$?

Problem 2.5 Jaden is taking a test. There are two questions he is stumped on and he decides to guess. Let A be the event that he gets the first question right; let B be the event he gets the second question right.

① Obtain an expression for p_1, the probability that he gets both questions right conditional on getting the first question right.

② Obtain an expression for p_2, the probability that he gets both questions right conditional on getting either of the two questions right (A or B).

③ Prove or disprove that $p_1 \leq p_2$.

Problem 2.6 Suppose $P(A) = 1/2$, $P(B|A \cap C) = 1/3$ and $P(C|A) = 1/4$. Find $P(A \cap B^c \cap C)$.

Problem 2.7 Consider flipping coins until either two tails ('TT') or tails then heads ('TH') first occurs. Find the probability that 'TH' occurs before 'TT'.

Problem 2.8 (A Card game called *Treize*) Consider a well-shuffled deck of n cards, labeled 1 through n. You flip over the cards one by one, saying the numbers 1 through n as you do so. You win the game if, at some point, the number you say aloud is the same as the number on the card being flipped over. Prove or disprove that the probability of winning is greater than the probability of losing as the number of cards increases, i.e. $n \to \infty$.

Problem 2.9 (A variation of the Monty Hall problem) Monty shows you three identical doors. One of them contains a luxurious car, and the other two contain goats. You choose one of the doors but do not open it, as given in Example 2.29. This time, however, Monty does not know the location of the car. He randomly choose one of the two doors different from your selection and opens it. The door turns out to conceal a goat. He now gives you the option either of sticking with your original door or switching to the other one. What should you do?

Problem 2.10 Lewis Carroll, the author of Alice's Adventures in Wonderland, is the pen name of Charles Lutwidge Dodgson, who was an Oxford mathematician and logician. Lewis Carroll's Pillow Problems (1958), is a collection of 72 challenging, and sometimes amusing, mathematical problems, several of which involve probability. Here is one probability question in Lewis Carroll's Pillow Problems :

- A bag contains one marble, known to be either white or black. A white marble is put in, the bag shaken, and a marble drawn out, which proves to be white. What is now the probability of drawing a white marble?

Problem 2.11 Suppose that A and B are events such that $P(A|B) = P(B|A)$ and $P(A \cup B) = 1$ and $P(A \cap B) > 0$. Show that $P(B) > 1/2$.

Problem 2.12 Prove or disprove

① If for the events A and B, $P(A)P(B) > 0$, then $P(A|B) > P(A)$ if and only if $P(B|A) > P(B)$.

② If for three events A, B and C, $P(A|B) < P(B|C)$ and $P(A|C^c) < P(B|C^c)$, then $P(A) < P(B)$.

③ If for three events A, B and C, $P(A|B \cap C) < P(A|B^c \cap C)$ and $P(A|B \cap C^c) < P(A|B^c \cap C^c)$, then $P(A|B) < P(A|B^c)$.

④ If for three events A, B and C, $P(A|C) \geq P(B|C)$ and $P(A|C^c) \geq P(B|C^c)$, then $P(A) < P(B)$.

Problem 2.13 Suppose that there are k boxes and that each of k boxes contains r red balls and g green balls. A ball is randomly chosen from box 1 and transferred to box 2, then a ball is randomly chosen from box 2 and transferred to box 3, and so on. Finally, a ball is randomly chosen from box k. Show that the probability that the last ball is red is $r/(r+g)$. In fact, the probability that the ball chosen from box i is $r/r+g$ for all $i = 1, \ldots, k$.

Problem 2.14 Jaden flips 3 coins and Joanne flips 2 coins. Jaden wins if the number of heads he gets is more than the number Joanne gets. What is the probability that Jaden will win?

3. Discrete random variables

In this chapter, we introduce *discrete random variables*. Some properties are discussed for discrete random variables, such as distributions, means, variances, moments, and joint distributions of such variables, with most of these properties actually hold for random variables in general.

3.1 Discrete random variables

A real-valued function X defined on the sample space (i.e., a possibly unknown real number) is a *random variable*. The *distribution* of X is the collection of all possible probability statements about X. Each random variable has a function that summarizes its distribution. The distribution of a random variable tells us how likely every possible value is to occur. Thus, the distribution is also called the "probability distribution".

Definition 3.1.1 A **random variable** is a function from the sample space Ω to the real numbers, $X : \Omega \to \mathbb{R}$ that assigns a real number $X(\omega)$ to each outcome $\omega \in \Omega$. We write

$$P(X \in B) = P(\{\omega \in \Omega : \ X(\omega) \in B\}),$$

and $X \sim P$ means that X has a distribution P.

Remark The terminology *random* variable would be appropriate because the domain of X is Ω, and we have a probability measure on subsets of Ω. A random variable has an associated probability distribution and can be regarded "a variable which takes its value according to a probability distribution".

- A random variable X needs to be defined as a function on Ω so that its probability distribution is explained with an underlying probability space.

- For example, we can consider an experiment of tossing two fair coins. Then, there are four possible outcomes (elements), $\Omega = \{\text{HH}, \text{HT}, \text{TH}, \text{TT}\}$, we can define X to be the number of heads, which maps the sample space $\Omega = \{\omega_1, \omega_2, \omega_3, \omega_4\}$ into the real line R, $X(\omega) : \Omega \to \{0, 1, 2\} \subset R$, as shown in Figure 3.1.

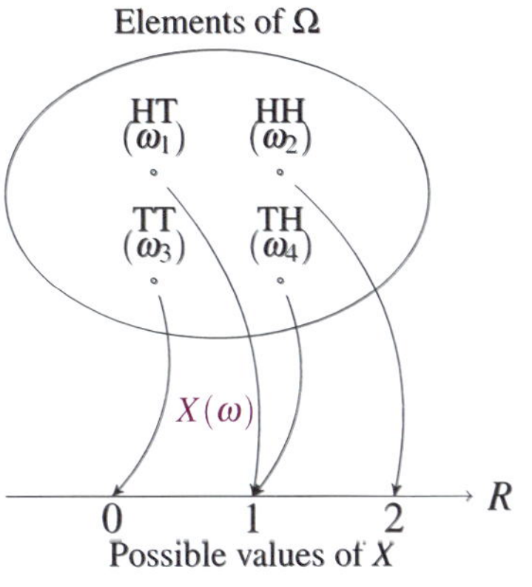

Figure 3.1: An example of random variable X that maps the sample space into the real line.

- We can compute $P(X \in B) = P(\{\omega : X(\omega) \in B\})$ for any $B \subset R$ such that $\{\omega : X(\omega) \in B\} = X^{-1}(B) \in \mathscr{F}$. Technically, a random variable is a measurablc function, i.c. $\{\omega \subset \Omega : X(\omega) \leq x\} \in \mathscr{F}$ for each $x \in \mathbb{R}$.

- By $\{X = x\}$ we mean the event $\{\omega \in \Omega | X(\omega) = x\}$. Similarly, $\{a < X < b\}$ means the event $\{\omega \in \Omega | a < X(\omega) < b\}$.

There are three types of random variables: discrete, continuous, and mixed. However, in most cases, we encounter either discrete or continuous random variable. In this chapter, we consider the discrete random variable.

Definition 3.1.2 A random variable that takes at most countably many different values is called a *discrete* random variable. Let $x_1, x_2, \ldots$ contain all of the possible values of a discrete random variable X. The *probability (mass) function* (p.f./p.m.f.) of X is

$$f_X(x) = P(X = x) = P(\{\omega : X(\omega) = x\}), \text{ for all real } x.$$

$$P(X \in B) = \sum_{\text{all } x_i \in B} f_X(x_i).$$

Remark We always use upper case letters (such as X, Y, Z, and others) to denote random variables and lower case letters (x, y, z, and so on) to denote their possible numerical values.

If X is discrete, there are at most countably many values of x such that $f_X(x) > 0$ and the corresponding values of $f(x)$ must add to 1. That is, let $x_1, x_2, \ldots$ include all the possible values of X. Then $\sum_{i=1}^{\infty} f_X(x_i) = 1,\ f_X(x_i) \geq 0.$

Definition 3.1.3 — Cumulative distribution function. Every random variable has a (cumulative) distribution function (c.d.f/d.f.) defined by $F_X(x) = P(X \leq x)$, for all $x \in R$. If X has a discrete distribution with p.m.f. f_X, then $F(x) = \sum_{t \leq x} f_X(t)$.

Example 3.1 Consider the experiment of tossing a coin three times. An outcome of this experiment must indicate the result of each tosses. For example, HHT could indicate that two heads and then a tail were observed. The sample space for this experiment has eight points, namely, {HHH, HHT, HTH, THH, TTH, THT, HTT, TTT}.

- Define the random variable X to be the number of heads in the three tosses.
- X can have values 0,1,2,3, and the probability mass function of X is given by

$$\begin{aligned}
f(0) &= P(X=0) = \frac{1}{8},\ f(1) = P(X=1) = \frac{3}{8}, \\
f(2) &= P(X=2) = \frac{3}{8},\ f(3) = P(X=3) = \frac{1}{8}, \\
f(x) &= P(X=x) = \binom{3}{x}\left(\frac{1}{2}\right)^x\left(\frac{1}{2}\right)^{3-x},\ x = 0,1,2,3.
\end{aligned}$$

- Then, the c.d.f. of X is $F(x) = \begin{cases} 0 & \text{if } -\infty < x < 0, \\ 1/8 & \text{if } 0 \leq x < 1, \\ 1/2 & \text{if } 1 \leq x < 2, \\ 7/8 & \text{if } 2 \leq x < 3, \\ 1 & \text{if } 3 \leq x < \infty. \end{cases}$

As shown in the following plot, the c.d.f of X is in between 0 and 1, and nondecreasing with jumps and flat regions.

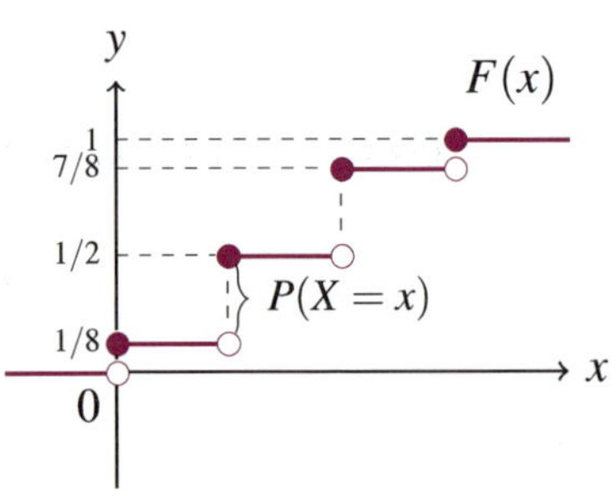

Figure 3.2: The graph of c.d.f. of X in Example 3.1

Note that for discrete distributions, the c.d.f. is a step function that jumps $P(X = x)$ (the p.f.) at each possible value x. If the c.d.f. of X jumps at a value x, the size of the jump is the probability (mass) that $X = x$. ■

Theorem 3.1.1 Every random variable has a c.d.f. that satisfies the following properties:

- $\lim_{x \to -\infty} F(x) = 0$, and $\lim_{x \to \infty} F(x) = 1$.
- F is continuous from the right (right-continuous) : for all x,

$$F(x^+) = \lim_{t \to x,\, t > x} F(t) = \lim_{t \downarrow x} F(t) = F(x).$$

- F is nondecreasing : $x_1 < x_2$ implies that $F(x_1) \le F(x_2)$.

Proof. We use the basic properties for probability given in Chapter 1 to establish three properties of $F(x) = P(X \le x)$ by rewriting $\{X \le x\}$ in terms of suitable events.

- Let a_n, $n = 1, 2, \ldots$, denote any increasing sequence diverging to ∞, say, $a_n = n$, and let A_n denote the event that $\{X \le a_n\}$. Then $P(A_n) = F(a_n)$ and $A_1 \subset A_2 \subset \cdots \subset \cup_{n=1}^{\infty} A_n = \{X < \infty\}$. Hence, we have

$$\lim_{x \to \infty} F(x) = \lim_{n \to \infty} F(a_n) = P\left(\lim_{n \to \infty} (A_n)\right) = P(\cup_{n=1}^{\infty} A_n) = P(X < \infty) = 1.$$

Similarly, we have $\lim_{x \to -\infty} F(x) = 0$.

- Let a_n, $n = 1, 2, \ldots$, denote any decreasing sequence converging to 0, say, $a_n = 1/n$, and let A_n denote the event that $\{X \le x + a_n\}$. Then, $A_1 \supset A_2 \supset \cdots \supset \cap_{n=1}^{\infty} A_n = \{X \le x\}$. Hence, we have

$$P(X \le x) = F(x) = P(\cap_{n=1}^{\infty} A_n) = \lim_{n \to \infty} P(A_n) = \lim_{n \to \infty} F(x + a_n) = F(x^+).$$

- Let $A_1 = \{X \le x_1\}$ and $A_2 = \{X \le x_2\}$. Then, $A_1 \subset A_2$, and hence

$$F(x_1) = P(A_1) \le P(A_2) = F(x_2).$$

■

Remark Every function that satisfies those properties in Theorem 3.1.1 is the c.d.f. of some random variable. If we define the event $B_n = \{X \le x - 1/n\}$, then we have

$$P(X < x) = P(\cup_{n=1}^{\infty} B_n) = P\left(\lim_{n \to \infty} B_n\right) = \lim_{n \to \infty} P(B_n) = \lim_{n \to \infty} F\left(x - \frac{1}{n}\right) = \lim_{t \uparrow x} F(t) = F(x^-).$$

- $f(x) = P(X = x) = P(X \le x) - P(X < x) = F(x) - F(x^-)$.
- $P(a < X \le b) = P(X \le b) - P(X \le a) = F(b) - F(a)$.

- $P(a < X < b) = P(X \le b) - P(X = b) - P(X \le a) = F(b^-) - F(a)$, where $F(b^-) = \lim_{t \to b,\, t<b} F(t) = \lim_{t \uparrow b} F(t) = \lim_{h \downarrow 0} F(x-h)$.

- We often use $F_X(x)$ and f_X to denote the c.d.f. and p.m.f. of a random variable X, respectively. If there is no confusion, we omit the subscript X.

- If X and Y have the same distribution, $F_X(t) = F_Y(t)$ for all $-\infty < t < \infty$, then we say that X and Y are equal in distribution and write $X \overset{d}{=} Y$.

■ **Example 3.2** Suppose that an urn contains n tickets with numbers from 1 to n and that we withdraw a ticket at random r times with replacement. Let X be the largest number withdrawn. Find $f_X(x)$ and $F_X(x)$. Let Y be the largest number withdrawn when the drawn tickets are not replaced (without replacement). Find $f_Y(y)$ and $F_Y(y)$.

- X can have possible values from $x = 1, \ldots, n$.

- Since the number of ways of choosing r numbers less than or equal to x is x^r and there are n^r outcomes, we have
$$P(X \le x) = F_X(x) = \left(\frac{x}{n}\right)^r, \ x = 1, \ldots, n.$$
For any real x, $F_x(x) = ([x]/n)^r$, where $[x]$ denotes the largest integer which is not greater than(equal or less than) x.

- Hence, for integer x and $1 \le x \le n$, we have
$$f_X(x) = F(x) - F(x^-) = \left(\frac{x}{n}\right)^r - \left(\frac{x-1}{n}\right)^r,$$
and elsewhere $f_X(x)$ is zero.

- Without replacement, it is easy to get $P(Y = y) = \dfrac{\binom{y-1}{r-1}}{\binom{n}{r}}$, $y = 1, \ldots, n$.

- Since the number of ways of choosing r different numbers less than or equal to y without replacement is $\binom{y}{r}$ for $y = 1, \ldots, n$, we have $F_Y(y) = \dfrac{\binom{y}{r}}{\binom{n}{r}}$. Alternatively, for integer $y = 1, \ldots, n$,
$$F_Y(y) = P(Y \le y) = \sum_{k=1}^{y} P(Y = k) = \sum_{k=1}^{y} \frac{\binom{k-1}{r-1}}{\binom{n}{r}} = \frac{\binom{y}{r}}{\binom{n}{r}},$$
which follows from Theorem 1.3.5.

■

■ **Example 3.3 — Bernoulli trial.** Now consider a simple experiment with two outcomes: success, with probability p, and failure, with probability $q = (1 - p)$. Such an experiment is called a Bernoulli trial.

- A Bernoulli trial is a random experiment in which there are precisely two possible outcomes, which we conveniently call 'failure' (F) and 'success' (S).

- We can define a random variable from the sample space $\{S, F\}$ into the set of real numbers as follows: $X(F) = 0$, and $X(S) = 1$.

- Let X be a random variable that takes only the values 0 and 1 with $P(X = 1) = p$. The distribution of X is called the *Bernoulli distribution with parameter* p, $X \sim \text{Bernoulli}(p)$ with p.m.f

$$f_X(x) = P(X = x) = p^x(1-p)^{1-x},\ x = 0, 1.$$

■

■ Example 3.4 — Binomial Random Variable. Perform a *fixed* number n of replicated trials, all of which are *independent* and have the *same probability* p of satisfying a particular criterion. For each trial, record a *binary* response (e.g., success or failure, yes or no) indicating whether or not the criterion was met. Then, the count of how many trials for which the criterion was met has what is called a Binomial(n, p) distribution.

- A random variable X is called a binomial random variable if it represents the total number of successes out of n independent Bernoulli trials.

- If we have x successes in n trials, then the probability of each n-tuple with x successes and $n-x$ failures is $p^x(1-p)^{n-x}$. However, there are $\binom{n}{x}$ tuples with x successes and $n-x$ failures in n trials. Hence, the probability mass function of X is

$$f(x) = P(X = x) = \binom{n}{x} p^x (1-p)^{n-x}, \qquad x = 0,\ 1, \ldots,\ n.$$

- Note that $f(x) \geq 0$ and $\sum_{x=0}^{n} f(x)$ is 1,

$$\sum_{x=0}^{n} f(x) = \sum_{x=0}^{n} \binom{n}{x} p^x (1-p)^{n-x} = (p+1-p)^n = 1,$$

which implies that $f(x)$ is a p.m.f.

- The Bernoulli distribution with parameter p, $X \sim \text{Ber}(p)$, is the binomial distribution with parameters 1 and p, $X \sim \text{Binomial}(1, p)$.

- X is the number of "successes" that we will achieve in n independent trials, where each trial can be either a success or a failure, each with the same probability p of success. We can also say that X is a sum of multiple independent Ber(p) random variables. Let $X \sim \text{Binomial}(n, p)$ and $Z_i \sim \text{Ber}(p)$, where all of the Bernoullis are independent. Then, we can express $X = \sum_{i=1} Z_i$.

- Suppose that $X \sim \text{Binomial}(n,p)$ and $Y \sim \text{Binomial}(n,q)$, where $p+q = 1$. Then $P(X \leq x) = 1 - P(Y \leq n-x-1)$ since $X+Y = n$ and $P(X \leq x) = P(Y \geq n-x)$.

We can also check by writing the corresponding c.d.f

$$P(Y \le n-x-1) = \sum_{i=0}^{n-x-1} \binom{n}{i} q^i p^{n-i} = \sum_{j=x+1}^{n} \binom{n}{j} p^j q^{n-j},\ j = n-i.$$

■

Theorem 3.1.2 If $X \sim \text{Bin}(n,p)$, then as k goes from 0 to n, $P\{X = k\}$ first increases monotonically and then decreases monotonically, reaching its largest value when k is the largest integer less than or equal to $(n+1)p$.

Proof. Consider $P(X=k)/P(X=k-1) \ge 1$ and determine for what values of k it holds. Then, $1 \le \dfrac{P(X=k)}{P(X=k-1)} = \dfrac{\binom{n}{k} p^k (1-p)^{n-k}}{\binom{n}{k-1} p^{k-1} (1-p)^{n-k+1}} = \dfrac{(n-k+1)p}{k(1-p)}$, which implies that $(n-k+1)p \ge k(1-p)$, equivalently, if and only if $k \le (n+1)p$. Figure 3.3 illustrates Theorem 3.1.2 with the graph of the p.m.f. of a binomial random variable with parameters $n = 10$ and $p = 0.5$.

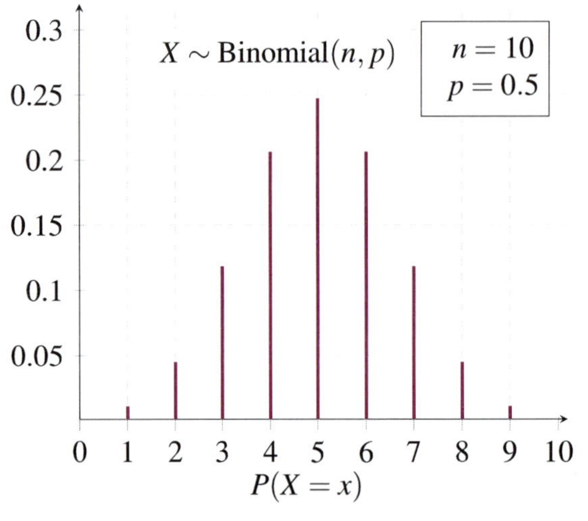

Figure 3.3: The p.m.f. of the binomial random variable, $X \sim \text{Binomial}(10, 0.5)$

■

■ **Example 3.5 — Best of Seven.** SK Wyverns play a "best of seven" Korean series with an opponent team, which means they play until one team wins four games. Assume Wyverns win every game independently with probability p. Find the probability Wyverns will win the series.

- The games are Bernoulli trials with success probability p, and the required probability is equivalent to the probability that $r = 4$ successes occurring before $s = 4$ failures in the independent Bernoulli trials, each resulting in a success with probability p.

- If the rth success occurs before or at the $(r+s-1)$ trial, it must have occurred before the sth failure. Thus, out of $n-1$ independent Bernoulli trials, there should be $r-1$ successes and $n-r$ failures, and the rth success should occur at the nth trial, $n = r, r+1, \ldots, r+s-1$. Hence the required probability is

$$P(\text{Wyverns win the series}) = \sum_{n=r}^{r+s-1} \binom{n-1}{r-1} p^{r-1}(1-p)^{n-r} p.$$

- On the other hand, if we assume that the games continue even after the winner of the series is decided, the required probability is the probability $P(X \geq r)$ when $X \sim \text{Binomial}(n,p)$, $n = r+s-1$, $P(X \geq r) = \sum_{k=r}^{n} \binom{n}{k} p^k (1-p)^{n-k}$.

■

3.2 Moments

There are several one-dimensional functionals of a distribution that play important roles in probability. The first of these functionals is the *mean* or *expectation* or *expected value*. Suppose X is a random variable. Then $\mathrm{E}(X)$ represents the long-term average value of X in an infinite number of replications of the random experiment. In some sense, the expected value is a "typical value" for the distribution of X. If we had to make one guess that would be close to the value X will actually take, $\mathrm{E}(X)$ would be our best guess.

Definition 3.2.1 The mean of a discrete random variable X with p.f. f is

$$\mathrm{E}(X) = \sum xP(X = x) = \sum xf(x),$$

if the sum is defined, i.e. $\sum_{\text{all } x} |x| f(x) < \infty$.

■ **Example 3.6** Let X be the indicator of an event A (also denoted I_A). That is

$$X(\omega) = I_A(\omega) = \begin{cases} 1 & \text{if } \omega \in A, \\ 0 & \text{if } \omega \notin A. \end{cases}$$

Since X takes only the values 0 or 1, it is a Bernoulli trial.

- The distribution of this random variable can be specified in four parts:

$$P(X \in B) = \begin{cases} 1 & \text{if } 0, 1 \in B, \\ P(A) & \text{if } 1 \in B \text{ and } 0 \notin B, \\ P(A^c) & \text{if } 0 \in B \text{ and } 1 \notin B, \\ 0 & \text{if } 0, 1 \notin B. \end{cases}$$

- Then, $X \sim \text{Ber}(p)$, $p = P(A)$, and $\mathrm{E}(X) = 1 \cdot p + 0 \cdot (1-p) = p = P(A)$.

■

Example 3.7 Suppose that X has p.m.f. $f_X(x) = \frac{c}{x(x+1)}$, $x = 1, 2, \ldots$. Determine the value (normalizing constant) c and the expected value of X if it exists.

- For $f_X(x)$ to be a p.m.f., we need $c > 0$ and $\sum_{x=1}^{\infty} f_X(x) = 1$. Note that

$$\sum_{x=1}^{\infty} \frac{c}{x(x+1)} = c \lim_{n\to\infty} \sum_{x=1}^{n} \left(\frac{1}{x} - \frac{1}{x+1}\right) = c\left(1 - \lim_{n\to\infty} \frac{1}{n+1}\right) = c.$$

Hence $c = 1$.

- Since $\sum_{x=1}^{\infty} |x| f_X(x) = \sum_{x=1}^{\infty} \frac{1}{x+1}$ is not finite, the expected value of X does not exist.

■

Example 3.8 If $X \sim \text{Bin}(n, p)$, then $\text{E}(X) = np$.

$$\begin{aligned}
\text{E}[X] &= \sum_{x=0}^{n} x \binom{n}{x} p^x (1-p)^{n-x} = \sum_{x=0}^{n} x \frac{n!}{x!(n-x)!} p^x (1-p)^{n-x} \\
&= np \sum_{x=1}^{n} \frac{(n-1)!}{(x-1)!(n-x)!} p^{x-1} (1-p)^{n-x} \\
&= np \sum_{r=0}^{n-1} \binom{n-1}{r} p^r (1-p)^{n-1-r} = np.
\end{aligned}$$

■

Theorem 3.2.1 — Law of the unconscious statistician. If X is a discrete random variable that takes on one of the values x_i, $i \geq 1$, with respective probabilities $f(x_i)$, then for any real-valued function g, $Y = g(X)$,

$$\text{E}[Y] = \text{E}[g(X)] = \sum_i g(x_i) f(x_i)$$

Proof. Note that $\text{E}[Y] = \sum_j y_j P(Y = y_j) = \sum_j y_j P(g(X) = y_j)$. Thus,

$$\begin{aligned}
\text{E}[g(X)] &= \sum_i g(x_i) f(x_i) = \sum_j \sum_{\{i\,:\,g(x_i)=y_j\}} g(x_i) f(x_i) \\
&= \sum_j \sum_{\{i\,:\,g(x_i)=y_j\}} y_j f(x_i) = \sum_j y_j \sum_{\{i\,:\,g(x_i)=y_j\}} f(x_i) \\
&= \sum_j y_j P(\{g(X) = y_j\}) = \text{E}[Y].
\end{aligned}$$

■

Theorem 3.2.2 Let X be a random variable with finite mean $\mathrm{E}(X)$, and let a and b be constants. Then:

- $\mathrm{E}(aX+b) = a\mathrm{E}(X)+b$.
- If $P(X=b)=1$, then $\mathrm{E}(X)=b$.
- If $P(a<X\le b)=1$, then $a<\mathrm{E}(X)\le b$.
- If $h_1(X)$ and $h_2(X)$ have finite mean, $\mathrm{E}[h_1(X)+h_2(X)] = \mathrm{E}[h_1(X)]+\mathrm{E}[h_2(X)]$.

Proof. We assume the necessary absolute convergence for finite expected values and use the law of the unconscious statistician in Theorem 3.2.1. Then,

- If $Y=aX+b$, that is, $g(x)=ax+b$, then $\mathrm{E}(Y)=a\mathrm{E}(X)+b$, which follows from

$$\mathrm{E}[aX+b] = \sum_i (ax_i+b)f(x_i) = a\sum_i x_i f(x_i) + b\sum_i f(x_i) = a\mathrm{E}(X)+b.$$

  BUT be advised that in general $\mathrm{E}[g(X)] \neq g[\mathrm{E}(X)]$.

- If X has a p.m.f. $f_X(b)=1$, then by definition $\mathrm{E}(X)=bf(b)=b$.
- Since $f_X(x)=0$ for $x\notin(a,b]$, we have

$$a=\sum_x a f_X(x) < \mathrm{E}(X) = \sum_x x f(x) \le \sum_x b f(x) = b.$$

- Let $g(X)=h_1(X)+h_2(X)$. Then,

$$\begin{aligned}\mathrm{E}[g(X)] &= \mathrm{E}[h_1(X)+h_2(X)] = \sum_x (h_1(x)+h_2(x))f(x)\\ &= \sum_x h_1(x)f(x) + \sum_x h_2(x)f(x) = \mathrm{E}[h_1(X)]+\mathrm{E}[h_2(X)].\end{aligned}$$

■

Remark Probability and expectation are closely related :

- For any event $A\subset\Omega$, $P(A)=\mathrm{E}(I_A)$, and $P(\Omega)=\mathrm{E}(I_\Omega)=\mathrm{E}(1)=1$.
- If A is an event, $0\le I_A$. Hence by the Expectation Monotonicity Rule,

$$0=\mathrm{E}(0)\le \mathrm{E}(I_A)=P(A).$$

■ **Example 3.9 — Expectation Monotonicity Rule.** If X and Y are random variables and $X\le Y$ (meaning that $X(\omega)\le Y(\omega)$ for all ω), then $\mathrm{E}(X)\le\mathrm{E}(Y)$.

- If $X \geq 0$, then $\mathrm{E}(X) \geq 0$. Since $X \leq |X|$, we have $\mathrm{E}(X) \leq \mathrm{E}(|X|)$.
- Since $0 \leq (X - \mathrm{E}(X))^2 = X^2 - 2X\mathrm{E}(X) + \mathrm{E}(X)^2$, we have $\mathrm{E}(X)^2 \leq \mathrm{E}(X^2)$.

■

■ **Example 3.10 — Chuck-a-luck.** Chuck-a-Luck is a casino game of chance and is played with three dice. To play this game, the player chooses one number from the numbers $1, \ldots, 6$. The three dice are then rolled. If the player's number does not come up at all, the player loses 10 dollars. If the chosen number comes up one, two, or three times, the player wins 10, 20, or 30 dollars, respectively. Is this game fair to the player?

- Let the random variable X denote the number of times the player's number comes up. Then, X has a binomial distribution with $n = 3$ and $p = 1/6$ with p.m.f

$$P(X = x) = \binom{3}{x}\left(\frac{1}{6}\right)^x\left(\frac{5}{6}\right)^{3-x}, \quad x = 0, 1, 2, 3.$$

- Let Y be the player's winnings (in dollars) in the game. Then $Y = -10$ if $X = 0$, $Y = 10X$ if $X = 1, 2, 3$. Thus, the expected winning is

$$\mathrm{E}(Y) = -10 \times \left(\frac{5}{6}\right)^3 + \sum_{x=1}^{3} 10x\binom{3}{x}\left(\frac{1}{6}\right)^x\left(\frac{5}{6}\right)^{3-x} \doteq -0.787.$$

- Hence, the game seems to be favorable for the player but it is actually not fair to the player.

■

R We can compute the expected winning by simulating rolls of three dice in the chuck-a-luck game, where you bet on a number "1".

```
n.sim=10^6; sides=6; rolls=3

simulation=replicate(n.sim,sample(sides,rolls,replace=TRUE))

chuck=colSums(simulation==1)

d0=-10; d1=10; d2=20; d3=30
wins=tabulate(chuck)

pay=wins[1]*d1+wins[2]*d2+wins[3]*d3+d0*(n.sim-sum(wins))
pay/n.sim

[1] -0.78984
```

Another common summary of a distribution is a measure of how much a distribution is spread out around its mean. The technical definition of *variance* is as follows.

Definition 3.2.2 Let $\mu = \mathrm{E}(X)$, then $\mathrm{Var}(X) = \mathrm{E}\left[(X-\mu)^2\right]$. The *standard deviation* of X (or of its distribution) is the square-root of its variance. Another formula for the variance is $\mathrm{Var}(X) = \mathrm{E}(X^2) - [\mathrm{E}(X)]^2$.

Theorem 3.2.3 There are some properties of variances that we can figure out immediately from what we already know:

- For any X, $\mathrm{Var}(X) \geq 0$.
- For any constant c, $\mathrm{Var}(c) = 0$ because $c = \mathrm{E}(c)$. In fact, $\mathrm{Var}(X) = 0$ if and only if $P[X = \mathrm{E}(X)] = 1$.
- For any constant a and b, $\mathrm{Var}(aX+b) = a^2\mathrm{Var}(X)$.

Example 3.11 Let $X \sim \mathrm{Ber}(p)$. Then $\mathrm{E}(X^2) = 1^2 \cdot p + 0^2 \cdot (1-p) = p$, and thus $\mathrm{Var}(X) = \mathrm{E}(X^2) - [\mathrm{E}(X)]^2 = p - p^2 = p(1-p)$. ■

Example 3.12 Suppose $Z_1, \ldots, Z_n$ are independent Bernouilli (p) random variables. Then we know that $X = \sum_{i=1}^n Z_i$ has a Binomial (n,p) distribution, and thus we have

$$\mathrm{E}(X) = np, \text{ and } \mathrm{Var}(X) = \mathrm{E}(X^2) - [\mathrm{E}(X)]^2 = \mathrm{E}(X^2) - (np)^2.$$

Thus, using the binomial p.m.f. and the law of the unconscious statistician, we have

$$\begin{aligned}
\mathrm{E}(X^2) &= \sum_{x=0}^{n} x^2 \binom{n}{x} p^x (1-p)^{n-x} = \sum_{x=1}^{n} x^2 \cdot \frac{n!}{x!(n-x)!} p^x (1-p)^{n-x} \\
&= \sum_{x=1}^{n} x \frac{n!}{(x-1)!(n-x)!} p^x (1-p)^{n-x} = \sum_{x=1}^{n} [(x-1)+1] \frac{n!}{(x-1)!(n-x)!} p^x (1-p)^{n-x} \\
&= \sum_{j=2}^{n} \frac{n!}{(j-2)!(n-j)!} p^j (1-p)^{n-j} + \sum_{x=1}^{n} \frac{n!}{(x-1)!(n-x)!} p^x (1-p)^{n-x} = S_2 + S_1
\end{aligned}$$

Since the sum consists of Binomial$(n-1,p)$ probabilities, it follows that

$$\begin{aligned}
S_1 &= \sum_{x=1}^{n} \frac{n!}{(x-1)!(n-x)!} p^x (1-p)^{n-x} = \sum_{k=0}^{n-1} \frac{n!}{k!(n-(k+1))!} p^{k+1} (1-p)^{n-(k+1)} \\
&= np \sum_{k=0}^{n-1} \frac{(n-1)!}{k!((n-1)-k)!} p^k (1-p)^{(n-1)-k} = np.
\end{aligned}$$

Similarly, letting $i = j-2$ and $m = n-2$,

$$S_2 = \sum_{j=2}^{n} \frac{n!}{(j-2)!} \frac{p^j (1-p)^{n-j}}{(n-j)!} = n(n-1)p^2 \sum_{i=0}^{m} \frac{m!}{i!(m-i)!} p^i (1-p)^{m-i} = n(n-1)p^2.$$

Then $\mathrm{E}(X^2) = S_1 + S_2 = np + n(n-1)p^2$. Hence, we have

$$\text{Var}(X) \;=\; \text{E}(X^2) - [\text{E}(X)]^2 = np + n(n-1)p^2 - (np)^2 = np(1-p).$$

Alternatively, we can compute

$$\begin{aligned} \text{E}[X(X-1)] &= \sum_{r=0}^{n} r(r-1)\frac{n!}{r!(n-r)!}p^r(1-p)^{n-r} \\ &= n(n-1)p^2 \sum_{r=2}^{n} \binom{n-2}{r-2} p^{r-2}(1-p)^{(n-2)-(r-2)} = n(n-1)p^2. \end{aligned}$$

Therefore, $\text{Var}(X) = \text{E}[X(X-1)] + \text{E}(X) - [\text{E}(X)]^2 = np(1-p)$. ■

The mean and variance are determined by the first two moments. Higher moments are sometimes useful, but not so much as the first two.

Definition 3.2.3 The *kth moment* of a random variable X is $\text{E}(X^k)$. Suppose that X is a random variable for which $\text{E}(X) = \mu$. For every integer k, the expectation $\text{E}[(X-\mu)^k]$ is called the *kth central moment of X* or kth moment of X about the mean.

Remark The term "moment" is borrowed from physics, and the central moments are not affected by adding a constant to a random variable.

- The 1st moment $\text{E}(X)$ can be viewed as *the center of gravity* or *the center of mass.*
- The 2nd central moment of X is *variance*, $\text{E}(X-\mu)^2 = \sigma^2$, a measure of how much a distribution is spread out around its mean μ, called the *moment of inertia about the center of mass.*
- The 3rd central (standardized) moment, $\dfrac{\text{E}(X-\mu)^3}{\sigma^3}$, is known as *skewness*, a measure of asymmetry.
- The 4th central (standardized) moment subtracted by 3, $\dfrac{\text{E}(X-\mu)^4}{\sigma^4} - 3$, is known as *kurtosis*, a measure of the thickness of the tails.

Definition 3.2.4 The moment generating function (m.g.f.) $M_X(t)$ of a discrete random variable X is defined for all real values of t by

$$M_X(t) = \text{E}\left(e^{tX}\right) = \sum_x e^{tx} f_X(x),$$

if this expected value exists for all t in the interval $-h < t < h$ for some $h > 0$, where $f_X(x)$ is the p.m.f. of X.

The following theorem tells us if we know the moment generating function of a random variable, then we can generate all the moments of X by taking derivatives of the moment generating function and then evaluating them at zero.

Theorem 3.2.4 If the m.g.f. of X is finite in an open interval around 0, then $\mathrm{E}(X^k)$ exists for all $k = 1, 2, \ldots$ and

$$\mathrm{E}(X^k) = \frac{d^k}{dt^k} M_X(t)\Big|_{t=0}.$$

Proof. Note that

$$\begin{aligned}
\frac{d}{dt} M_X(t) &= \frac{d}{dt}\mathrm{E}\left(e^{tX}\right) = \mathrm{E}\left(\frac{d}{dt}e^{tX}\right) = \mathrm{E}\left(X e^{tX}\right), \\
\frac{d^2}{dt^2} M_X(t) &= \frac{d^2}{dt^2}\mathrm{E}\left(e^{tX}\right) = \mathrm{E}\left(\frac{d^2}{dt^2}e^{tX}\right) = \mathrm{E}\left(X^2 e^{tX}\right).
\end{aligned}$$

Thus, in general we get $\frac{d^k}{dt^k} M_X(t) = \frac{d^k}{dt^k}\mathrm{E}\left(e^{tX}\right) = \mathrm{E}\left(\frac{d^k}{dt^k}e^{tX}\right) = \mathrm{E}\left(X^k e^{tX}\right)$. If we set $t = 0$ in the kth derivative, we get $\frac{d^k}{dt^k} M_X(t)\Big|_{t=0} = \mathrm{E}\left(X^k e^{tX}\right)\Big|_{t=0} = \mathrm{E}\left(X^k\right)$. Hence the kth derivative of the moment generating function of X evaluated at $t = 0$ is the kth moment of X about the origin.

■

■ **Example 3.13** Suppose X has a discrete uniform distribution on $\{1, 2, \ldots, N\}$ with p.m.f. $P(X = x) = f_X(x) = \frac{1}{N}$ for $x = 1, \ldots, N$. Then

- The moment generating function is given by

$$M_X(t) = \sum_{x=1}^{N} \frac{1}{N} e^{tx} = \frac{1}{N}(e^t + e^{2t} + \cdots + e^{Nt}) = \frac{e^t(e^{Nt} - 1)}{N(e^t - 1)}.$$

- It is to see that

$$\begin{aligned}
\mathrm{E}(X) &= M_X'(0) = \frac{1}{N}(1 + 2 + 3 + \cdots + N) = \frac{N+1}{2}, \\
\mathrm{E}(X^2) &= M_X''(0) = \frac{1}{N}(1 + 2^2 + 3^2 + \cdots + N^2) = \frac{(N+1)(2N+1)}{6}, \\
\mathrm{Var}(X) &= \mathrm{E}(X^2) - \mathrm{E}(X)^2 = \frac{N^2 - 1}{12}.
\end{aligned}$$

■

■ **Example 3.14** For Binomial Distribution, $X \sim \mathrm{Binomial}(n, p)$,

$$M_X(t) = \mathrm{E}(e^{tX}) = \sum_{x=0}^{n} e^{t \cdot x} \binom{n}{x} p^x (1-p)^{n-x} = (1 - p + pe^t)^n,$$

which follows from Binomial theorem.

- Thus, when $n = 1$, we have $1 - p + pe^t$, m.g.f. of Bernoulli distribution, which is also obtained as $M_X(t) = \mathrm{E}(e^{tX}) = e^{t\cdot 0}(1-p) + e^{t\cdot 1}p = 1 - p + pe^t$.

- In addition, we have

$$\begin{aligned}
\mathrm{E}(X) &= M_X'(0) = n(pe^t + 1 - p)^{n-1}pe^t\big|_{t=0} = np, \\
\mathrm{E}(X^2) &= M_X''(0) = n(n-1)(pe^t + 1 - p)^{n-2}p^2e^t + n(pe^t + 1 - p)^{n-1}pe^t\big|_{t=0} \\
&= n(n-1)p^2 + np, \\
\mathrm{Var}(X) &= \mathrm{E}(X^2) - (np)^2 = np(1-p).
\end{aligned}$$

■

3.3 Joint distributions

In most interesting problems, there are several random variables of interest. We will begin by considering two discrete random variables at a time.

Definition 3.3.1 Let X and Y be two discrete random variables taking values, $x_1, x_2, \ldots,$ and $y_1, y_2, \ldots$, respectively. Their joint (bivariate) distribution can be specified through the *joint probability mass function*, $f_{X,Y}(x,y)$ defined

$$f_{X,Y}(x,y) = P(\{X = x, Y = y\}),\ x = x_1, \ldots, \text{ and } y = y_1, \ldots,$$

Here, the event $\{X = x,\ Y = y\}$ means $\{X = x\} \cap \{Y = y\}$.

Remark Every joint p.m.f. $f_{X,Y}(x,y)$ satisfies the following properties :

- $f_{X,Y}(x,y) \geq 0$ for all x, y, and $\sum_{\text{all }(x,y)} f_{X,Y}(x,y) = 1$.
- There are at most countably many (x,y) such that $f_{X,Y}(x,y) > 0$.
- For each set A, $P((X,Y) \in A) = \sum_{(x,y)\in A} f_{X,Y}(x,y)$.

Definition 3.3.2 If X and Y have the joint p.m.f. $f_{X,Y}$, then the p.m.f. of X and p.m.f. of Y are simply $f_X(x) = P(X = x) = \sum_{\text{all }y} f_{X,Y}(x,y)$, and $f_Y(y) = P(Y = y) = \sum_{\text{all }x} f_{X,Y}(x,y)$. These are called the *marginal* p.m.f. of X and the marginal p.m.f. of Y respectively.

■ **Example 3.15** Suppose we roll a pair of dice. Let X denote the smaller and Y denote the larger outcome on the dice, then what is the joint probability mass function of X and Y?

- For example, $f_{X,Y}(3,4) = P(X=3, Y=4) = \dfrac{2}{36}$ since there are two outcomes $(3,4)$ and $(4,3)$. Thus, the joint p.m.f can be written as

$$f_{X,Y}(x,y) = \begin{cases} 1/36, & \text{if } 1 \le x = y \le 6, \\ 2/36, & \text{if } 1 \le x < y \le 6, \\ 0 & \text{otherwise.} \end{cases}$$

- The following table shows joint and marginal probability mass functions :

	$f_{X,Y}(x,y)$	Y 1	2	3	4	5	6	$f_X(x)$
X	1	1/36	2/36	2/36	2/36	2/36	2/36	11/36
	2	0	1/36	2/36	2/36	2/36	2/36	1/4
	3	0	0	1/36	2/36	2/36	2/36	7/36
	4	0	0	0	1/36	2/36	2/36	5/36
	5	0	0	0	0	1/36	2/36	1/12
	6	0	0	0	0	0	1/36	1/36
	$f_Y(y)$	1/36	1/12	5/36	7/36	1/4	11/36	1

■

In addition to joint and marginal p.m.f.'s, we can define a conditional p.m.f. of a discrete random variable X given some event $B \subset \Omega$ with $P(B) > 0$, by extending the concept of conditional probability of one event given the other event to random variables. Specifically, we write $P(X=x|B) = P(A_x|B) = \dfrac{P(A_x \cap B)}{P(B)} \ge 0$, where $A_x = \{\omega : X(\omega) = x\} = \{X = x\}$.

Definition 3.3.3 Let X be a discrete random variable. Define $f_X(x|B)$ by

$$f_{X|B}(x|B) = P(X=x|B) = P(A_x|B) = \frac{P(A_x \cap B)}{P(B)},$$

which is called the *conditional p.m.f.* of X given B.

- More generally, we define the conditional p.m.f. of X given Y, $f_{X|Y}(x|y)$ by

$$f_{X|Y}(x|y) = P(X=x|Y=y) = P(A_x|B_y) = \frac{P(A_x \cap B_y)}{P(B_y)} = \frac{P(\{X=x, Y=y\})}{P(Y=y)},$$

for each y such that $P(B_y) > 0$, where $B_y = \{Y = y\}$.

- Thus, $P(X \in A|Y=y) = \dfrac{P(A \cap B_y)}{P(B_y)} = \dfrac{\sum_{x \in A} f_{X,Y}(x,y)}{f_Y(y)} = \displaystyle\sum_{x \in A} \frac{f_{X,Y}(x,y)}{f_Y(y)} = \sum_{x \in A} f_{X|Y}(x|y).$

Remark Since X is a discrete random variable with countably many values with $\bigcup_x A_x = \Omega$ and $A_{x_1} \cap A_{x_2} = \emptyset$ if $x_1 \ne x_2$, we have $\sum_x P(A_x|B) = \sum_x \frac{P(A_x \cap B)}{P(B)} = \frac{P(\Omega \cap B)}{P(B)} = 1$.

- $f_{X|Y}(x|y)$ is a p.m.f., in that $f_{X|Y}(x|y) \geq 0$ and $\sum_x f_{X|Y}(x|y) = 1$, if $f_Y(y) > 0$.
- By multiplication rule, we have $P(X = x, Y = y) = P(X = x|Y = y)P(Y = y)$, that is, $f_{X,Y}(x,y) = f_{X|Y}(x|y)f_Y(y)$.

▪ **Example 3.16** Consider the discrete uniform random distribution given in Example 3.13 defined on $\{1,2,\ldots,N\}$. Let B the event that $1 \leq X \leq n$, where $n < N$. Then, the conditional p.m.f. of X given B is

$$f_{X|B} = \frac{P(A_x \cap B)}{P(B)} = \frac{P(\{X = x\} \cap B)}{P(B)} = \begin{cases} \frac{1/N}{n/N} = \frac{1}{n}, & \text{for } x \in B = \{1,2,\ldots,n\}, \\ 0, & \text{otherwise,} \end{cases}$$

which implies that the conditional distribution of X given B is a discrete uniform defined on $B = \{1,2,\ldots,n\}$. ▪

Theorem 3.3.1 — Expectation of Sums. If X and Y have a joint p.m.f. $f_{X,Y}(x,y)$, then $\mathrm{E}[g(X,Y)] = \sum_y \sum_x g(x,y) f_{X,Y}(x,y)$.

- If $\mathrm{E}(X)$ and $\mathrm{E}(Y)$ exist, then $\mathrm{E}(X+Y) = \mathrm{E}(X) + \mathrm{E}(Y)$.
- If $\mathrm{E}(X_i)$ exists for each $i = 1,\ldots,n$, and if $a_1,\ldots,a_n,b$ are constants, then

$$\mathrm{E}(a_1X_1 + \cdots + a_nX_n + b) = a_1\mathrm{E}(X_1) + \cdots + a_n\mathrm{E}(X_n) + b.$$

Proof. Let the sample spaces of X and Y be denoted by Ω_X and Ω_Y, and suppose that $\Omega_X = \{x_1, x_2, \ldots\}$ and $\Omega_Y = \{y_1, y_2, \ldots\}$. Then we can consider the random variable $g(X,Y) = X + Y$. Then, we have

$$\begin{aligned} E(X+Y) &= \sum_j \sum_k (x_j + y_k) P(X = x_j,\ Y = y_k) \\ &= \sum_j \sum_k x_j P(X = x_j,\ Y = y_k) + \sum_j \sum_k y_k P(X = x_j,\ Y = y_k) \\ &= \sum_j x_j P(X = x_j) + \sum_k y_k P(Y = y_k)\ . \end{aligned}$$

■

▪ **Example 3.17** Let S_n be the number of successes in n Bernoulli trials with probability p for success on each trial. Then the expected number of successes is np. That is, $\mathrm{E}(S_n) = np$.

- Let X_j be a random variable which has the value 1 if the jth outcome is a success and 0 if it is a failure, $X_j \sim \text{Bernoulli}(p)$. Then, $\mathrm{E}(Z_j) = p$.

- Since $S_n = X_1 + X_2 + \cdots + X_n$ and the expected value of the sum is the sum of the expected values, we have $\mathrm{E}(S_n) = \mathrm{E}\left(\sum_{i=1}^{n} X_i\right) = \sum_{i=1}^{n} \mathrm{E}(X_i) = np$.

■

■ **Example 3.18** Consider a class of n students in the birthday problem. We would like to know if there is at least one pair of students with the same birthday. Let X be the number of different pairs of students who share a common birthday. Find the expected value of X.

- Define $I_{i,j}$ as the indicator of the event that students i and j have the same birthday. Then, $I_{i,j} \sim \text{Bernoulli}\left(\frac{1}{365}\right)$, $\mathrm{E}(I_{i,j}) = P(I_{i,j} = 1) = \frac{1}{365}$, $i, j = 1, \ldots, n$, $i \neq j$.
- X is represented in terms of $I_{i,j}$'s, $i, j = 1, \ldots, n$, $X = \sum_{i<j} I_{ij}$. There are $\binom{n}{2}$ possible ways of getting different pairs of students. Thus, $\mathrm{E}(X) = \sum_{i<j} \mathrm{E}(I_{i,j}) = \binom{n}{2}/365$.

# of People	# of Pair	Average # of Matches	Probability of Match
4	6	0.016	0.0164
20	190	0.520	0.411
23	253	0.693	0.507
30	435	1.191	0.706
40	780	2.136	0.891
50	1,225	3.356	0.970

■

■ **Example 3.19 — Expected value of matching.** In the matching problem, let X be the number of students who get the correct phone. Find the mean of X.

- Suppose $n = 3$. There are six possible outcomes of X, and we assign to each of them the probability $1/6$. Then we can calculate $\mathrm{E}(X)$ as
$$3\left(\frac{1}{6}\right) + 1\left(\frac{1}{6}\right) + 1\left(\frac{1}{6}\right) + 0\left(\frac{1}{6}\right) + 0\left(\frac{1}{6}\right) + 1\left(\frac{1}{6}\right) = 1\,.$$
- For general n, let $X_i = 1$ if student i gets the correct phone and $X_i = 0$ if not. Then $X = \sum_{i=1}^{n} X_i$ and $\mathrm{E}(X) = \sum_{i=1}^{n} \mathrm{E}(X_i)$. Clearly $\mathrm{E}(X_i)$ is the same for all i. Indeed, $P(X_i = 1) = 1/n$ and $P(X_i = 0) = 1 - 1/n$ so that $\mathrm{E}(X_i) = 1/n$. Therefore $\mathrm{E}(X) = 1$, no matter what n is!

■

■ **Example 3.20** The indicator function I_A of an event $A \subset \Omega$ is defined by $I_A(\omega) = \begin{cases} 1, & \text{if } \omega \in A, \\ 0, & \text{if } \omega \notin A, \end{cases}$ and has a Bernoulli distribution with parameter $P(A)$ as discussed in Example 3.6. In addition, there are several properties with indicator functions.

- $\mathrm{E}(I_A) = P(A)$, $I_{A^c} = 1 - I_A$, $I_{A\cap B} = I_A \cdot I_B$, and $I_{A\cup B} = I_A + I_B - I_A \cdot I_B$.

- Let I_j be an indicator variable for the event A_j. Define

$$S_r = \sum_{i_1<i_2<\ldots<i_r} I_{i_1} I_{i_2} \ldots I_{i_r},\ s_r = \mathrm{E}(S_r) = \sum_{i_1<i_2<\ldots<i_r} P(A_{i_1} \cap \ldots \cap A_{i_r}).$$

Let $U = \cup_{j=1}^n A_j$. Then, since $U^c = \cap_{j=1}^n A_j^c$, it follows that

$$\begin{aligned} I_{U^c} &= 1 - I_U = (1 - I_{A_1})(1 - I_{A_2}) \cdots (1 - I_{A_n}) \\ &= 1 - \sum_{j=1}^n I_{A_j} + \sum_j \sum_{j<k} I_{A_j} I_{A_k} - \cdots + (-1)^n I_{A_1} I_{A_2} \cdots I_{A_n}. \end{aligned}$$

Hence, taking expectations, we obtain

$$\begin{aligned} 1 - \prod_{j=1}^n (1 - I_j) &= S_1 - S_2 + \ldots + (-1)^{n-1} S_n \\ P\left(\bigcup_{j=1}^n A_j\right) &= \mathrm{E}\left[1 - \prod_{j=1}^n (1 - I_j)\right] = s_1 - s_2 + \ldots + (-1)^{n-1} s_n, \end{aligned}$$

which establishes the Inclusion-exclusion principle.

■

Definition 3.3.4 — Hypergeometric Distribution. Suppose that a population contains A successes and B failures. Let X be the number of successes in n draws without replacement from this population. The distribution of X is called the *hypergeometric distribution with parameters A, B, and n*, denoted as $X \sim \text{Hypergeo}(n, A, B)$. It is required that $n \le A + B$. The p.m.f. of X is

$$P(X = x) = f(x) = \frac{\binom{A}{x}\binom{B}{n-x}}{\binom{A+B}{n}}, \text{ for } x = \max\{0, n-B\}, \ldots, \min\{n, A\}.$$

Remark The p.m.f. of $X \sim \text{Hypergeo}(n, A, B)$ sums to 1 by Vandermonde's identity. Also, we note that $P(X = x)$ remains the same if the roles of A and n are interchanged.

■ **Example 3.21** Consider an urn with r red balls and b black balls. Suppose n balls are drawn without replacement, $n \le r + b$. The probability of drawing exactly x red balls is given by the hypergeometric distribution.

- Think of $r + b$ balls with r of them labelled success (red) and the other b labeled failure (black).

- Now, we pick n trials to be called the first n trials and see how many successes appear in these first trials. There are $\binom{r+b}{n}$ ways to choose the n first trials. There are $\binom{r}{x}\binom{b}{n-x}$ ways to pick which x trials have the successes and which $n-x$ have the failures.

- Notice that $\dfrac{\binom{r}{x}\binom{b}{n-x}}{\binom{r+b}{n}} = \dfrac{\binom{n}{x}\binom{r+b-n}{r-x}}{\binom{r+b}{r}}$, which implies that $P(X=x) = P(Y=x)$, where $X \sim \text{Hypergeo}(n,r,b)$ and $Y \sim \text{Hypergeo}(r,n,r+b-n)$. In other words, X and Y have the same distribution.

■

■ **Example 3.22** Consider an urn with r red balls and b blue balls. Suppose that we draw three marbles one at a time from the urn without replacement, $r+b \geq 3$. We see that the probability of drawing exactly x blue balls is given by the hypergeometric distribution.

- Let X_i for $i = 1,2,3$ be the indicator that the ith ball is blue. Let X be the number of blue balls in the sample of three. Note that the each of X_i's is a Bernoulli trial but are not independent and that $X = X_1 + X_2 + X_3$.

- Since all $r+b$ balls in the urn at the first drawing are equally likely to be selected, each ball has probability $\frac{1}{r+b}$ of being selected. Thus, we have $P(X_1 = 1) = \frac{b}{r+b}$ and $P(X_1 = 0) = \frac{r}{r+b}$ because there are b blue balls and r red balls.

- Consider the value of X_3, which depends on the balls selected beforehand. However, when computing the (marginal) distribution of X_3 we are using only the information available at the beginning of the experiment. To compute $P(X_3 = 1)$, we see that there are four different possibilities for the colors of the balls on the first two draws. For example, there are $rb(b-1)$ ways to choose the three balls such the first is red and the last two are blue. There are a total of $(r+b)(r+b-1)(r+b-2)$ ways to choose three balls. We can also choose blue-red-blue, blue-blue-blue, red-red-blue. In each of these cases, the total number of draws is the same but the number of draws that match the pattern differs. Thus,

$$P(X_3 = 1) = \frac{b(b-1)(b-2) + 2b(b-1)r + r(r-1)b}{(r+b)(r+b-1)(r+b-2)} = \frac{b((b-1)(b-2) + 2(b-1)r + r(r-1))}{(r+b)(r+b-1)(r+b-2)} = \frac{b}{r+b},$$

which implies that X_1 and X_3 have the same distribution and that X_1, X_2 and X_3 have the same distribution as well.

- Then, for $j,k,\ell = 0,1$, we can compute the probability that $X_1 = j$, $X_2 = k$ and $X_3 = \ell$ that $P(X_1 = j, X_2 = k, X_3 = \ell) = \dfrac{\prod_{s=0}^{2-j-k-\ell}(r-s)\prod_{t=0}^{j+k+\ell-1}(b-t)}{(r+b)(r+b-1)(r+b-2)}$. Thus, for example, we can have $P(X_2 = k)$ by adding up four possible cases of $X_1 = j$,

$X_2 = k$, and $X_3 = \ell$, $j,\ell = 0,1$, $P(X_2 = k) = \sum_{j,\ell=0,1} P(X_1 = j, X_2 = k, X_3 = \ell) = \left(\frac{r}{r+b}\right)^k \left(\frac{b}{r+b}\right)^{1-k}$, for $k = 0,1$, which implies that $X_2 \sim \text{Bernoulli}\left(\frac{r}{r+b}\right)$.

- Now, we find the distribution of X. First, note that X can take values from 0 to 3. Since $X = X_1 + X_2 + X_3$, we have

$$\{X = x\} = \{X_1 + X_2 + X_3 = x\} = \bigcup_{j,k,\ell : j+k+\ell = x} \{X_1 = j, X_2 = k, X_3 = \ell\}$$

Second, note that $P(X_1 = j, X_2 = k, X_3 = \ell)$ depends only on the sum of its arguments $j+k+\ell$ and that for each $X = x$, all the events in the union above have the same probability. Thus, it follows that

$$\begin{aligned}
P(X=0) &= \binom{3}{0} P(X_1 = 0, X_2 = 0, X_3 = 0) = \frac{r(r-1)(r-2)}{(r+b)(r+b-1)(r+b-2)} \\
P(X=1) &= \binom{3}{1} P(X_1 = 1, X_2 = 0, X_3 = 0) = 3\frac{r(r-1)b}{(r+b)(r+b-1)(r+b-2)} \\
P(X=2) &= \binom{3}{2} P(X_1 = 1, X_2 = 1, X_3 = 0) = 3\frac{rb(b-1)}{(r+b)(r+b-1)(r+b-2)} \\
P(X=3) &= \binom{3}{3} P(X_1 = 1, X_2 = 1, X_3 = 1) = \frac{b(b-1)(b-2)}{(r+b)(r+b-1)(r+b-2)},
\end{aligned}$$

which can be rewritten as $P(X = x) = \frac{\binom{b}{x}\binom{r}{3-x}}{\binom{r+b}{3}}$, a hypergeometric distribution with parameters 3, b, and r. Hence, $X \sim \text{Hypergeo}(3, b, r)$.

■

■ **Example 3.23** Suppose that Jaden chooses a subset S of size s $(0 \le s \le n)$ at random from the set $\{1,2,\ldots,n\}$, each of the $\binom{n}{s}$ subsets of size s having probability $1/\binom{n}{s}$ of being chosen. Suppose that, independently of Jaden, Joanne chooses a subset T of size $t(0 \le t \le n)$ at random from the set $\{1,2,\ldots,n\}$, each of the $\binom{n}{t}$ subsets of size t having probability $1/\binom{n}{t}$ of being chosen. Let X be the number of elements in the set $S \cap T$. We see that the p.m.f of X can be expressed in the three equivalent ways :

$$P(X = k) = \frac{\binom{s}{k}\binom{n-s}{t-k}}{\binom{n}{t}} = \frac{\binom{t}{k}\binom{n-t}{s-k}}{\binom{n}{s}} = \frac{\binom{n}{k}\binom{n-k}{s-k}\binom{n-s}{t-k}}{\binom{n}{s}\binom{n}{t}}.$$

- Let $|S \cap T|$ denote the number of elements in the set $S \cap T$. Then, it is clear that $\max(0, s+t-n) \le |S \cap T| \le \min(s,t)$. Then, we show that $|S \cap T|$ has a hypergeometric distribution.

- Firstly, consider partitioning the set $\{1,2,\ldots,n\}$ into 4 disjoint sets, $S \cap T$, $S \cap T^c$,

$S^c \cap T$, and $S^c \cap T^c$ with k, $s-k$, $t-k$ and $n-s-t+k$ elements, respectively.

	T	T^c	subtotal	#(possible choices)
S	k	$s-k$	s	$\binom{s}{k}$
S^c	$t-k$	$n-s-t+k$	$n-s$	$\binom{n-s}{t-k}$
subtotal	t	$n-t$	n	$\binom{n}{t}$

- Joanne's subset T is chosen by random, with $\binom{n}{t}$ possible ways. Suppose that Jaden has chosen the subset $\{1,2,\ldots,s\}$. Since $|S\cap T| = k$ and $|S^c \cap T| = t-k$, Joanne's subset T would consist of a subset of size k in $\{1,2,\ldots,s\}$ and a subset of size $t-k$ in $\{s+1, s+2, \ldots, n\}$. Thus, there are $\binom{s}{k} \times \binom{n-s}{t-k}$ possible choices. Hence, the required probability is

$$P(X=k) = \frac{\binom{s}{k}\binom{n-s}{t-k}}{\binom{n}{t}}, \quad k = \max(0, s+t-n), \ldots, \min(s,t).$$

- Secondly, the same partition can be viewed in a column-wise way.

	T	T^c	subtotal
S	k	$s-k$	s
S^c	$t-k$	$n-s-t+k$	$n-s$
subtotal	t	$n-t$	n
#(possible choices)	$\binom{t}{k}$	$\binom{n-t}{s-k}$	$\binom{n}{s}$

- Then, the set S is chosen by random, with $\binom{n}{s}$ possible outcomes. When $|S\cap T| = k$ and $|S\cap T^c| = s-k$, there are $\binom{t}{k} \times \binom{n-t}{s-k}$ possible choices, using a similar argument to the first approach. Hence, the required probability is

$$P(X=k) = \frac{\binom{t}{k}\binom{n-t}{s-k}}{\binom{n}{s}}, \quad k = \max(0, s+t-n), \ldots, \min(s,t).$$

- Lastly, consider two sets S and T with s and t elements, respectively. Then, there are $\binom{n}{s} \cdot \binom{n}{t}$ possible choices of two sets S and T. The event that $S\cap T$ has size (the number of elements in $S\cap T$) k corresponds to all the cases of partitioning $\{1,2,\ldots,n\}$ into four subsets, $S\cap T$, $S\cap T^c$, $S^c\cap T$, and $S^c \cap T^c$ of sizes k, $s-k$, $t-k$, and $n-s-t+k$, respectively. Thus,

$$\#\{X=k\} = \binom{n}{k,\, s-k,\, t-k,\, n-s-t+k} = \frac{n!}{k!(s-k)!(t-k)!(n-s-t+k)!}$$

$$= \frac{n!}{k!(n-k)!} \cdot \frac{(n-k)!}{(s-k)!(n-s)!} \cdot \frac{(n-s)!}{(t-k)!(n-s-t+k)!} = \binom{n}{k}\binom{n-k}{s-k}\binom{n-s}{t-k}.$$

Hence, the p.m.f. of X is

$$P(X=k) = \frac{\#(X=k)}{\binom{n}{s}\binom{n}{t}} = \frac{\binom{n}{k}\binom{n-k}{s-k}\binom{n-s}{t-k}}{\binom{n}{s}\binom{n}{t}}, \quad k = \max(0, s+t-n), \ldots, \min(s,t).$$

- Alternatively, we can check algebraically that they all represent $P(X=k)$ in the three equivalent ways.

$$\begin{aligned}
\frac{\binom{n}{k}\binom{n-k}{s-k}\binom{n-s}{t-k}}{\binom{n}{s}\binom{n}{t}} &= \frac{\binom{n-s}{t-k}}{\binom{n}{t}} \times \frac{\binom{n}{k}\binom{n-k}{s-k}}{\binom{n}{s}} = \frac{\binom{n-s}{t-k}}{\binom{n}{t}} \times \frac{\frac{n!}{k!(n-k)!}\frac{(n-k)!}{(s-k)!(n-s)!}}{\frac{n!}{s!(n-s)!}} \\
&= \frac{\binom{n-s}{t-k}}{\binom{n}{t}} \times \frac{s!}{k!(s-k)!} = \frac{\binom{s}{k}\binom{n-s}{t-k}}{\binom{n}{t}} \\
&= \frac{1}{\binom{n}{s}} \times \frac{t!(n-t)!}{k!(s-k)!(t-k)!(n-s-t+k)!} = \frac{\binom{t}{k}\binom{n-t}{s-k}}{\binom{n}{s}}.
\end{aligned}$$

■

■ **Example 3.24 — Capture-recapture.** Capture–recapture methods have been used in ecology, public health, social science, and other disciplines to count rare and elusive populations. In order to estimate the size N of a population, like fish in a lake or animals in wildlife, researchers "capture" a sample of size k. Subjects are "tagged" and released to the general population. Researchers then "recapture" a second sample of size n and count the number X which are found tagged. If the first sample is sufficiently mixed up in the general population then the proportion of those tagged in the second sample should be approximately equal to the proportion tagged in the population, which gives an estimate of the population size.

- The probability $P(X=x)$ that among these n recaptured samples, x are tagged is

$$P(X=x) = \frac{\binom{k}{x}\binom{N-k}{n-x}}{\binom{N}{n}}$$

since there are $\binom{N}{n}$ ways of drawing n animals, $\binom{k}{x}$ ways of drawing x tagged animals, and $\binom{N-k}{n-x}$ ways of drawing $n-x$ untagged ones. That is, X has a hypergeometric distribution with parameters n, k and $N-k$.

- Suppose instead that researchers were to carry on recapturing animals until they gets r tagged animals. Let the number of animals to be recaptured denote Y. For the event $\{Y=y\}$ to occur, it is necessary that the yth animal is tagged, which can occur in $\binom{k}{1}$ ways, and the preceding $y-1$ animals include exactly $r-1$ tagged and $(y-1)-(r-1)$ untagged animals, which may occur in $\binom{k-1}{r-1}\binom{N-k}{y-r}$ ways. In addition, the total number of ways of first selecting a tagged animal to fill the yth place, and then choosing $y-1$ animals to fill the remaining $y-1$ places is $N \cdot \binom{N-1}{y-1}$. Hence, the required probability is

$$P(Y=y) = \frac{k}{N} \times \frac{\binom{k-1}{r-1}\binom{N-k}{y-r}}{\binom{N-1}{y-1}}, \quad r \le y \le N-k+r.$$

■

■ **Example 3.25** A bowl contains k red and b blue balls. You randomly select without replacement one ball at a time until you have r red balls. Find the probability mass function of the number of trials needed.

- Let X be the number of trials needed to obtain r red balls.
- Then, the event $\{X = x\}$ is equivalent to $A \cap B$, where A is the event that $r-1$ red balls are selected in the first $x-1$ trials and B is the event that a red ball is selected at the xth trial.
- Note that $P(B) = \dfrac{k}{k+b}$ since it is a Bernoulli trial marginally with $p = k/(k+b)$ and that $P(A|B)$ is the probability that $r-1$ red balls and $x-1-(r-1)$ blue balls are chosen when $x-1$ balls are chosen from the urn containing $k+b-1$ balls. Thus, the required probability is

$$\begin{aligned} P(X=x) &= P(A|B)P(B) = \frac{\binom{k-1}{r-1}\binom{b}{x-1-(r-1)}}{\binom{k+b-1}{x-1}} \times \frac{k}{k+b} \\ &= P(A)P(B|A) = \frac{\binom{k}{r-1}\binom{b}{x-1-(r-1)}}{\binom{k+b}{x-1}} \frac{k-(r-1)}{k+b-(x-1)}. \end{aligned}$$

■

Theorem 3.3.2 Let $X \sim \text{Hypergeo}(n, A, B)$. Then,

$$\mathrm{E}(X) = \frac{nA}{A+B}, \quad \mathrm{Var}(X) = \frac{nAB}{(A+B)^2} \cdot \frac{A+B-n}{A+B-1}.$$

Proof. We compute the mean and variance of X by computing the first and the second factorial moments of X. First, we compute the first factorial moment, i.e. $\mathrm{E}(X)$:

$$\begin{aligned} \mathrm{E}(X) &= \sum_{x=0}^{n} x f(x) = \sum_{x=0}^{n} x \frac{\binom{A}{x}\binom{B}{n-x}}{\binom{A+B}{n}} = A \sum_{x=1}^{n} \frac{(A-1)!}{(x-1)!(A-x)!} \frac{\binom{B}{n-x}}{\binom{A+B}{n}} \\ &= A \sum_{x=1}^{n} \frac{\binom{A-1}{x-1}\binom{B}{n-x}}{\frac{A+B}{n}\binom{A+B-1}{n-1}} = n\frac{A}{A+B} \sum_{y=0}^{n-1} \frac{\binom{A-1}{y}\binom{B}{n-1-y}}{\binom{A+B-1}{n-1}} = n\frac{A}{A+B}, \end{aligned}$$

where the last equality is obtained since $\sum_{y=0}^{n-1} \frac{\binom{A-1}{y}\binom{B}{n-1-y}}{\binom{A+B-1}{n-1}} = 1$. Similarly, we find the second factorial moment of X to be $\mathrm{E}(X(X-1)) = \dfrac{n(n-1)A(A-1)}{(A+B)(A+B-1)}$. Hence, the variance of X is obtained as

$$\begin{aligned}
\mathrm{Var}(X) &= \mathrm{E}[X(X-1)] + \mathrm{E}(X) - [\mathrm{E}(X)]^2 \\
&= \frac{n(n-1)A(A-1)}{(A+B)(A+B-1)} + n\,\frac{A}{A+B} - \left(n\,\frac{A}{A+B}\right)^2 \\
&= n\left(\frac{A}{A+B}\right)\left(\frac{B}{A+B}\right)\left(\frac{A+B-n}{A+B-1}\right).
\end{aligned}$$

■

Remark If the large portion of the Bernoulli process is sufficiently large compared to the smaller portion, then this is the same as sampling without replacement from a very large population. That is, it should not make much difference whether the selection is being done with or without replacement.

- If A and B go to infinity with $A/(A+B) = p$ fixed, and n, x fixed, then we find

$$\begin{aligned}
P(X=x) &= \binom{A}{x}\binom{B}{n-x}\Big/\binom{A+B}{n} \\
&= \binom{n}{x}\frac{A}{A+B}\frac{A-1}{A+B-1}\cdots\frac{A-x+1}{A+B-x+1} \\
&\quad\times\frac{B}{A+B-x}\frac{B-1}{A+B-x-1}\cdots\frac{B-n+x+1}{A+B-n+1} \approx \binom{n}{x}p^x(1-p)^{n-x}.
\end{aligned}$$

■ **Example 3.26** A box contains 10 balls, some red and some green. We believe that it is equally likely to contain any number of red balls from 0 to 10. We will draw three balls at random without replacement and see what colors they are.

- Let X be the number of red balls that we observe in our sample of 3. Let Y be the number of red balls in the box. If we condition on the value of Y, say $Y = y$, then the distribution of X (conditional on $Y = y$) is computed precisely,

$$P(X = x|Y = y) = \frac{\binom{y}{x}\binom{10-y}{3-x}}{\binom{10}{3}}, \quad 0 \le x \le 3 \text{ and } y-7 \le x \le y.$$

  Thus, the conditional distribution of X given $Y = y$ is a hypergeometric distribution with parameters 10, y and $10-y$, that is, $X|Y = y \sim \mathrm{Hypergeo}(10, y, 10-y)$.

- Now, we can combine this conditional probability with the marginal probability of $\{Y = y\}$ to get the joint p.f. of X and Y for x and y integers, using multiplication rule

$$f_{X,Y}(x,y) = \frac{\binom{y}{x}\binom{10-y}{3-x}}{\binom{10}{3}}\frac{1}{11}, \quad 0 \le x \le 3,\ 0 \le y \le 10, \text{ and } y-7 \le x \le y.$$

■

■ **Example 3.27** Suppose that a box contains balls labeled 0 and 1. Suppose that the proportion of balls with 1 is p. We will sample n balls without replacement and let X_i be the number on the ith ball drawn. Afterward, we will sample n balls with replacement and let Y_i be the number on the ith ball drawn. Let X be the number of 1's drawn without replacement (first set of draws), and let Y be the number of 1's drawn with replacement (second set of draws). Find the means of X and Y.

- The distributions of X and Y are different in general. (Just imagine the case in which n is larger than the number of 1's in the box.)

- However, $X = \sum_{i=1}^{n} X_i$ and $Y = \sum_{i=1}^{n} Y_i$ are both true.

- Also, $\mathrm{E}(X_i) = p = \mathrm{E}(Y_i)$ for each i. Hence, $\mathrm{E}(X) = np = \mathrm{E}(Y)$.

- X_1 and X_2 have the same marginal distribution : We know that $P(X_1 = 1) = p$. Use the law of total probability to write

$$\begin{aligned} P(X_2 = 1) &= P(X_2 = 1|X_1 = 1)P(X_1 = 1) + P(X_2 = 1|X_1 = 0)P(X_1 = 0) \\ &= \frac{np-1}{n-1}p + \frac{np}{n-1}(1-p) = p. \end{aligned}$$

  A generalization of this argument can be used to show that all X_i have the same distribution as $X_i \sim \mathrm{Bernoulli}(p)$, $i = 1, \ldots, n$. However, note that they are not independent, that is, the ith ball drawn is obviously affected by the previous balls drawn since the sampling is without replacement, as also discussed in Example 3.22.

- Indeed, we can think of drawing a ball one by one from the box that contains 100 balls, of which r are red, that is, "labeled 1", as we have seen in Example 1.22. Then, the proportion of red balls is $\frac{r}{100}$, which can be regarded as p, and $\{X_1 = 1\}$ indicates the event that the first ball drawn is red, $P(X_1 = 1) = \frac{r}{100} = p$. As argued in Example 1.22, the probability that the ith ball draw is red, $P(X_i = 1)$ is all the same as $\frac{99!r}{100!} = \frac{r}{100} = p$, which implies that $X_i \sim \mathrm{Bernoulli}(p)$, $i = 1, \ldots, n$.

■

■ **Example 3.28** An airport limousine bus has N passengers. Each passenger is as likely to get off at any stop as at any other (each passenger is equally likely to choose any of stops), and the passengers act independently of one another. The bus makes a stop only if someone wants to get off. Suppose that there are 10 stops and that each passenger gets off at one of stops. Find the expected number of stops that the airport limousine bus makes.

- Let X be the number of stops the airport limousine bus makes and let W_i denote the indicator that the at least one passenger gets off at the ith stop, $i = 1, \ldots, 10$. If no passenger gets off at the ith stop, then $W_i = 0$ and otherwise $W_i = 1$. Then, $X = \sum_{i=1}^{10} W_i$.

- Note that $P(W_1 = 0) = P(\text{no passenger gets off at the first stop}) = 0.9^N$ and that

$E(W_1) = P(W_1 = 1)$. Thus, by symmetry $E(X) = \sum_{i=1}^{10} E(W_i) = 10\left(1 - 0.9^N\right)$.

■

■ **Example 3.29** Suppose that n different numbers are picked at random from the numbers $1, 2, 3, \ldots, N$, where $n < N$. Let the random variable X be the sum of the n numbers picked. Find the expected value of X, $E(X)$.

- Let I_i be an indicator that has the value 1 if the number i is one of n different numbers picked out of N different numbers, and 0 otherwise, and let $X_i = i \times I_i$, $i = 1, 2, \ldots, N$. Then, we can write $X = X_1 + \cdots + X_N$. Note that $E(I_i) = \dfrac{\binom{N-1}{n-1}}{\binom{N}{n}} = \dfrac{n}{N}$ since the event that the number i is one of n different picked out of N different numbers is equivalent to the event that we choose $n-1$ different numbers out of $N-1$ numbers other than i. Hence, we have $E(X) = \sum_{i=1}^N E(X_i) = \sum_{i=1}^N i \times E(I_i) = \sum_{i=1}^N i \times \frac{n}{N} = \frac{n(N+1)}{2}$.
- Alternative, let W_i be the ith number in n different numbers picked when we pick a number one by one without replacement. Then $X = W_1 + \cdots + W_n$ and $E(X) = \sum_{i=1}^n E(W_i)$. Note that $P(W_1 = k) = \frac{1}{N}$ and that $E(W_1) = \frac{1}{N}\sum_{i=1}^N i = \frac{N+1}{2}$. Next, consider the joint p.m.f. of W_1 and W_2, which is given $P(W_1 = k, W_2 = \ell) = \dfrac{1}{N} \times \dfrac{1}{N-1}$, $k, \ell = 1, \ldots, N$ and $k \neq \ell$. Thus, the marginal p.m.f. of W_2 is given by $P(W_2 = \ell) = \sum_{k=1, k\neq\ell}^N P(W_1 = k, W_2 = \ell) = \sum_{k=1,k\neq\ell}^N \frac{1}{N} \times \frac{1}{N-1} = \frac{1}{N}$, which is the same of the p.m.f. of W_1. Indeed, $W_1, \ldots, W_n$ have all the same distribution. Hence, it follows that $E(W_1) = E(W_2) = \cdots = E(W_n) = \frac{N+1}{2}$ and that $E(X) = \sum_{i=1}^n \frac{N+1}{2} = \frac{1}{2}n(N+1)$.

■

■ **Example 3.30 — Variance of hypergeometric random variable.** Let X have a hypergeometric distribution with parameters A, B, and n. Think of X as being the sum $X_1 + \cdots + X_n$ where $X_i = 1$ if the ith draw is success and $X_i = 0$ otherwise. Then all X_i have the same mean and all X_iX_j for $j \neq i$ have the same mean.

- The variance of X is $E(X^2) - [E(X)]^2$, and we have

$$E[X^2] = E\left[\sum_{i=1}^n X_i^2 + \sum_{i=1}^n \sum_{j\neq i} X_iX_j\right] = \sum_{i=1}^n E[X_i^2] + \sum_{i=1}^n \sum_{j\neq i} E[X_iX_j],$$

- We compute

$$\begin{aligned} E(X_1^2) &= E(X_1) = \frac{A}{A+B}, \\ E(X_1X_2) &= P(\text{first two draws are both success}) \\ &= P(X_1 = 1)P(X_2 = 1|X_1 = 1) = \frac{A(A-1)}{(A+B)(A+B-1)}. \end{aligned}$$

- Then, we have

$$\mathrm{E}(X) = \frac{nA}{A+B}, \quad \mathrm{Var}(X) = \frac{nAB}{(A+B)^2} \cdot \frac{A+B-n}{A+B-1}.$$

- Let $T = A+B$ denote the total number of balls in the box, and let $p = A/T$ denote the proportion of red balls in the box. Then,

$$\mathrm{Var}(X) = np(1-p)\alpha, \quad \alpha = \frac{T-n}{T-1},$$

where α is called the *finite population correction*. As $T \to \infty$, $\alpha \to 1$ for each fixed sample size n.

■

■ **Example 3.31 — Variance of matching.** In the matching problem, let X be the number of students who get the correct phone. We have computed the mean of X, $\mathrm{E}(X) = 1$ in Example 3.19. Here we compute $\mathrm{Var}(X)$

- Let I_i be the indicator of the event that the ith student gets the correct phone. Then,

$$X = \sum_{i=1}^{n} I_i, \quad \mathrm{Var}(X) = \mathrm{E}(X^2) - [\mathrm{E}(X)]^2 = \mathrm{E}(X^2) - 1.$$

- By symmetric $P(I_i = 1) = 1/n$ for all i, and for $i \neq j$, $P(I_i = 1, I_j = 1) = \dfrac{1}{n(n-1)}$.
- Then, since $I_i^2 = I_i$, we have

$$\mathrm{E}(X^2) = \mathrm{E}\left(\sum_{i=1}^{n} I_i^2 + \sum_{i \neq j} I_i I_j\right) = 1 + 2\binom{n}{2}\frac{1}{n(n-1)} = 1 + 1.$$

Hence, $\mathrm{Var}(X) = 1$.

■

Definition 3.3.5 — Independence of two random variables. Two random variables X and Y are *independent* if $P(X \in A, Y \in B) = P(X \in A)P(Y \in B)$, for all sets A and B. Thus, for two discrete random variables X and Y, if X and Y are independent, then

$$P(X = x, Y = y) = P(X = x)P(Y = y), \quad f_{X,Y}(x,y) = f_X(x) \cdot f_Y(y).$$

Equivalently, X and Y are *independent* if $P(X \in A | Y \in B) = P(X \in A)$, for all sets A and B such that $P(Y \in B) > 0$, which implies that $f_{X|Y}(x|y) = f_X(x)$.

■ **Example 3.32** Suppose that 3 balls are drawn at random from an urn that contains 2 red, 5 blue, and 3 green balls. Note that 3 balls are selected at one time, equivalently, it is selected one by one without replacement. Let X be the number of red balls and Y be the number of blue balls.

- The joint p.m.f. of X and Y is given by

$$f_{X,Y}(x,y) = P(X=x, Y=y) = \frac{\binom{2}{x}\binom{5}{y}\binom{3}{3-x-y}}{\binom{10}{3}}, \; x=0,1,2, \; y=0,1,2,3.$$

- Then, the marginal p.m.f. of X is obtained by adding up the joint p.m.f. with respect to y, $f_X(x) = \sum_{y=0}^{3} f_{X,Y}(x,y)$, and the same with the marginal p.m.f. of Y, $f_Y(y) = \sum_{x=0}^{2} f_{X,Y}(x,y)$, which are summarized in the following tale :

	Y $f_{X,Y}(x,y)$	0	1	2	3	$f_X(x)$
X	0	1/120	15/120	30/120	10/120	56/120
	1	6/120	30/120	20/120	0	56/120
	2	3/120	5/120	0	0	8/120
	$f_Y(y)$	10/120	50/120	50/120	10/120	1

Thus, we see that X and Y are not independent. For example, $P(X=2, Y=2) = f_{X,Y}(2,2) = 0 \neq \frac{8}{120} \cdot \frac{10}{120} = f_X(2) f_Y(2) = P(X=2)P(Y=2)$. Indeed, it is intuitively true that X and Y are not independent since the sampling is without replacement.

■

Theorem 3.3.3 If the nonnegative, integer-valued (discrete) random variables X and Y are independent, then

$$P(X+Y=m) = \sum_{x=0}^{m} P(X=x)P(Y=m-x), \; m=0,1,\ldots,$$

which is called the *convolution* for discrete random variables.

More generally, let X and Y be two independent integer-valued discrete random variables, with p.m.f $f_X(x)$ and $g_Y(y)$ respectively. Then the convolution of $f_X(x)$ and $g_Y(y)$ is the p.m.f h_Z of $Z = X+Y$ given by

$$h_Z(z) = \sum_{n=-\infty}^{\infty} f_X(n)\, g_Y(z-n),$$

for $z = -\infty, \ldots, -2, -1, 0, 1, 2, \ldots, \infty$.

Proof. Let A be the event $\{X+Y=m\}$ and B_x be the event $\{X=x\}$. Then $A = \cup_{x=0}^{\infty} A \cap B_x$ and $A \cap B_x$ are all disjoint. Thus,

$$P(X+Y=m) = \sum_{x=0}^{\infty} P(A \cap B_x) = \sum_{x=0}^{\infty} P(X+Y=m, X=x)$$

$$= \sum_{x=0}^{\infty} P(X=x, Y=m-x) = \sum_{x=0}^{\infty} P(X=x)P(Y=m-x), \; 0 \leq x \leq m,$$

if X and Y are two independent, nonnegative integer-valued random variables. ■

■ **Example 3.33** Let $X \sim \text{Binomial}(n_1, p)$ and $Y \sim \text{Binomial}(n_2, p)$, which are independent. Then, the conditional distribution of X given that $X+Y=m$ is the hypergeometric distribution.

- $X+Y \sim \text{Binomial}(n_1+n_2, p)$ since we can write $X = \sum_{i=1}^{n_1} X_i$ and $Y = \sum_{j=1}^{n_2} Y_j$, where $X_1, \dots, X_{n_1} \overset{i.i.d}{\sim} \text{Bernoulli}(p)$ and $Y_1, \dots, Y_{n_2} \overset{i.i.d}{\sim} \text{Bernoulli}(p)$.

- Alternatively, by Theorem 3.3.3 and Vandermonde's identity, we have

$$P(X+Y=m) = \textstyle\sum_{x=0}^{m} \binom{n_1}{x} p^x (1-p)^{n_1-x} \binom{n_2}{m-x} p^{m-x} (1-p)^{n_2-(m-x)}$$
$$= p^m (1-p)^{n_1+n_2-m} \textstyle\sum_{x=0}^{m} \binom{n_1}{x}\binom{n_2}{m-x} = \binom{n_1+n_2}{m} p^m (1-p)^{n_1+n_2-m}.$$

Note that X is the number of successes out of n_1 independent trials and Y is the number of successes out of n_2 additional independent trials with the same success probability. Thus $X+Y$ is the total number of successes out of n_1+n_2 trials.

- The conditional p.m.f. of X given $X+Y=m$ is

$$P(X=x|X+Y=m) = \frac{P(X=x, X+Y=m)}{P(X+Y=m)} = \frac{P(X=x)P(Y=m-x)}{P(X+Y=m)}$$
$$= \frac{\binom{n_1}{x} \cancel{p^x (1-p)^{n_1-x}} \binom{n_2}{m-x} \cancel{p^{m-x}(1-p)^{n_2-(m-x)}}}{\binom{n_1+n_2}{m} \cancel{p^m (1-p)^{n_1+n_2-m}}} = \frac{\binom{n_1}{x}\binom{n_2}{m-x}}{\binom{n_1+n_2}{m}},$$

which is a hypergeometric p.m.f with parameters m, n_1 and n_2.

■

■ **Example 3.34 — Capture-recapture (revisited).** In the capture-recapture problem of Example 3.24, we can compute the expected value of Y, the number of animals to be recaptured until r tagged animals are recaptured, whose p.m.f. is $P(Y=y) = \frac{k}{N} \times \frac{\binom{k-1}{r-1}\binom{N-k}{y-r}}{\binom{N-1}{y-1}}$, $r \le y \le N-k+r$.

- Note that $\sum_{y=r}^{N-k+r} \frac{k}{N} \times \frac{\binom{k-1}{r-1}\binom{N-k}{y-r}}{\binom{N-1}{y-1}} = 1$ since $P(Y=y)$ is the p.m.f. Thus, we have

$$E(Y) = \sum_{y=r}^{N-k+r} yP(Y=y) = \sum_{y=r}^{N-k+r} r\binom{k}{r}\binom{N-k}{y-r} \Big/ \binom{N}{y}$$
$$= r\left(\frac{N+1}{k+1}\right) \sum_{y=r}^{N-k+r} \frac{k+1}{N+1} \binom{k}{r}\binom{N-k}{y-r} \Big/ \binom{N}{y}$$
$$= r\left(\frac{N+1}{k+1}\right) \sum_{t=r+1}^{(N+1)-(k+1)+(r+1)} \frac{\frac{k+1}{N+1}\binom{(k+1)-1}{(r+1)-1}\binom{(N+1)-(k+1)}{t-(r+1)}}{\binom{(N+1)-1}{t-1}} = r\left(\frac{N+1}{k+1}\right),$$

where the last equality holds since it is the p.m.f. of a random variable T, the number of animals to be captured until $r+1$ tagged animals are recaptured, whose sum equals one.

- Alternatively, suppose that you were to capture them all, and let X_0 be the number of untagged animals captured before the first tagged animal, X_r the number of untagged animals captured between the rth and the $(r+1)$st tagged animals and X_k the number captured after the last tagged animal. Then, note that $\sum_{i=0}^{k} X_k = N-k$, and $\mathrm{E}(X_i) = \mathrm{E}(X_j)$ for all i and j since they have the same marginal distribution by symmetry. Hence, by taking expectations of both sides, $\sum_{i=0}^{k} \mathrm{E}(X_k) = N-k$, we have $\mathrm{E}(X_i) = \frac{N-k}{k+1}$. Finally, the number of animals recaptured until the rth tagged animal is captured can be expressed as $Y = (X_0+1)+(X_1+1)+\ldots+(X_{r-1}+1)$, we have $\mathrm{E}(Y) = r + \sum_{i=0}^{r-1} \mathrm{E}(X_i) = r\left(\frac{N+1}{k+1}\right)$.

■

Remark The distribution of Y in Example 3.24 is called the *negative hypergeometric distribution.*

- Under the dependent Bernoulli trials, the waiting time has the *negative hypergeometric distribution*, similar to the negative binomial distribution for the waiting time based on independent Bernoulli trials, which we discuss in the next section.

■ **Example 3.35 — Quick sort algorithm.** A sorting algorithm is an algorithm that puts elements of a list in a certain order. The output of a sorting algorithm is in nondecreasing order, and it is a permutation (reordering) of a list of the input values. The quick sort algorithm is (on average) one of the fastest sorting algorithms currently available and is widely used. The input is a list of n numbers $x_1, x_2, \ldots, x_n$. For convenience, we will assume that the numbers are distinct. Here, we would like to consider a simple version of the algorithm, proceeds as follows.

- When $n = 2$, the algorithm compares the two numbers and puts them in order. When $n > 2$, the algorithm chooses a number at random (all numbers equally likely to be chosen) and partitions the numbers based on how they compare to the chosen number. Specifically, suppose the chosen number is x_i. Every number in the list is compared to x_i. Those less than x_i are put in a "bracket" to the left of x_i, and those greater than x_i are put in a bracket to the right of x_i.

- The remainder of the algorithm is recursive in that we now choose the next bracket *with more than one element in it*. We apply the same procedure to the numbers in that bracket, and we continue this way until all brackets have only one number.

- That is, we start the algorithm with a single bracket containing all the numbers. Then we do the following:

① Choose a number x (called the *pivot* element) at random in the current bracket.
② Partition the current bracket by comparing every element in that bracket with x as described above.
③ If every bracket has only one number, stop, the numbers are sorted. Otherwise, choose any bracket with more than one number and go to step ①.

Note that any bracket is equally likely to be chosen and that if the comparisons are performed in the natural order, from left to right, then the order of the elements in each bracket(sublist) is the same as in the initial list.

Let N be the number of comparisons made in executing the quick sort. We compute the expected value of N.

- Let $x_{(1)} \le x_{(2)}, \dots, x_{(n)}$ be the same values as the input values $x_1, x_2, \dots, x_n$ but sorted in increasing order. For $i < j$, let X_{ij} be a random variable that takes on the value 1 if $x_{(i)}$ and $x_{(j)}$ are compared at any time over the course of the algorithm, and 0 otherwise. Then the total number of comparisons X satisfies $N = \sum_{i=1}^{n-1} \sum_{j=i+1}^{n} X_{ij}$ and that $\mathrm{E}(N) = \sum_{i=1}^{n-1} \sum_{j=i+1}^{n} \mathrm{E}(X_{ij})$.

- Now, $x_{(i)}$ and $x_{(j)}$ are compared if and only if either $x_{(i)}$ or $x_{(j)}$ is the first pivot selected from the set $A_{ij} = \{x_{(i)}, x_{(i+1)}, \dots, x_{(j-1)}, x_{(j)}\}$. This is because if $x_{(i)}$ (or $x_{(j)}$) is the first pivot selected from this set, then $x_{(i)}$ and $x_{(j)}$ must still be in the same bracket, and hence they will be compared. Similarly, if neither is the first pivot from this set, then $x_{(i)}$ and $x_{(j)}$ will be separated into distinct brackets, and so will not be compared. Since our pivots are chosen independently and uniformly at random from each bracket, it follows that, the first time a pivot is chosen from A_{ij}, it is equally likely to be any element from this set. Thus, the probability that $x_{(i)}$ or $x_{(j)}$ is the first pivot selected from A_{ij}, which is the probability that $X_{ij} = 1$, is $\dfrac{2}{j-i+1}$. Here, $P(X_{ij} = 1)$ is computed as the probability of $P(\{X \in \{x_{(i)}, x_{(j)}\})$, where X has a discrete uniform distribution on A_{ij} as discussed in Example 3.13.

- Alternatively, we can get $P(X_{ij} = 1)$ in the following way. Two numbers $x_{(i)}$ and $x_{(j)}$ (with $i < j$) will be compared only if $x_{(i)}$ or $x_{(j)}$ is selected before any of $x_{(i+1)}, \dots, x_{(j-1)}$ is selected. Thus, we can compute the probability that the event C_{ij} that $x_{(i)}$ or $x_{(j)}$ is chosen occurs before the event D_{ij} that $x_{(i+1)}, \dots, x_{(j-1)}$ is chosen, which is $P(C_{ij})/(P(C_{ij}) + P(D_{ij}))$. Specifically, we compute

$$
\begin{aligned}
&P(X_{ij} = 1) = P(C_{ij} | C_{ij} \cup D_{ij}) \\
= \;& P\left(x_{(i)} \text{ or } x_{(j)} \text{ is chosen} \,\middle|\, \text{one of } x_{(i)}, \dots, x_{(j)} \text{ is chosen} \right) \\
= \;& \frac{P\left(x_{(i)} \text{ or } x_{(j)} \text{ is chosen}\right)}{P\left(\text{one of } x_{(i)}, \dots, x_{(j)} \text{ is chosen}\right)} = \frac{2/\cancel{\#(\Omega)}}{(j-i+1)/\cancel{\#(\Omega)}} = \frac{2}{j-i+1},
\end{aligned}
$$

where we do not care about the value $\#(\Omega)$, all the possible ways of choosing a pivot from any bracket in the process of the algorithm, because it is the same in both numerator and denominator and thus cancels.

- Using the substitution $k = j - i + 1$, we then have

$$\begin{aligned} \mathrm{E}(N) &= \sum_{i=1}^{n-1}\sum_{j=i+1}^{n}\frac{2}{j-i+1} = \sum_{i=1}^{n-1}\sum_{k=2}^{n-i+1}\frac{2}{k} = \sum_{k=2}^{n}\sum_{i=1}^{n+1-k}\frac{2}{k} \\ &= \sum_{k=2}^{n}(n+1-k)\frac{2}{k} = (2n+2)\sum_{k=1}^{n}\frac{1}{k} - 4n \\ &\approx (2n+2)\int_1^n \frac{1}{x}dx - 4n = (2n+2)\log n - 4n. \end{aligned}$$

■

R Suppose that we have the following sequence of numbers

$$5, 9, 3, 10, 11, 14, 8, 4, 17, 6$$

and the second value '9' was chosen at random.

- The first stage of the sorting would bracket the numbers as follows

$$\{5,3,8,4,6\}\{9\}\{10,11,14,17\}.$$

- The remainder of the algorithm is recursive in that we now choose the next bracket with more than one element in it. We apply the same procedure to the numbers in that bracket, and we continue this way until all brackets have only one number.
- Suppose we choose the left bracket above at the next stage and that the chosen number within that bracket is the second value '3'. After partitioning the left bracket our list looks as follows:

$$\{3\}\{5,8,4,6\}\{9\}\{10,11,14,17\}.$$

- Next we choose the first bracket with more than one number $\{5,8,4,6\}$. Suppose the fourth value '6' is the chosen number within that bracket, the next stage of our sort looks as follows:

$$\{3\}\{5,4\}\{6\}\{8\}\{9\}\{10,11,14,17\}.$$

This continues until all brackets have only one element. Try it.

```
rand.quick.sort = function(x) {
  if (length(x) <= 1) return(x)
  initIdx = sample(seq(length(x)), size=1)
  pivot = x[initIdx]
  therest = x[-initIdx]
  lefts = therest[therest < pivot]
  rights = therest[therest >= pivot]
  print(c(lefts, pivot, rights))
  lefts = rand.quick.sort(lefts)
  rights = rand.quick.sort(rights)
  return(c(lefts, pivot, rights))
}
```

```
x <- c( 5, 9, 3, 10, 11, 14, 8, 4, 17, 6)

rand.quick.sort(x)
[1]  5  3  8  4  6  9 10 11 14 17

[1] 3 5 8 4 6

[1] 5 4 6 8

[1] 4 5

[1] 10 11 14 17

[1] 14 17

[1]  3  4  5  6  8  9 10 11 14 17
```

3.4 Waiting time distributions

The Bernoulli process can be understood as a sequence of independent coin tosses, where the probability of heads in each toss is a fixed number p in the range $0 < p < 1$. In general, the Bernoulli process consists of a sequence of Bernoulli trials, where each trial produces a 1 (a success) with probability p, and a 0 (a failure) with probability $1-p$, independently of what happens in other trials.

■ **Example 3.36** Suppose that you toss a coin until you get a head. Let X be the number of trials until the first head is obtained. Suppose that the coin tosses are independent Bernoulli trials with parameter p. Find the p.m.f. of X.

- The p.m.f. of X is for $x = 1, 2, \ldots,$

  $$P(X = x) = f_X(x) = P(\text{first } x-1 \text{ tosses are tail and next is head}) = (1-p)^{x-1}p.$$

- Note that $f_X(x) > 0$ and $\sum_{x=1}^{\infty} f_X(x) = \sum_{x=1}^{\infty} (1-p)^{x-1}p = \frac{p}{1-(1-p)} = 1.$

■

Definition 3.4.1 If a discrete random variable X whose p.m.f. is given by

$$f_X(x) = \begin{cases} (1-p)^{x-1}p, & \text{if } x = 1, 2, \ldots, \\ 0, & \text{otherwise.} \end{cases}$$

is said to have a *geometric distribution* with parameter p, denoted by $X \sim \text{Geo}(p)$.

Remark Let X be the number of failures until the first success in a sequence of i.i.d. Bernoulli trials with probability p of success. The p.m.f. of X:

$$f_X(x) = \begin{cases} (1-p)^x p, & \text{for } x = 0, 1, \ldots, \\ 0, & \text{otherwise.} \end{cases}$$

The distribution of X, the number of failures, can also be called the Geometric distribution, excluding the success.

Theorem 3.4.1 Suppose that $X \sim \text{Geo}(p)$. Then,

- $\text{E}(X) = 1/p$, and $\text{Var}(X) = \dfrac{q}{p^2}$, $q = 1-p$,
- $P(X > x-1) = P(X \geq x) = q^{x-1}$ for $x = 1, 2, \ldots$.

Proof. A geometric series $\sum_{n=1}^{\infty} ar^{n-1} = \dfrac{a}{1-r}$ for $|r| < 1$ can be used to obtain properties for a geometric random variable.

- The expected value of X is

$$\begin{aligned} \text{E}(X) &= \sum_{x=1}^{\infty} xP(x=x) = \sum_{x=1}^{\infty} x(1-p)^{x-1}p = p\sum_{x=0}^{\infty} -\frac{d}{dp}(1-p)^x \\ &= -p\frac{d}{dp}\sum_{x=0}^{\infty}(1-p)^x = -p\frac{d}{dp}\left(\frac{1}{p}\right) = p\frac{1}{p^2} = \frac{1}{p}. \end{aligned}$$

- Alternatively, the m.g.f. of the geometric distribution with parameter p can be evaluated

$$M_X(t) = \text{E}(e^{tX}) = \sum_{x=1}^{\infty} p(1-p)^{x-1}e^{tx} = \frac{pe^t}{1-(1-p)e^t}, \; t < -\log(1-p).$$

  Then, by differentiating $M_X(t)$ and plugging $t = 0$, we get $\text{E}(X) = M_X'(0)$

$$\text{E}(X) = M_X'(0) = M_X'(t)|_{t=0} = \left.\frac{pe^t}{[1-(1-p)e^t]^2}\right|_{t=0} = \frac{1}{p}.$$

  Hence the expected value of X is the reciprocal of the parameter p.

- Consider $\text{E}[X(X-1)] = \sum_{x=1}^{\infty} x(x-1)P(X=x) = \sum_{x=1}^{\infty} x(x-1)q^{x-1}p$, $q = 1-p$. Then, since $x(x-1)q^{x-2}$ is a second derivative of q^x with respect to q, we get

$$\begin{aligned} \text{E}[X(X-1)] &= \sum_{x=1}^{\infty} x(x-1)q^{x-1}p, \; q = 1-p = pq\sum_{x=1}^{\infty} x(x-1)q^{x-2} \\ &= pq\sum_{x=1}^{\infty}\frac{d^2}{dq^2}q^x = pq\frac{d^2}{dq^2}\sum_{x=1}^{\infty} q^x = pq\frac{d^2}{dq^2}\frac{q}{1-q}, \end{aligned}$$

which yields $\mathrm{E}[X(X-1)] = pq\dfrac{2}{(1-q)^3} = \dfrac{2q}{p^2}$. Thus, we have

$$\mathrm{Var}(X) = \mathrm{E}[X(X-1)] + \mathrm{E}(X) - (\mathrm{E}[X])^2 = \frac{2q}{p^2} + \frac{1}{p} - \frac{1}{p^2} = \frac{q}{p^2}.$$

Alternatively, the mean of X^2 is what we get by plugging $t = 0$ into this the second derivative of $M_X(t)$,

$$M_X''(t) = \frac{[1-(1-p)e^t]^2 pe^t + pe^t 2[1-(1-p)e^t](1-p)e^t}{[1-(1-p)e^t]^4}.$$

Hence, the variance of X is

$$\mathrm{Var}(X) = M_X''(0) - [M_X'(0)]^2 = \frac{p^3 + 2p^2(1-p)}{p^4} - \frac{1}{p^2} = \frac{1-p}{p^2}.$$

- Note that

$$P(X > x-1) = P(X \geq x) = \sum_{n=x}^{\infty} P(X = n) = \sum_{n=x}^{\infty} q^{n-1}p = q^{x-1},\ x = 1, 2, \ldots.$$

■

■ **Example 3.37** An urn contains N white and M black balls. Balls are randomly selected, one at a time with replacement, until a black one is obtained. Let X be the number of draws to get one black ball.

- What is the probability that exactly n draws are needed?

$$P(X = n) = (1-p)^{n-1}p,\ X \sim \mathrm{Geo}(p),\ p = \left(\frac{M}{N+M}\right).$$

- What is the probability that at least k draws are needed? $P(X \geq k) = (1-p)^{k-1}$ as shown in Theorem 3.4.1.

■

Theorem 3.4.2 Let X be a discrete random variable with positive integers. Then,

$$\mathrm{E}(X) = \sum_{n=1}^{\infty} P(X \geq n) = 1 + \sum_{n=1}^{\infty} P(X > n).$$

Proof. Let F_n denote the event $\{\omega : X(\omega) \geq n\}$. Then, X can be written as $X(\omega) = \sum_{n\geq 1} I_{F_n}(\omega)$. Thus, it follows that $\mathrm{E}(X) = \sum_{n\geq 1} \mathrm{E}(I_{F_n}) = \sum_{n\geq 1} P(F_n) = \sum_{n\geq 1} P(X \geq n)$.

- Alternatively, note that $\sum_{i=1}^{\infty} P(X \geq i) = P(X \geq 1) + P(X \geq 2) + P(X \geq 3) + \cdots$ Thus, it is rewritten as

$$
\begin{aligned}
\sum_{i=1}^{\infty} P(X \geq i) &= P(X \geq 1) + P(X \geq 2) + P(X \geq 3) + \cdots \\
&= P(X = 1) + P(X = 2) + P(X = 3) + \cdots \\
&\qquad\qquad\quad + P(X = 2) + P(X = 3) + \cdots \\
&\qquad\qquad\qquad\qquad\quad + P(X = 3) + \cdots \\
&\qquad\qquad\qquad\qquad\qquad\qquad \vdots \\
&= 1 \cdot P(X = 1) + 2 \cdot P(X = 2) + 3 \cdot P(X = 3) + \cdots
\end{aligned}
$$

Hence, $\sum_{i=1}^{\infty} P(X \geq i) = \sum_{i=1}^{\infty} iP(X = i) = \mathrm{E}(X)$. On the other hand, it can also be understood as follows :

$$
\begin{aligned}
\sum_{i=1}^{\infty} iP(X = i) &= 1 \cdot P(X = 1) + 2 \cdot P(X = 2) + 3 \cdot P(X = 3) + \cdots \\
&= P(X = 1) \\
&+ P(X = 2) + P(X = 2) \\
&+ P(X = 3) + P(X = 3) + P(X = 3) + \cdots \\
&= P(X \geq 1) + P(X \geq 2) + P(X \geq 3) + P(X \geq 4) + \cdots
\end{aligned}
$$

■

Remark If X is a discrete random variable that can take only the values $0, 1, 2, \ldots$, then we also have $\mathrm{E}(X) = \sum_{n=1}^{\infty} P(X \geq n)$.

■ **Example 3.38 — Expected waiting time.** Suppose an event occurs at the kth trial with probability p_k, with $\sum_1^{\infty} p_k = 1$. Let X be the waiting time that has a Geometric distribution, $X \sim \mathrm{Geo}(p)$ which implies that $p_k = P(X = k) = (1-p)^{k-1}p$. Then the expected waiting time to see the event is

$$
\mathrm{E}(X) = \sum_{1}^{\infty} kp_k = \sum_{k=1}^{\infty} P(X \geq k) = 1 + \sum_{k=1}^{\infty} q_k = \frac{1}{p}.
$$

where q_k is the probability of no success through the kth trial, that is, $q_k = P(X > k) = P(X \geq k+1) = (1-p)^k$. ■

Remark A Bernoulli process can also be used to model systems involving "arrival" of customers or jobs at service centers. We can model the *interarrival times* (the times between successive arrivals) as independent variables.

- Time is *discretized* into periods, and a "success" at the kth trial is associated with the arrival of at least one customer at the service center during the kth period. We can use the term "arrival" in place of "success" given the context of modeling.
- If we consider the case where arrivals occur in discrete time with Bernoulli process, then the interarrival times are geometrically distributed. In other words, given an arrival process with independent Bernoulli trials, the time until the first arrival has a geometric distribution.

Definition 3.4.2 — Memoryless property. A discrete distribution has the memoryless property if a random variable X has a distribution satisfying

$$P(X > m+n|X > m) = P(X > n) \text{ for all nonnegative integers } m,n.$$

Theorem 3.4.3 The geometric random variable has the memoryless property. That is, if X has a geometric distribution with parameter p, then

- The conditional distribution of $X-k$ given that $X > k$ is also a geometric distribution with parameter p

$$\begin{aligned} P(X = k+x|X > k) &= P(X = x), \\ P(X > s|X > t) &= P(X > s-t), \text{ for positive integers } s > t. \end{aligned}$$

Proof. If X has a geometric distribution with p.m.f $P(X = x) = (1-p)^{x-1}p$, then we have

$$\begin{aligned} P(X = k+x|X > k) &= \frac{P(X = k+x,\ X > k)}{P(X > k)} = \frac{(1-p)^{k+x-1}p}{(1-p)^k} = P(X = x), \\ P(X > n+m) &= (1-p)^{n+m} = (1-p)^n(1-p)^m = P(X > n)P(X > m). \end{aligned}$$

Hence the geometric distribution has the lack of memory property. ■

■ **Example 3.39** A baseball player's batting average is 0.300. That is, when the player comes up to bat there is a 30% chance of getting a hit. Given that after three times at bat, the player has not had a hit, what is the probability the player will not get a hit after five times at bat? The number of times at bat until getting a hit is a geometric random variable with $p = 0.3$. By the memoryless property, the desired probability is $P(X > 5|X > 3) = P(X > 2) = (1-p)^2 = (0.70)^2 = 0.49$. ■

■ **Example 3.40** People who play the lottery often can be heard saying that they are due to win because their number has not been drawn for a long time. Let X be the number of failures (their number is not drawn) before the first (or next) time that their number is drawn. If it has been $k = 365$ days since their number was last drawn (or since we started counting), then $X \geq 365$ and the conditional distribution of $X - 365$ is still a geometric distribution. They should expect to wait just as long from now as if they were starting over! ■

Remark The independence assumption underlying the Bernoulli process has important implications with the memoryless property.

- Whatever has happened in past trials provides no information on the outcomes of future trials. That is, future trials are independent from the past ones.
- Starting from any given point in time, the future is also modeled by a Bernoulli process, which is independent of the past(*fresh-start* or *starting over* property of the Bernoulli process).
- For any given time n, the sequence of random variables $X_{n+1}, X_{n+2}, \ldots$ (the future of the process) is also a Bernoulli process, and is independent from $X_1, \ldots, X_n$ (the past of the process).

■ **Example 3.41 — Coupon collector's problem(expected waiting time).** In the coupon collector's problem discussed in Example 2.41, find the expected number of packages until Joanne obtains all 12 different animal pictures. That is, compute $\mathrm{E}(T)$, where T, is the number of packages that needs to be collected until one obtains a complete set of at least one of each animal picture.

- As shown in Example 2.41, we have $P(T > n) = \sum_{j=1}^{n-1} (-1)^{j+1} \binom{n}{j} \left(1 - \frac{j}{n}\right)^k$. Thus, we can compute $\mathrm{E}(T)$ from its definition or the result in Theorem 3.4.2

$$\begin{aligned}
\mathrm{E}(T) &= \sum_{n=1}^{\infty} nP(T = n) = \sum_{n=1}^{\infty} n[P(T > n-1) - P(T > n)] \\
&= \sum_{n=1}^{\infty} P(T \geq n) = \sum_{n=1}^{\infty} P(T > n-1),
\end{aligned}$$

  which might be hard to evaluate.

- Alternatively, let T_i be the number of packages of Cheetos after Joanne has first seen $i-1$ new pictures until we see the ith new pictures. Then T is written as a sum of T_i, $T = \sum_{i=1}^{r} T_i$, where T_1 is the number of packages until the 1st new picture you have not seen before, which is always 1. Thus, $T_2 \sim \mathrm{Geo}\left(\frac{r-1}{r}\right)$ since after collecting the 1st animal picture, there is a $1/r$ change of getting the same animal picture you already had and an $(r-1)/r$ chance you will get something new. In general, $T_i \sim \mathrm{Geo}\left(\frac{r-i+1}{r}\right)$.

- Hence, by the linearity of expectation, we have

$$E(T) = \sum_{i=1}^{n} E(T_i) = \sum_{i=1}^{n} \frac{n}{n-i+1} = n \sum_{i=1}^{n} \frac{1}{i}.$$

■

■ **Example 3.42** Suppose that X_1 and X_2 are two independent and identically distributed(i.i.d) random variables with the geometric distribution with parameter p. Find the p.m.f of $Y = X_1 + X_2$.

- Using the convolution in Theorem 3.3.3, we have

$$\begin{aligned} P(Y=y) &= \sum_{x=1}^{y-1} P(X_1 = x)P(X_2 = y-x),\ y = 2,3,\ldots,\infty, \\ &= \sum_{x=1}^{y-1} (1-p)^{x-1} p (1-p)^{y-x-1} p \\ &= p^2 \sum_{x=1}^{y-1} (1-p)^{y-2} = (y-1)p^2(1-p)^{y-2},\ y = 2,3,\ldots, \end{aligned}$$

where Y is actually the number of trials until two successes (or the second success) occur, which is known as the *negative binomial distribution*.

■

Definition 3.4.3 — Negative binomial distribution. Let X denote the number of trials until the rth success occurs with success probability p. Here r is a positive integer greater than or equal to one. Then, $X = x$ if and only if the first $x+r-1$ trials contain exactly $r-1$ successes and the rth trial is success as shown in Figure 3.4. Thus, the p.m.f of X is given by

$$P(X=x) = \binom{x-1}{r-1} p^r (1-p)^{x-r}, \qquad x = r, r+1, \ldots.$$

- Notice that this p.m.f $f_X(x)$ can also be expressed as

$$f_X(x) = \binom{x+r-1}{r-1} p^r (1-p)^x, \qquad x = 0, 1, \ldots,$$

where X denotes the number of failures before the rth success in a Bernoulli process.

- When $r = 1$, X is geometrically distributed. For a general r, we say that X has a negative binomial distribution, $X \sim \text{Negbin}(r, p)$.

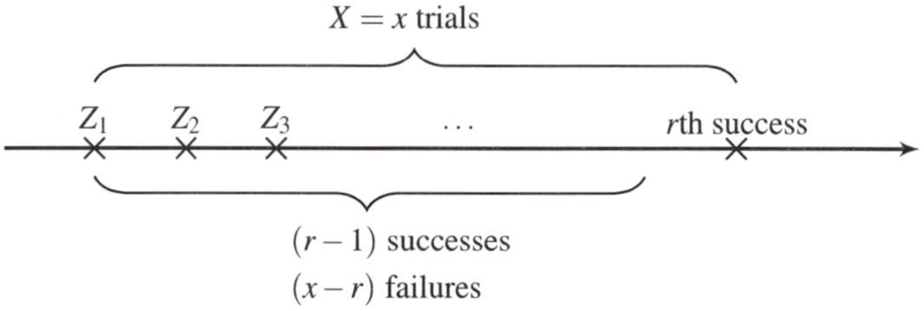

Figure 3.4: Arrival of the rth success at the xth Bernoulli trial

Theorem 3.4.4 Let r be a nonzero positive integer. Then

$$(1-y)^{-r} = \sum_{x=r}^{\infty} \binom{x-1}{r-1} y^{x-r}, \ |y| < 1.$$

Proof. Define $h(y) = (1-y)^{-r}$. By Taylor series expansion of $h(y)$ about $y = 0$, we get $(1-y)^{-r} = \sum_{k=0}^{\infty} \frac{h^{(k)}(0)}{k!} y^k$, where $h^{(k)}(0)$ is the kth derivative of $h(y)$ evaluated at $y = 0$, $h^{(k)}(0) = r(r+1)(r+2)\cdots(r+k-1) = \frac{(r+k-1)!}{(r-1)!}$. Then, plugging this into the Taylor's expansion of $h(y)$, we obtain

$$\begin{aligned}(1-y)^{-r} &= \sum_{k=0}^{\infty} \frac{(r+k-1)!}{(r-1)!k!} y^k = \sum_{k=0}^{\infty} \binom{r+k-1}{r-1} y^k \\ &= \sum_{x=r}^{\infty} \binom{x-1}{r-1} y^{x-r}, \ (\because x = r+k),\end{aligned}$$

which completes the proof of the theorem. ∎

Remark For $-1 < x < 1$, $(1+x)^m = \sum_{k=0}^{\infty} \binom{m}{k} x^k$, $\binom{m}{k} = \frac{m(m-1)(m-2)\cdots(m-k+1)}{k!}$. where we define

- If m is a positive integer, the series stops after $(m+1)$ terms.
- If $m = -1$,

$$\begin{aligned}\binom{-1}{1} &= -1, \ \binom{-1}{2} = \frac{-1(-2)}{2!} = 1, \\ \binom{-1}{k} &= \frac{-1(-2)(-3)\cdots(-1-k+1)}{k!} = (-1)^k \left(\frac{k!}{k!}\right) = (-1)^k.\end{aligned}$$

- Thus, $(1+x)^{-1} = \sum_{k=0}^{\infty} \binom{-1}{k} x^k = \sum_{k=0}^{\infty} (-1)^k x^k = 1 - x + \cdots + (-1)^k x^k + \cdots$.
- When m is a negative integer, we call the Binomial series **negative binomial expansion**. When $m = -n$, $n \in \mathbb{N}$,

$$
\begin{aligned}
(1+x)^m &= \sum_{k=0}^{\infty} \binom{-n}{k} x^k = \sum_{k=0}^{\infty} (-1)^k \frac{n(n+1)\cdots(n+k-1)}{k!} x^k \\
&= \sum_{k=0}^{\infty} (-1)^k \binom{n+k-1}{k} x^k \\
&= \sum_{\ell=n}^{\infty} \binom{\ell-1}{n-1} (-x)^{\ell-n}.
\end{aligned}
$$

- Let $X \sim \text{Negbin}(r,p)$. Then, we can check that $\sum_{x=r}^{\infty} P(X=x)$ is equal to one,

$$
\begin{aligned}
&\sum_{x=r}^{\infty} P(X=x) = \sum_{x=r}^{\infty} \binom{x-1}{r-1} p^r (1-p)^{x-r} \\
&= p^r \sum_{x=r}^{\infty} \binom{x-1}{r-1} (1-p)^{x-r} = p^r (1-(1-p))^{-r} = 1.
\end{aligned}
$$

Theorem 3.4.5 Let $X \sim \text{Negbin}(r,p)$. Then, the moment generating function is

$$
M_X(t) = \left(\frac{pe^t}{1-(1-p)e^t} \right)^r, \quad \text{for } t < -\log(1-p).
$$

Proof. By Theorem 3.4.4, we obtain

$$
\begin{aligned}
M(t) &= \sum_{x=r}^{\infty} e^{tx} f_X(x) = \sum_{x=r}^{\infty} e^{tx} \binom{x-1}{r-1} p^r (1-p)^{x-r} \\
&= p^r e^{tr} \sum_{x=r}^{\infty} \binom{x-1}{r-1} \left[e^t (1-p) \right]^{x-r}, \\
&= p^r e^{tr} \left[1-(1-p)e^t \right]^{-r}, \text{ if } |e^t(1-p)| < 1 \\
&= \left(\frac{pe^t}{1-(1-p)e^t} \right)^r, \text{ for } t < -\log(1-p).
\end{aligned}
$$

■

Example 3.43 — Interarrival times. In the Bernoulli process, consider the time (discretized time, i.e. the number of trials) of the rth success, which we denote by W_r. A related random variable is the r th interarrival time, denoted by T_r. It is defined by $T_1 = W_1$, $T_r = W_r - W_{r-1}$, $r = 2, 3, \ldots$, and represents the number of trials following the $(r-1)$st success until the next (the rth) success as shown in Figure 3.5

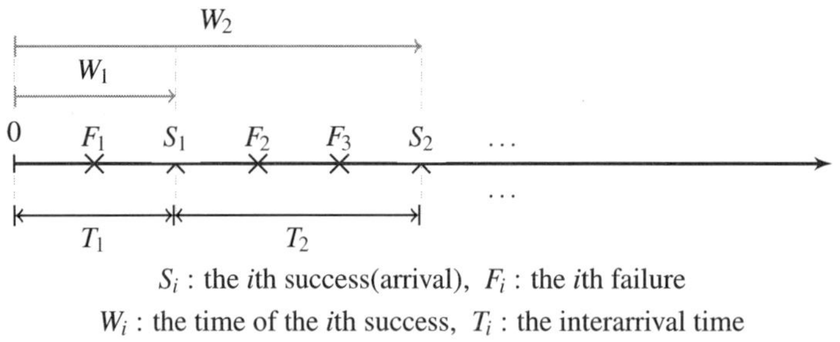

Figure 3.5: Interarrival time in the Bernoulli process

- Note that $W_r = T_1 + T_2 + \cdots + T_r$. In fact, W_r is the number of trials until the rth success, which has a negative binomial distribution with parameters r and p, $W_r \stackrel{d}{=} X \sim \text{Negbin}(r, p)$.

- Each T_i has a geometric distribution with parameter p, $T_i \sim \text{Geo}(p)$, and $T_1, T_2, T_3, \ldots$ are independent since the number of trials T_i after the $(i-1)$st success until the ith success has the number of trials T_1 until the first success with a fresh-starting Bernoulli process. Hence, we have $\text{E}(X) = \sum_{i=1}^{r} \text{E}(T_i) = \frac{r}{p}$.

■

■ **Example 3.44** Suppose that $X_1, X_2, \ldots$ are independent random variables, each having the geometric distribution with a p.m.f. $P(X_i = k) = (1-p)^k p$, $k = 0, 1, \ldots$. We can find the p.m.f. of $Y_n = \sum_{i=1}^{n} X_i$ and its expected value.

- X_i denotes the number of failures until the first success, in other words, $Y_1 = X_1$ is the number of tails until a head first appears.

- Let $Y_2 = X_1 + X_2$. Then, using the convolution formula in Theorem 3.3.3,

$$P(Y_2 = y) = P(X_1 + X_2 = y) = \sum_{x=0}^{y} (1-p)^x (1-p)^{y-x} p^2$$
$$= (y+1)(1-p)^y p^2 = \binom{y+1}{1} (1-p)^y p^2, \ \ y = 0, 1, 2, \ldots.$$

Thus, Y is the number of failures shown until the second head appears.

- Let $Y_n = \sum_{i=1}^{n} X_i$. Then, we see that Y_n is the number of tails until the nth head first appears. Thus, the event $\{Y_n = y\}$ is the event that $n-1$ heads and y tails appear in the first $y+n-1$ tosses and the nth head is obtained in the nth toss. Hence the required probability is $P(Y_n = y) = \binom{y+n-1}{n-1} (1-p)^y p^n$ for $y = 0, 1, 2, \ldots$, which is the p.m.f. of a negative binomial distribution with parameters n and p.

- We can also use the convolution formula to obtain $P(Y_n = y)$ by mathematical induction. Assume that $Y_n = \sum_{i=1}^{n} X_i$ has the p.m.f. above. Then, the p.m.f. of $Y_{n+1} = \sum_{i=1}^{n+1} X_i$ is

$$\begin{aligned} P(Y_{n+1} = y) &= P(X_{n+1} = x,\ Y_n = y - x) \\ &= \sum_{x=0}^{y} p(1-p)^x \binom{y-x+n-1}{n-1} (1-p)^{y-x} p^n \\ &= \sum_{x=0}^{y} p^{n+1}(1-p)^y \binom{y-x+n-1}{n-1} = p^{n+1}(1-p)^y \binom{y+n}{n}. \end{aligned}$$

Since for $n = 1$, it holds for $P(Y_1 = y)$, we get the result by mathematical induction.

- Here, we have used the identity $\sum_{x=0}^{y} \binom{n-1+x}{n-1} = \binom{n+y}{n}$, which follows from Theorem 1.3.5, $\binom{n}{k} = \sum_{\ell=k-1}^{n-1} \binom{\ell}{k-1}$. Alternatively, it can be derived by equating the coefficient of t^y on both sides $\frac{1}{1+t} \cdot \frac{1}{(1+t)^n} = \frac{1}{(1+t)^{n+1}}$ with negative binomial expansion.
- $\mathrm{E}(Y_n) = \sum_{i=1}^{n} \mathrm{E}(X_i) = n \sum_{k=0}^{\infty} k(1-p)^k p = n\frac{1-p}{p}$.

■

■ **Example 3.45** Suppose that X and Y are independent with X having a negative binomial distribution with parameters r and p and Y having a negative binomial distribution with parameters s and p. Then X has the distribution of the number of failures until we observe the rth success and Y has the distribution of the number of additional failures until we observe s additional successes. So $X+Y$ has the distribution of the number of failures until we observe $r+s$ successes. That is, $X+Y$ has a negative binomial distribution with parameters $r+s$ and p. Indeed, the same reasoning as in Example 3.44 can be applied to derive the distribution of $X+Y = \sum_{i=1}^{r} X_i + \sum_{j=1}^{s} Y_j$, where $X_1, \ldots, X_r, Y_1, \ldots, Y_s$ are all independent and identically distributed as geometric distribution with parameter p. ■

Remark Each toss of a coin results in heads with probability p. If m_r is the mean number of tosses up to and including the rth head, then we have

$$m_r = p(1+m_{r-1}) + (1-p)(1+m_r),\ r = 1,2,\ldots, \text{with } m(0) = 0,$$

which can be explained by the memoryless property of a geometric distribution or in terms of the conditional expectation. Indeed, we have the following result about the relationship between expectation and conditional expectation.

Definition 3.4.4 — Conditional expectation of discrete random variables. Let X be a discrete random variable and let B be an event such that $P(B) > 0$. Then, the *conditional expectation* of X given B denoted by $\mathrm{E}(X|B)$ is defined by

$$\mathrm{E}(X|B) = \sum_x xP(X = x|B) = \sum_x x f_{X|B}(x|B), \text{ if } \textstyle\sum_x |x| f_{X|B}(x|B) < \infty.$$

Suppose that X and Y have a joint p.m.f. $f_{X,Y}(x,y)$. For each y with $P(Y=y)>0$, the conditional expectation of X given $Y=y$ is defined by

$$\mathrm{E}(X|Y=y)=\sum_x xP(X=x|Y=y)=\sum_x xf_{X|Y}(x|y),\ \text{if } \textstyle\sum_x |x|f_{X|Y}(x|y)<\infty.$$

Theorem 3.4.6 Let X be a discrete random variable with mean $\mathrm{E}(X)$, and let B be an event such that $P(B)P(B^c)>0$. Then, $\mathrm{E}(X)=\mathrm{E}(X|B)P(B)+\mathrm{E}(X|B^c)P(B^c)$.

- Suppose that $\{B_i : i\geq 1\}$ is a partition of the sample space Ω such that $\cup_i B_i=\Omega$, $B_i\cap B_j=\emptyset$, $i\neq j$, and $P(B_i)>0$, $\forall i$. Then $\mathrm{E}(X)=\sum_i \mathrm{E}(X|B_i)P(B_i)$.
- Suppose that X and Y have a joint p.m.f. $f_{X,Y}(x,y)$. Then, it follows that

$$\mathrm{E}(X)=\sum_y \mathrm{E}(X|Y=y)P(Y=y).$$

Proof. By the definition of conditional p.m.f. and the law of total probability, we have

$$f_X(x)=P(X=x)=P(\{X=x\}\cap B)+P(\{X=x\}\cap B^c)=f_X(x|B)P(B)+f_X(x|B^c)P(B^c).$$

Hence, it follows that

$$\begin{aligned}\mathrm{E}(X) &= \sum_x xf_X(x)=P(B)\sum_x xf(x|B)+P(B^c)\sum_x xf(x|B^c)\\ &= P(B)\mathrm{E}(X|B)+P(B^c)\mathrm{E}(X|B^c).\end{aligned}$$

- If $\{B_i : i\geq 1\}$ is a partition of the sample space Ω, then by the law of total probability, $f_X(x)=\sum_i f_X(x|B_i)P(B_i)$. Thus, we have

$$\mathrm{E}(X)=\sum_x xf_X(x)=\sum_x x\sum_i f_X(x|B_i)P(B_i)=\sum_i \mathrm{E}(X|B_i)P(B_i).$$

- Suppose that X and Y have a joint p.m.f. $f_{X,Y}(x,y)$. Then, we have

$$\begin{aligned}\mathrm{E}(X) &= \sum_x xf_X(x)=\sum_x x\sum_y f_{X,Y}(x,y)=\sum_x\sum_y f_{X|Y}(x|y)f_Y(y)\\ &= \sum_y\left[\sum_x xf_{X|Y}(x|y)\right]f_Y(y)=\sum_y \mathrm{E}(X|Y=y)P(Y=y),\end{aligned}$$

where the interchange of the order of summation in the second equality holds since the summand is nonnegative, and the third equality holds by the multiplication rule and the fact $f_Y(y)>0$.

■

■ **Example 3.46** Each toss of a coin results in heads with probability p. Let X be the number of tosses before the first head (success) is obtained. Then, $P(X=x)=(1-p)^{x-1}p$, $x\geq 1$. To compute the expectation of X, we can use Theorem 3.4.6.

- By Theorem 3.4.6, it follows that $\mathrm{E}(X) = p\mathrm{E}(X|H_1) + (1-p)\mathrm{E}(X|T_1)$, where H_1 is the event that a head is obtained in the first toss and $T_1 = H_1^c$.

- Note that $\mathrm{E}(X|H_1) = 1$ given that a head is obtained on the first toss, whereas $\mathrm{E}(X|T_1) = 1 + \mathrm{E}(X)$ given that a tail is obtained on the first toss. Thus, we have $\mathrm{E}(X) = p + (1-p)[1 + \mathrm{E}(X)]$ and $\mathrm{E}(X) = \dfrac{1}{p}$, which is also given in Theorem 3.4.1.

- More generally, let T_r be the number of tosses up to and including the rth head. Then, we have the following recursive equation as before.

$$\begin{aligned} \mathrm{E}(T_r) &= \mathrm{E}(T_r|H_1)P(H_1) + \mathrm{E}(T_r|T_1)P(T_1) \\ &= p[1 + \mathrm{E}(T_{r-1})] + (1-p)[1 + \mathrm{E}(T_r)],\ r = 1, 2, \ldots,\ \mathrm{E}(T_0) = 0. \end{aligned}$$

■

■ **Example 3.47** Suppose that each printed character in your probability textbook is a typo independently with probability p, or is correct with probability $1-p$. Let n be the number of characters in the book and let X be the number of typos. Find the expected value of X.

- Let B be the event that the first character is a typo. Then, it follows that $\mathrm{E}(X) = P(B)\mathrm{E}(X|B) + P(B^c)\mathrm{E}(X|B^c) = p\mathrm{E}(X|B) + (1-p)\mathrm{E}(X|B^c)$.

- Let N_n be the expected number of typos in n characters, which means $N_n = \mathrm{E}(X)$. Then, $\mathrm{E}(X|B) = 1 + N_{n-1}$ and $\mathrm{E}(X|B^c) = N_{n-1}$. Thus, we have the following difference equation $N_n = p(1 + N_{n-1}) + (1-p)N_{n-1} = N_{n-1} + p$ with $N_0 = 0$. Hence, the solution of the difference equation is $N_n = np = \mathrm{E}(X)$.

- Indeed, we see that X has a binomial distribution with parameters n and p with $\mathrm{E}(X) = np$, which is the same as the solution of the difference equation.

■

■ **Example 3.48** Jaden and Joanne are playing the leisure activity of escaping the maze with three different paths. Each time they try to find the exit with three paths in the maze, they have a 30% chance of choosing the path 1, from which they can get out of the maze in a hour, a 40% of choosing the path 2, from which they come back to where they started after an hour, and a 30% of choosing the path 3, from which they return back to where they started after three hours. Find the expected amount of time they spend in the maze.

- Let X be the amount of time they spend in the maze. Let B_i be the event that they choose the path i, $i = 1, 2, 3$. Then, $\{B_1, B_2, B_3\}$ is a partition of the sample space. Thus, by Theorem 3.4.6, we have that $\mathrm{E}(X) = \sum_{i=1}^{3} P(B_i)\mathrm{E}(X|B_i)$.

- Note that $P(B_1) = 0.3$, $\mathrm{E}(X|B_1) = 1$, $P(B_2) = 0.4$, $\mathrm{E}(X|B_2) = 1 + \mathrm{E}(X)$, $P(B_3) = 0.3$, and $\mathrm{E}(X|B_3) = 3 + \mathrm{E}(X)$. Hence, $\mathrm{E}(X) = 0.3 \cdot 1 + 0.4 \cdot [1 + \mathrm{E}(X)] + 0.3 \cdot [3 + \mathrm{E}(X)]$, which implies that $\mathrm{E}(X) = \frac{16}{3}$.

■

■ **Example 3.49 — Gambler's ruin problem (revisited).** Suppose you have \$1000 to gamble with picking at random a number from 1 to 10. You bet \$1 on an odd number every time. That is, if you get an odd number, you will get \$1 and lose \$1 otherwise. Also, suppose that you quit the game once you achieve \$1050 or you are bankrupt. Find the expected length of the game.

- Let X be the length of the game (number of plays, i.e. pickings at random a number from 1 to 10) with $\$n$ until you achieve \$1050 or you are bankrupt and let B be the event that you get an odd number in the first picking. Then, we have $\mathrm{E}(X) = P(B)\mathrm{E}(X|B) + P(B^c)\mathrm{E}(X|B^c)$.

- Let T_n be the expected length of the game, that is, $T_n = \mathrm{E}(X)$. Then, $\mathrm{E}(X|B) = T_{n+1}$ and $\mathrm{E}(X|B^c) = T_{n-1}$. Thus, we have the following difference equation

$$T_n = 1 + (T_{n-1} + T_{n+1})/2, \; T_0 = T_{1050} = 0.$$

- To solve the difference equation, we rewrite it as $T_{n+1} - T_n = T_n - T_{n-1} - 2$,

$$\begin{aligned} T_2 - T_1 &= T_1 - T_0 - 2, \\ T_3 - T_2 &= T_2 - T_1 - 2, \\ &\vdots \\ T_{n-1} - T_{n-2} &= T_{n-2} - T_{n-1} - 2, \\ T_n - T_{n-1} &= T_{n-1} - T_{n-2} - 2. \end{aligned}$$

  Adding both sides of these equations, we have $T_n = T_{n-1} + (T_1 - T_0) - 2(n-1)$.

- Likewise, we rewrite the equation,

$$\begin{aligned} T_2 &= T_1 + (T_1 - T_0) - 2, \\ T_3 &= T_2 + (T_1 - T_0) - 2 \cdot 2, \\ &\vdots \\ T_{n-1} &= T_{n-2} + (T_1 - T_0) - 2(n-2), \\ T_n &= T_{n-1} + (T_1 - T_0) - 2(n-1). \end{aligned}$$

  Once again, adding both sides of these equations, we have $T_n = nT_1 - n(n-1)$.

- Further, with the boundary condition $T_{1050} = 0$, we have the equation $0 = T_{1050} = 1050T_1 - 1050(1050-1)$. Hence, we have the solution to the difference equation $T_n = n(1050 - n)$.

■

3.5 Poisson distribution

A Bernoulli process is a sequence of i.i.d. Bernoulli trials. Sometimes, we do not have that kind of control over how much data we collect or how many trials we consider. On the other hand, you may know the average rate of events, i.e., number of events that occur in a period. You count the actual number of events that occur in such a time period. For instance, you measure the number of calls at the help desk with two hours, the number of cars that pass in a given time period, the number of times that cell phones in a class room are ringing, and so on.

Example 3.50 — Help Desk. You are answering calls as a user consultant at the help desk. You may get 3 calls during five minutes, or it might take two hours before you get 3 calls. You can't just observe 3 calls the way you observe 3 patients in a clinical trial or 3 parts produced by a machine running on a schedule. ■

The family of Poisson distributions is commonly used to model count data. These are used for counts of occurrences that happen at random times without being the result of a fixed number of trials. For example, we might count how many customers arrive at a queue during a fixed time period, or how many times a machine breaks down during a shift, or how many typos are found in a given page of a book, etc.

Definition 3.5.1 A random variable X is said to have a Poisson distribution with parameter λ, denoted as $X \sim \text{Poisson}(\lambda)$, if its p.m.f is given by

$$P(X=x) = f_X(x) = \frac{e^{-\lambda}\lambda^x}{x!}, \; x = 0, 1, 2, \cdots, \;\; \lambda > 0.$$

Remark The probability mass function f_X is called the Poisson distribution after Siméon D. Poisson (1781-1840). It is easy to see that $\sum_{x=0}^{\infty} f_X(x)$ is equal to one,

$$\sum_{x=0}^{\infty} f_X(x) = \sum_{x=0}^{\infty} \frac{e^{-\lambda}\lambda^x}{x!} = e^{-\lambda}\sum_{x=0}^{\infty}\frac{\lambda^x}{x!} = e^{-\lambda}e^{\lambda} = 1.$$

Theorem 3.5.1 The expected value and variance of a Poisson random variable, $X \sim \text{Poisson}(\lambda)$ are both equal to its parameter λ, $\text{E}(X) = \lambda$, and $\text{Var}(X) = \lambda$. In addition,

- $\text{E}[X(X-1)(X-2)\cdots(X-k)] = \lambda^{k+1}$ for $k = 0, 1, 2, \ldots$.
- The m.g.f. is $\text{E}(e^{tX}) = M_X(t) = \exp(\lambda[\exp(t)-1])$. From this, we can also compute the mean and variance.

Proof. Note that

$$\begin{aligned} E(X) &= \sum_{x=0}^{\infty} x\frac{\lambda^x}{x!}e^{-\lambda} = \lambda e^{-\lambda}\sum_{x=1}^{\infty}\frac{\lambda^{x-1}}{(x-1)!} = \lambda e^{-\lambda}e^{\lambda} = \lambda, \\ E[X(X-1)] &= \sum_{x=0}^{\infty} x(x-1)\frac{\lambda^x}{x!}e^{-\lambda} = \lambda^2 e^{-\lambda}\sum_{x=2}^{\infty}\frac{\lambda^{x-2}}{(x-2)!} = \lambda^2 e^{-\lambda}e^{\lambda} = \lambda^2, \\ \text{Var}(X) &= E[X(X-1)] + E(X) - [E(X)]^2 = \lambda. \end{aligned}$$

- In general, we have

$$\begin{aligned} E[X(X-1)(X-2)\cdots(X-k)] &= \sum_{x=0}^{\infty} x(x-1)\cdots(x-k)\frac{\lambda^x}{x!}e^{-\lambda} \\ &= \lambda^k e^{-\lambda}\sum_{x=k+1}^{\infty}\frac{\lambda^{x-k-1}}{(x-k-1)!} = \lambda^k. \end{aligned}$$

- Also, we find the moment generating function of X.

$$M_X(t) = \sum_{x=0}^{\infty} e^{tx}\frac{e^{-\lambda}\lambda^x}{x!} = e^{-\lambda}\sum_{x=0}^{\infty} e^{tx}\frac{\lambda^x}{x!} = e^{-\lambda}e^{\lambda e^t} = e^{\lambda(e^t-1)}.$$

Thus, $M_X'(t) = \lambda e^t e^{\lambda(e^t-1)}$, and $E(X) = M_X'(0) = \lambda$. Similarly, $M_X''(t) = \lambda e^t e^{\lambda(e^t-1)} + (\lambda e^t)^2 e^{\lambda(e^t-1)}$. and $M_X''(0) = E(X^2) = \lambda^2 + \lambda$. Hence we get

$$\text{Var}(X) = E(X^2) - [E(X)]^2 = \lambda^2 + \lambda - \lambda^2 = \lambda = E(X).$$

■

Remark The binomial distribution with parameters n and p is nearly the same as the Poisson distribution with mean np if n is very large and p is very small while np is neither large nor small.

- That is, suppose that $X \sim \text{Binomial}(n,p)$ and $np = \lambda > 0$, then we have $P(X = 0) = (1-p)^n = \left(1-\frac{\lambda}{n}\right)^n \overset{n\to\infty}{\longrightarrow} e^{-\lambda}$.
- Also, $\frac{P(X=1)}{P(X=0)} = \frac{np(1-p)^{n-1}}{(1-p)^n} = \frac{np}{1-p} \approx \lambda$, which implies that $P(X=1) \approx \lambda e^{-\lambda}$. In general, we have $P(X=x) \approx \frac{\lambda^x}{x!}e^{-\lambda}$.
- More generally, we have the Poisson approximation to binomial in the following theorem.

Theorem 3.5.2 — Poisson approximation to binomial. Suppose that X has a binomial distribution with very large n and very small p and let $np = \lambda$, so that $p = \lambda/n$. We can write the p.m.f. as

$$f_X(x) = \frac{n(n-1)\cdots(n-x+1)}{x!}\left(\frac{\lambda}{n}\right)^x\left(1-\frac{\lambda}{n}\right)^{n-x} \approx \frac{\lambda^x}{x!}\exp(-\lambda).$$

Proof. Recall that $\left(1+\frac{a}{n}\right)^n \to e^a$ as $n \to \infty$ for $a \in R$. Then,

$$\begin{aligned} f_X(x) &= \binom{n}{x}p^x(1-p)^{n-x} = \frac{1}{x!}\frac{n(n-1)\ldots(n-x+1)}{n^x}(np)^x\left(1-\frac{np}{n}\right)^{n-x} \\ &\to \frac{1}{x!}\lambda^x e^{-\lambda},\ x = 0, 1, \ldots \end{aligned}$$

Hence, the Poisson distribution with mean λ is nearly the same as the binomial distribution with parameters n and λ/n if n is very large. ■

Remark To deal with $P(X = x) = \frac{n\cdot(n-1)\cdots(n-x+1)}{n\cdot n\cdots n}\frac{\lambda^x}{x!}\frac{(1-\lambda/n)^n}{(1-\lambda/n)^x}$,

- Take a limit as $n \to \infty$ and $p \to 0$ in such a way that $\lambda = np$ is fixed.
- Since x is fixed, so $\frac{n\cdot n-1\cdots n-x+1}{n\cdot n\cdots n} \to 1$ and $(1-\lambda/n)^x \to 1$.
- To deal with the $(1-\lambda/n)^n$ term, consider its logarithm $\log[(1-\lambda/n)^n] = n\log(1-\lambda/n) \approx n(-\lambda/n) = -\lambda$, so $(1-\lambda/n)^n \to e^{-\lambda}$.
- Therefore $P(X = x) \approx \frac{\lambda^x}{x!}e^{-\lambda}$.
- This indicates that in the limit as p becomes small and n large with np constant, the Binomial distribution converges to another distribution, which is a Poisson distribution with parameter λ.

■ **Example 3.51 — Poisson approximation to matching probability.** Suppose that there are 160 students in P&E college who take two courses A and B. Both courses are held in the same lecture hall (which has 160 seats), and the seating is random and independent for the two courses. Let N be the number of students in the same seat for both classes.

- We can find the (exact) probability that no one has the same seat for both courses by recalling matching problem,

$$P(N = 0) = P(\text{there is no match out of } n \text{ students}) = \sum_{j=0}^{n}\frac{(-1)^j}{j!},\ n = 160.$$

- Define I_i to be the indicator for student i having the same seat in both courses, so that $N = \sum_{i=1}^{160} I_i$. Then, $I_i \sim \text{Bernoulli}(p)$, $p = 1/160$, $i = 1, \dots,$ but they are not independent, so N does not have a binomial distribution. If N is a binomial random variable, then $np = 1$ and $n = 160$ is large enough for Poisson approximation of N. However, I_i and I_j can be considered approximately independent because

$$P(I_i = 1, I_j = 1) = \left(\frac{1}{160}\right)\left(\frac{1}{159}\right) \approx \left(\frac{1}{160}\right)^2 = P(I_i = 1)P(I_j = 1).$$

- We can find an approximation to the probability that no one has the same seat for both courses, using Poisson approximation,

$$P(N = 0) \approx e^{-\lambda} = e^{-1} \approx \sum_{j=0}^{n} \frac{(-1)^j}{j!}, \; n \to \infty, \; \lambda = np = 1.$$

- We can find an approximation to the probability that at least one student has the same seat for both courses, $P(N \geq 1) = 1 - P(N = 0) \approx 1 - e^{-1}$.

- Further, the probability that exactly k students have the same seat for both courses,

$$P(N = k) = \frac{1}{k!} \sum_{j=0}^{n-k} \frac{(-1)^j}{j!} \approx \frac{e^{-1}}{k!}, \text{ as } n \to \infty,$$

which is the Poisson probability $P(X = k)$ with parameter $\lambda = 1$, $X \sim \text{Poisson}(1)$.

- We can find the expected value of the number of students in the same seat for both classes, that is, $\text{E}(N) = 1$. Note that N has an approximate mean $\lambda = 1$ from Poisson approximation but actually has the exact mean $\text{E}(N) = \sum_{i=1}^{n} \text{E}(I_i) = 1 = \lambda$.

■

■ **Example 3.52 — Balls-and-bins problem.** Suppose we have m balls, labeled $i = 1, 2, \dots, m$ and n bins, labeled $j = 1, 2, \dots, n$. Each ball is thrown into one of the bin independently and uniformly at random from the n possibilities. Find the expected value (mean) and variance of the number of empty bins.

- Let X be the number of empty bins. Then, $X = X_1 + \cdots + X_n$, where X_i is the indicator variable defined as $X_i = \begin{cases} 1, & \text{if the } i\text{th bin is empty,} \\ 0, & \text{otherwise.} \end{cases}$ Then,

$$\text{E}(X_i) = P(X_i = 1) = \left(1 - \frac{1}{n}\right)^m, \; i = 1, \dots, n.$$

Hence, $\text{E}(X) = n\left(1 - \frac{1}{n}\right)^m \approx ne^{-m/n}$ since $\left(1 - \frac{x}{n}\right)^n \approx e^{-x}$ for large n. We may interpret that the expected proportion of empty bins is approximately $e^{-m/n}$.

- To find $\mathrm{Var}(X) = \mathrm{E}(X^2) - [\mathrm{E}(X)]^2$, note that $\mathrm{E}(X^2) = \sum_{i=1}^n \mathrm{E}(X_i^2) + \sum_{i \neq j} \mathrm{E}(X_i X_j)$, $\mathrm{E}(X_i^2) = \mathrm{E}(X_i)$ and that $\mathrm{E}(X_i X_j) = P(X_i = 1, X_j = 1)$ since X_i's are indicator variables. Thus, we have

$$\mathrm{E}(X_i X_j) = P(X_i = 1, X_j = 1) = \left(\frac{n-2}{n}\right)^m = \left(1 - \frac{2}{n}\right)^m,$$

$$\mathrm{E}(X^2) = n\left(1 - \frac{1}{n}\right)^m + n(n-1)\left(1 - \frac{2}{n}\right)^m,$$

$$\mathrm{Var}(X) = n(n-1)\left(1 - \frac{2}{n}\right)^m + n\left(1 - \frac{1}{n}\right)^m \left[1 - n\left(1 - \frac{1}{n}\right)^m\right]$$

$$\approx n(n-1)e^{-\frac{2m}{n}} + ne^{-\frac{m}{n}}\left[1 - ne^{-\frac{m}{n}}\right] = n\left(e^{-\frac{m}{n}} - e^{-\frac{2m}{n}}\right) \text{ for large } n.$$

- We can generalize the result of $\mathrm{E}(X)$ to find the expected fraction of bins with k balls. Let Y be the number of bins with k balls. Then, $Y = \sum_{i=1}^n Y_i$, where Y_i is the indicator variable defined as $Y_i = \begin{cases} 1, & \text{if the } i\text{th bin has } k \text{ balls,} \\ 0, & \text{otherwise.} \end{cases}$ Thus, we have

$$P(Y_i = 1) = \binom{m}{k}\left(\frac{1}{n}\right)^k \left(1 - \frac{1}{n}\right)^{m-k} = \frac{1}{k!}\frac{m(m-1)\cdots(m-k+1)}{n^k}\left(1 - \frac{1}{n}\right)^{m-k}.$$

Hence, when m and n large with fixed k, we have

$$\mathrm{E}(Y) = n\mathrm{E}(Y_1) = n\frac{1}{r!}\frac{m(m-1)\cdots(m-k+1)}{n^k}\left(1 - \frac{1}{n}\right)^{m-k} \approx n\frac{e^{-\frac{m}{n}}\left(\frac{m}{n}\right)^k}{k!}.$$

In other words, we see that the number of balls in a specific bin has a Binomial distribution with parameters m and $\frac{1}{n}$ and it has approximately Poisson distribution with parameters λ as $m, n \to \infty$ and $m/n \to \lambda$.

■

■ **Example 3.53** Suppose that $X \sim \mathrm{Poisson}(\lambda_1)$, $Y \sim \mathrm{Poisson}(\lambda_2)$ and they are independent. Then, $X + Y$ has a Poisson distribution with parameter $\lambda_1 + \lambda_2$, and the conditional distribution of X given that $X + Y = n$ is binomial distribution.

- To find the distribution of $X + Y$, we also use the convolution of two distributions of X and Y given in Theorem 3.3.3, $h_Z(z) = \sum_{n=-\infty}^{\infty} f_X(n) g_Y(z - n)$. Thus, we have

$$P(X + Y = n) = \sum_{x=0}^{n} P(X = x)P(Y = n - x),\ n = 0, 1, \ldots,$$

$$= \sum_{x=0}^{n} e^{-\lambda_1}\frac{\lambda_1^x}{x!} e^{-\lambda_2}\frac{\lambda_2^{n-x}}{(n-x)!} = e^{-(\lambda_1+\lambda_2)}\frac{1}{n!}\sum_{x=0}^{n}\frac{n!}{x!(n-x)!}\lambda_1^x \lambda_2^{n-x},\ n = 0, 1, \ldots,$$

$$= e^{-(\lambda_1+\lambda_2)}\frac{(\lambda_1 + \lambda_2)^n}{n!},\ n = 0, 1, \ldots,$$

where the last equality holds from the binomial theorem.

- The conditional p.m.f. of X given $X+Y=n$ is

$$P(X=x|X+Y=n)=\frac{P(X=x,X+Y=n)}{P(X+Y=n)}=\frac{P(X=x)P(Y=n-x)}{P(X+Y=n)}$$

$$=\frac{\cancel{e^{-\lambda_1}}\frac{\lambda_1^x}{x!}\cancel{e^{-\lambda_2}}\frac{\lambda_2^{n-x}}{(n-x)!}}{\cancel{e^{-(\lambda_1+\lambda_2)}}\frac{(\lambda_1+\lambda_2)^n}{n!}}=\binom{n}{x}\left(\frac{\lambda_1}{\lambda_1+\lambda_2}\right)^x\left(\frac{\lambda_2}{\lambda_1+\lambda_2}\right)^{n-x},$$

which is a binomial p.m.f with parameters n and p, $p=\lambda_1/(\lambda_1+\lambda_2)$.

■

Remark The sum of two independent Poisson random variables is also Poisson, and the parameter of the sum is the sum of the parameters.

- We can also understand this result in terms of Poisson approximation of binomial distribution.
- The sum of two independent Binomial random variables with an identical p, $X_1 \sim \text{Binomial}(n_1,p)$ and $X_2 \sim \text{Binomial}(n_2,p)$ has a Binomial distribution, $X_1+X_2 \sim \text{Binomial}(n_1+n_2,p)$. Then, they are all approximately distributed as Poisson distributions $X_1 \sim \text{Poisson}(\lambda_1)$, $X_2 \sim \text{Poisson}(\lambda_2)$, and $X_1+X_2 \sim \text{Poisson}(\lambda_1+\lambda_2)$, as $n_1,n_2 \to \infty$, $n_1p \to \lambda_1$ and $n_2p \to \lambda_2$.
- In general, let $X_1,\ldots,X_k$ be independent and let $X_i \sim \text{Poisson}(\lambda_i)$, $i=1,\ldots,k$. Then $\sum_{i=1}^k X_i \sim \text{Poisson}(\sum_{i=1}^k \lambda_i)$.

■ **Example 3.54 — Consulting Requests.** Suppose that we have a situation in which a certain kind of occurrence happens at random over a period of time. For example, the occurrences that we are interested in might be incoming telephone calls to help desk for consulting requests. We want to model this situation so that we can consider the probabilities of events such as more than 10 phone calls occurring in an hour. There could be more incoming calls between 4:00 and 5:00 p.m. than between 8:00 and 9:00 a.m., and this fact would certainly affect the above probability. Thus, we need to assume that the average rate, i.e., the average number of occurrences per an hour, is a constant. This rate will be denoted as λ.

- Let X be the number of consulting requests that arrive at a help desk during two hours. Suppose that the overall rate of arrivals is 3 per hour. Then it would be common to model X as having a Poisson distribution with mean $2\times 3=6$, $X \sim \text{Poisson}(6)$.
- Consider what happens if we break our 2-hour period into little pieces as shown in Figure 3.6.

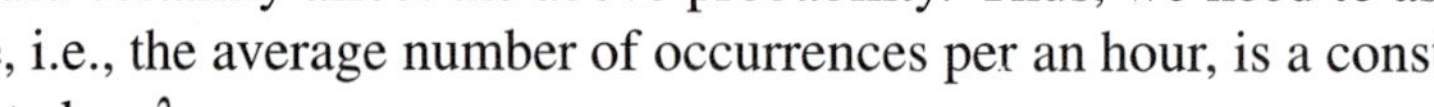

Figure 3.6: Dividing an interval with equally-spaced subintervals

- Each interval is of length $2/n$ hours. Suppose that we model the number of requests X_i in the ith interval as a Poisson random variable, $X_i \sim \text{Poisson}(6/n)$. Then, we see that $\sum_{i=1}^n X_i = X$, the number of requests in the whole 2-hour period.
- We can use the binomial distribution to model this situation. We imagine that a given time interval is broken up into n subintervals of equal length. If the subintervals are sufficiently short, we can assume that two or more occurrences happen in one subinterval with a probability which is negligible in comparison with the probability of at most one occurrence.

- Thus, in each subinterval of Figure 3.6, we are assuming that there is either 0 or 1 occurrence. This means that the sequence of subintervals can be thought of as a sequence of Bernoulli trials, with a success corresponding to an occurrence in the subinterval.

- Suppose that the intervals are so small that $P(X_i \geq 2) \ll P(X_i = 1)$, so small even that $nP(X_i \geq 2)$ is essentially 0. Then $\sum_{i=1}^n X_i \approx \sum_{i=1}^n Y_i$, where $Y_i = 1$ if $X_i \geq 1$ and $Y_i = 0$ if $X_i = 0$. But $\sum_{i=1}^n Y_i$ has a binomial distribution with parameters n and $p - 1 - \exp(-6/n) \approx 6/n$, which illustrates Poisson approximation to Binomial.

- To decide upon the proper value of p, the probability of an occurrence in a given subinterval, we reason as follows. On the average, there are λt occurrences in a time interval of length t. If this time interval is divided into n subintervals, then we would expect, using the Bernoulli trials interpretation, that there should be np occurrences. Thus, we want $\lambda t = np$, which implies that $p = \dfrac{\lambda t}{n}$.

■

Remark Stirling's formula gives an approximation to $n!$ that is valid for large n

$$n! \approx \sqrt{2\pi} n^{n+\frac{1}{2}} e^{-n}, \text{ as } n \to \infty.$$

- "$a_n \approx b_n$" indicates that the ratio of the two quantities, a_n/b_n converges to 1.

- From the Taylor expansion of $e^n = 1 + n + \ldots + n^n/n! + \ldots$, we have $1 \leq n^n/n! \leq e^n$. Note that $\log n! = \sum_{k=1}^n \log k$ and that $\int_1^n \log x dx \leq \sum_{k=1}^n \log k \leq \int_1^{n+1} \log x dx$. Thus, since $\int_1^z \log x dx = z \log z - z + 1$, we have

$$n \log n - n + 1 \leq \log n! \leq (n+1) \log (n+1) - n.$$

Divide by $n \log n$ and let $n \to \infty$, $\dfrac{\log n!}{n \log n}$ converges to 1. Hence, $\log n! \sim n \log n$, which is known as the weak form of Stirling's formula.

- We can get the "strong form" of Stirling's approximation $n! \sim \sqrt{2\pi} n^{n+\frac{1}{2}} e^{-n}$ using the Gamma function $\Gamma(x)$ that will be discussed in Chapter 4.

Example 3.55 Suppose that a fair coin is flipped 100 times. Calculate the probability of obtaining exactly 50 Heads.

- In 100 flips, there are $\binom{100}{50}$ ways to get exactly 50 heads. Thus, the required probability is $\frac{1}{2^{100}}\binom{100}{50} = \frac{1}{2^{100}}\frac{100!}{50!50!}$.
- If a fair coin is flipped $2n$ times. The probability of obtaining exactly n heads can be obtained using Stirling's approximation

$$\frac{\binom{2n}{n}}{2^{2n}} = \frac{(2n)!}{[2^n(n!)]^2}. \approx \frac{\sqrt{2\pi}(2n)^{2n+\frac{1}{2}}e^{-2n}}{\left[2^n\sqrt{2\pi}n^{n+\frac{1}{2}}e^{-n}\right]^2} = \frac{1}{\sqrt{\pi n}}$$

- For $n = 13$ this is 0.156478. The exact answer is 0.154981.

■

Example 3.56 The number N of customers entering KU cafe on a given day is supposed to have a Poisson distribution with parameter λ, $N \sim \text{Poisson}(\lambda)$. Each customer orders "Cafe Latte" with probability p, independently of the other customers. Let X be the number of customers who order "Cafe Latte" on a given day and Y be the number of customers who do not. Then, X and Y are independent and have Poisson distributions with parameters λp and $\lambda(1-p)$, respectively.

- Consider the joint p.m.f. of X and Y, $P(X = j, Y = k)$, which is the same as $P(X = j, Y = k, N = j+k)$ since $X+Y = N$. Then,

$$\begin{aligned} P(X = j, Y = k) &= P(X = j, Y = k | N = j+k)P(N = j+k) \\ &= \binom{j+k}{j} p^j (1-p)^k e^{-\lambda} \frac{\lambda^{j+k}}{(j+k)!} \\ &= e^{-\lambda p}\frac{(\lambda p)^j}{j!} \times e^{-\lambda(1-p)}\frac{[\lambda(1-p)]^k}{k!}, \quad j,k \geq 0. \end{aligned}$$

- Thus, the marginal p.m.f.'s of X and Y are obtained by

$$\begin{aligned} P(X = j) &= \sum_{k=0}^{\infty} P(X = j, Y = k) = e^{-\lambda p}\frac{(\lambda p)^j}{j!}, \\ P(Y = k) &= \sum_{j=0}^{\infty} P(X = j, Y = k) = e^{-\lambda(1-p)}\frac{[\lambda(1-p)]^k}{k!}. \end{aligned}$$

Thus, X and Y have Poisson distributions with parameters λp and $\lambda(1-p)$, respectively. In addition, it follows that $P(X = j, Y = k) = P(X = j)P(Y = k)$, which implies that X and Y are independent.

- Alternatively, we can compute the p.m.f. of X from the joint p.m.f. of X and N with the multiplication rule

$$
\begin{aligned}
P(X=x) &= \sum_{n=x}^{\infty} P(X=x|N=n)P(N=n) \\
&= \sum_{n=x}^{\infty} \binom{n}{x} p^x(1-p)^{n-x}\frac{\lambda^n e^{-\lambda}}{n!} = e^{-\lambda}\sum_{n=x}^{\infty}\frac{n!}{x!(n-x)!}p^x(1-p)^{n-x}\frac{\lambda^n}{n!} \\
&= \frac{e^{-\lambda}p^x\lambda^x}{x!}\sum_{n=x}^{\infty}\frac{(1-p)^{n-x}\lambda^{n-x}}{(n-x)!} = \frac{e^{-\lambda}(p\lambda)^x}{x!}\sum_{\ell=0}^{\infty}\frac{((1-p)\lambda)^\ell}{\ell!} \\
&= \frac{e^{-\lambda}(p\lambda)^x}{x!}e^{(1-p)\lambda} = \frac{e^{-\lambda p}(\lambda p)^x}{x!}.
\end{aligned}
$$

- Imagine that among the sufficiently large number ($M \to \infty$) of people in a town, each of them visits KU cafe with a probabilty $\varepsilon \approx 0$. Thus, the number of customers N entering KU cafe would have a binomial distribution with parameters M and ε. By the Poisson approximation with $\lambda = M\varepsilon$, N has a Poisson distribution with $\lambda - M\varepsilon$. In addition, the number of customers who order "Cafe Latte" would have a binomial distribution with parameters M and $p\varepsilon$, which has a Poisson distribution with $\lambda = Mp\varepsilon$ by the Poisson approximation as well.

■

We revisit the coupon collector's problem for illustration by running simulations. Suppose that each package of Cheetos snack contains the picture of one of twelve zodiac animal characters. The picture of each animal character is equally likely to be placed in any given package of Cheetos snack, and that pictures are placed in different packages independently of each other.

- Suppose that Joanne buys $k = 20$ packages of Cheetos. Does she get a complete set of 12 animals in 20 packages she purchases? How many different animals does she collect in this case?

- Suppose that Joanne buys $k = 20$ packages of Cheetos. What is the probability that a complete set of 12 animals is obtained?

```
animals= c("rat","Ox","Tiger","Rabbit","Dragon","Snake",
  "Horse", "Sheep", "Monkey", "Rooster", "Dog", "Pig")
samp.animals=sample(animals,size=20,replace=TRUE)
unique(samp.animals) # extracts unique elements
length(unique(samp.animals)); table(samp.animals)

[1] 9
samp.animals
    Dog   Horse  Monkey      Ox     Pig  Rabbit Rooster
      1       2       2       1       3       1       2
  Sheep   Tiger
      3       5
```

```
collector=function(n,m) {
 samp.animals=sample(animals, size=m, replace=TRUE)
 ifelse(length(unique(samp.animals))==n, "yes", "no")
}
collector(12,20)

table(replicate(1000,collector(12,20)))/1000

  no   yes
0.953 0.047
# the required probability is approximately obtained as 0.047
```

We can find the expected number of packages until Joanne obtains all 12 different animal pictures.

- As given in Example 3.41, the expected number of packages until Joanne obtains all 12 different animal pictures is $\mathrm{E}(T) = 12\sum_{i=1}^{12}\frac{1}{i} \doteq 37.2385$.
- We can compute the expected number of packages by running simulations.

```
avg_collect=function(n=length(animal)){
full.animals=sample(animals,1,replace=TRUE)
count <- 1
while (length(unique(full.animals))<n) {
full.animals=c(full.animals,sample(animals,1,replace=TRUE))
count <- count + 1
 }
return(count)
}

Nsim <- 10000
count_collect <- replicate(Nsim,avg_collect(12))

mean(count_collect) #by simulations
[1] 37.2116

12*sum(1/seq(1,12)) #numerical calculation
[1] 37.23853
```

3.6 Exercises

Problem 3.1 Let X have a binomial distribution, $X \sim \text{Binomial}(n, p)$. Compute $\text{Var}(Xe^X)$.

Problem 3.2 For a midterm in STAT 221, your professor gives you a list of 15 problems to study. He indicates that he will choose eight problems for the actual midterm. You will be required to answer correctly at least five of those in order to pass the midterm. You decide to study 10 of the 15 problems. What is the probability that you will pass the exam? How many problems would you expect to answer correctly?

Problem 3.3 A graph is a set of vertices (or nodes), with edges joining them. Start with a graph on n vertices and no edges. For each pair of vertices, flip a coin with heads probability p. If the coin lands heads, place an edge between that pair of vertices. If tails, do not place an edge. The parameter p is called the edge probability.

① Let X be the number of edges in a random graph of n vertices. Compute the expected value of $\text{E}[X(X-1)]$.

② The degree of a vertex is defined to be the number of edges that contain that vertex, also described as the number of edges incident to that vertex. Let Y be the degree of vertex v in a random graph. Determine the p.m.f of Y.

Problem 3.4 Suppose a fair die is rolled 15 times. Let X be the number of faces that never show up in these 15 rolls.

① Find the probability mass function of X.

② Determine the expected value of X.

Problem 3.5 Suppose X has a Poisson distribution and $P(X=2) = P(X=1)$.

① Find $\text{E}[1/(X+1)]$.

② Compute $P(X=k|X \neq 0)$.

Problem 3.6 Suppose that there are 59 students in the classroom for STAT 221. Consider their birth month, the month each student was born in, and assume that twelve months $\{$January, February, ..., November, December$\}$ are all equally likely for their birth month.

① Let X be the number of students in the classroom who was born in October. Show that $P(X=3)$ is smaller than $P(X=4)$.

② Each student in the classroom is supposed to call out one by one which month the student was born in. Record how many months are called out. On average, how many students call out their birth month until 12 months are all recorded?

Problem 3.7 Each time, you simultaneously toss a fair coin and roll a fair die until a head is tossed or a six is rolled, or both. What is the probability mass function of the number of trials? Also, find the expected value of the number of trials until a head is tossed or a six is rolled, or both. More generally, what is the distribution of $\min\{X, Y\}$, where X and Y are independent and geometrically distributed with parameters p_1 and p_2, respectively?

Problem 3.8 Consider a well-shuffled deck of n cards, labeled 1 through n. You flip over the cards one by one, saying the numbers 1 through n as you do so (or you can say the numbers 1 through n in a different order but the same number is said not more than once). Let X be the number of times you say the correct number labeled in the card you flipped over.

① Compute the probability that X is greater or equal to 2, $P(X \geq 2)$.

② Determine the expected value of X.

Problem 3.9 An urn contains 5 blue balls, 5 red balls, and 5 white balls.

① Suppose that you randomly pick ***without replacement*** one ball at a time until you have 1 red ball. What is the probability mass function of the number of picks needed?

② Suppose that you randomly pick ***with replacement*** one ball at a time until you have 1 blue ball. Let X be the number of picks until you have 1 blue ball. Find the moment generating function of X, $M_X(t) = \mathrm{E}(e^{tX})$.

③ Suppose that you randomly pick ***with replacement*** one ball at a time until you have either 1 blue ball or 1 red ball. Compute the probability that the blue ball is obtained before the red ball is obtained.

Problem 3.10 Suppose that you have a box with N keys, exactly one of which is your house key. Further suppose that you get home after dark and can't see the keys on the key box. You randomly try one key at a time until you get the correct key. Let X denote the number of times you try until you get your house key. Determine the probability mass function of the random variable X and its expectation if you choose one key randomly without replacement.

Problem 3.11 A box contains N smartphones with d defective phones. A random sample of size n, the first sample, is taken without replacement from the box. Then, a random sample of size k, the second sample, is taken without replacement from the first sample. Let p_i denote the probability of exactly i defective phones in the second sample. If there are no defective phones in the second sample, what is the probability that there were no defective phones in the first sample?

II Continuous Distributions

4. Continuous random variables

4.1 Continuous random variables

Until now, we have assumed that a random variable is discrete, taking values in some (possibly infinite but countable) set of numbers that we can explicitly enumerate (e.g., the integers). However, it is often inconvenient and unreasonable to require that random variables are countable such as times, distances, weights, heights, amplitudes, angles, and so forth. Thus, we introduce another type of random variables such that the sample space Ω is uncountable, and $X(w)$ may take any one of uncountable real values.

A random variable X is said to be continuous if its space is either an interval or a union of intervals. The following definition formally defines a continuous random variable.

Definition 4.1.1 A random variable X is said to be *continuous* or have a *continuous distribution* if there exists a nonnegative function f_X such that for all $a \leq b$,

$$P(a < X \leq b) = \int_a^b f_X(x)dx.$$

- If X has a continuous distribution, the function f_X above is called the *probability density function* (p.d.f.) or just the *density* of X.

- Every continuous random variable has a p.d.f. that satisfies

  ① $f_X(x) \geq 0$ for all x.

  ② $\int_{-\infty}^{\infty} f_X(x)dx = 1$.

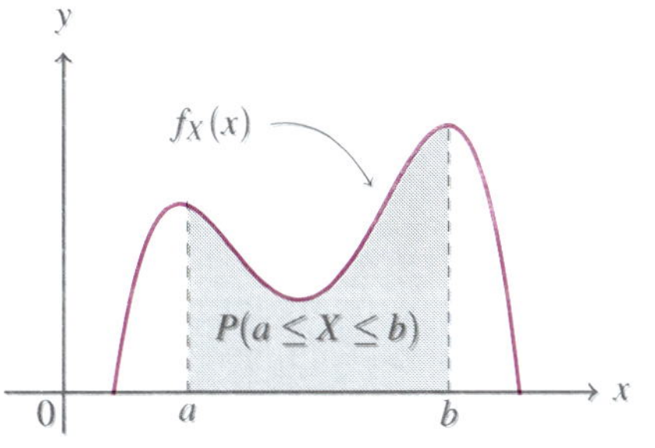

Figure 4.1: The shaded area below $f_X(x)$ and above the x axis equals $P(a \le X \le b)$

Remark Continuous distributions assign probability 0 to individual values.

- Let X have a continuous distribution with p.d.f. f_X. For all $a < b$,
$$P(X = b) \quad \le \quad P(a < X \le b) = \int_a^b f_X(x)dx, \;\; \lim_{a \uparrow b} \int_a^b f_X(x)dx = 0,$$
which means that $P(X = b) = 0$ for all b.

- This does not mean that $X = x$ is impossible. For each $\varepsilon > 0$, $P(b - \varepsilon \le X \le b + \varepsilon) > 0$ if f is positive in an interval around b, but the probability is spread so thinly that there is no positive probability on individual values. When we say that $X = b$ we mean that X is very close to b. We can think that our 1 unit of probability is being spread out over too many values - too many to explicitly assign probabilities to each value.

- Instead, we look at $\lim_{\Delta x \to 0} \dfrac{P(x \le X \le x + \Delta x)}{\Delta x} = f_X(x)$. The fraction here tells us the amount of probability "mass" in the interval ("volume") of length Δx. (Mass/Volume is "density".) When we take the limit, we are asking "what is the density of probability for values infinitesimally close to x.

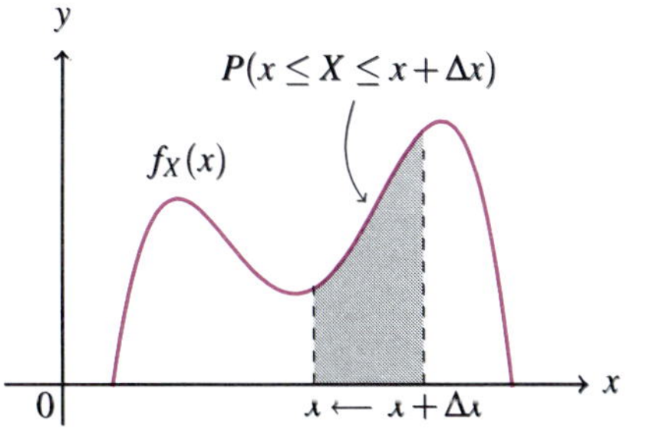

Figure 4.2: The density of probability for values infinitesimally close to x

- We can say that $f_X(x)$ is the density of probability in an *infinitesimal* neighborhood around x. It follows that $f_X(x)\,dt$ (think of this as density $\times$ length) is the infinitesimal probability mass around x. To add up infinitesimal quantities for $P(a < X < b)$, we integrate $P(a < X < b) = \int_a^b f_X(x)dx$.

■ **Example 4.1 — Normalizing Constants.** It is common to have an idea what the shape of a p.d.f. should be, such as "proportional to x for $0 < x < 2$" or "goes down like $1/x^2$ for $x > 1$". If these functions have finite integral, then there exist constants that we can multiply to make them densities.

- Find all values of c and a such that $f_X(x) = c/x^a$ for $x > 1$ is a p.d.f. We must have $c > 0$, and then $f \geq 0$. To check the integral,

$$\int_1^\infty \frac{c}{x^a}dx = \begin{cases} \frac{c}{-a+1}[\infty^{-a+1} - 1^{-a+1}], & \text{if } a \neq 1, \\ c[\log(\infty) - \log(1)], & \text{if } a = 1. \end{cases}$$

This is finite only for $a > 1$. If $a > 1$, then $c = a - 1$ makes the integral equal to 1.

■

Remark Often it is necessary to integrate over the entire real line, or an unbounded interval of the form $[a, \infty)$ or $(-\infty, b]$ as in Example 4.1. Also, we may wish to integrate unbounded functions defined on an open bounded interval (a, b). These integrals are called *improper integral*, and are limits of integrals rather than integrals themselves.

- Suppose $f : [a, b) \to R$ is a function (not necessarily bounded) that is Riemann integrable on $[a, c]$ for all $c < b$. We define $\int_a^b f(x)dx = \lim_{c \to b^-} \int_a^c f(x)dx$ if the limit exists.
- Suppose $f : [a, \infty) \to R$ is a function such that f is Riemann integrable on $[a, c]$ for all $c < \infty$. We define $\int_a^\infty f(x)dx = \lim_{c \to \infty} \int_a^c f(x)dx$, if the limit exists.
- If the limit exists, we say the improper integral *converges*. If the limit does not exist, we say the improper integral *diverges*. We similarly define improper integrals for the left-hand endpoint, $\int_{-\infty}^a f(x)dx = \lim_{c \to -\infty} \int_c^a f(x)dx$.

■ **Example 4.2 — Unbounded p.d.f.** It is necessary that a p.d.f. have finite integral, but it is not necessary that a p.d.f. be bounded. As an example, the function

$$f_X(x) = \begin{cases} \frac{1}{2\sqrt{x}}, & \text{for } 0 < x < 1, \\ 0, & \text{otherwise.} \end{cases} \quad \text{is a p.d.f.}$$

■

Remark The set $\{x : f_X(x) > 0\}$ is called the *support* of (the distribution of) X. The support of X can be either bounded or unbounded.

Definition 4.1.2 Every random variable (discrete, continuous, mixed) has a *(cumulative) distribution function* (c.d.f/d.f.) defined by $F_X(x) = P(X \leq x)$ for all real x. Let $f_X(x)$ be the probability density function of a continuous random variable X. The cumulative distribution function $F_X(x)$ of X is defined as

$$F_X(x) = P(X \leq x) = \int_{-\infty}^{x} f_X(t)\,dt.$$

Theorem 4.1.1 If $F_X(x)$ is the cumulative distribution function of a continuous random variable X, the probability density function $f(x)$ of X is the derivative of $F_X(x)$, that is

$$\frac{d}{dx}F_X(x) = f_X(x).$$

Proof. By Fundamental Theorem of Calculus, we get

$$\frac{d}{dx}(F_X(x)) = \frac{d}{dx}\left(\int_{-\infty}^{x} f_X(t)\,dt\right) = f_X(x)\frac{dx}{dx} = f_X(x).$$ ■

Remark *Fundamental Theorem of Calculus* is a basic theorem that explains the relationship between differentiation and integration.

- Let $F\colon [a,b] \to R$ be a continuous function, differentiable on (a,b). Let $f(x)$ be such that $f(x) = F'(x)$ for $x \in (a,b)$. Then $\int_a^b f(x)dx = F(b) - F(a)$.
- Let $f\colon [a,b] \to R$ be a Riemann integrable function. Define $F(x) = \int_a^x f(t)dt$. Then, $F(x)$ is continuous on $[a,b]$, and if $f(x)$ is continuous at $c \in [a,b]$, then $F(x)$ is differentiable at c with $F'(c) = f(c)$.

■ **Example 4.3** A continuous random variable X has the following c.d.f

$$F_X(x) = \begin{cases} 0, & \text{for } x < 1, \\ c(x - 2 + 1/x), & \text{for } 1 \leq x < 2, \\ 1, & \text{for } x \geq 2. \end{cases}$$

Since $F_X(x)$ is continuous, we must have $F_X(1) = 0$ and $F_X(2) = 1$, which implies that $c = 2$. The p.d.f. $f_X(x)$ is obtained by differentiating $F_X(x)$, which gives

$$f_X(x) = \frac{d}{dx}F_X(x) = \begin{cases} 2(1 - 1/x^2), & \text{for } 1 \leq x < 2, \\ 0, & \text{otherwise.} \end{cases}$$

■

Remark Every random variable has a c.d.f. $F_X(x)$ that satisfies the following properties as shown in Theorem 3.1.1, and every function that satisfies these properties is the c.d.f. of some random variable.

① $F_X(x)$ is non-decreasing : $x_1 < x_2$ implies that $F_X(x_1) \leq F_X(x_2)$.

② $F_X(x)$ is normalized : $\lim_{x \to -\infty} F_X(x) = 0$, $\lim_{x \to \infty} F_X(x) = 1$.

③ $F_X(x)$ is right-continuous : $F_X(x) = F_X(x^+)$ for all x, where

$$F_X(x^+) = \lim_{y \to x,\ y > x} F_X(y) \equiv \lim_{y \downarrow x} F_X(y) = F_X(x).$$

Note that given a function $F_X(x)$ satisfying ① – ③, we can define a random variable X by $X(\omega) = \inf\{x \in R : F_X(x) \geq \omega\}$, whose c.d.f. is $F_X(x)$.

Example 4.4 — Uniform Distribution. An important example of a continuous distribution is the Uniform distribution. This represents the symmetric case where all values are "equally likely". What can be the density of probability if every value is "equally likely"?

- We say that X has a Uniform(a,b) distribution for $a < b$ denoted as $X \sim$ Uniform(a,b), if the p.d.f of X is a constant on the interval (a,b), and the c.d.f of X is given by

$$f_X(x) = \begin{cases} \frac{1}{b-a}, & a < x < b, \\ 0, & \text{otherwise}, \end{cases} \qquad F_X(x) = \begin{cases} 0, & x \leq a, \\ \frac{x-a}{b-a}, & a < x < b, \\ 1, & x \geq b. \end{cases}$$

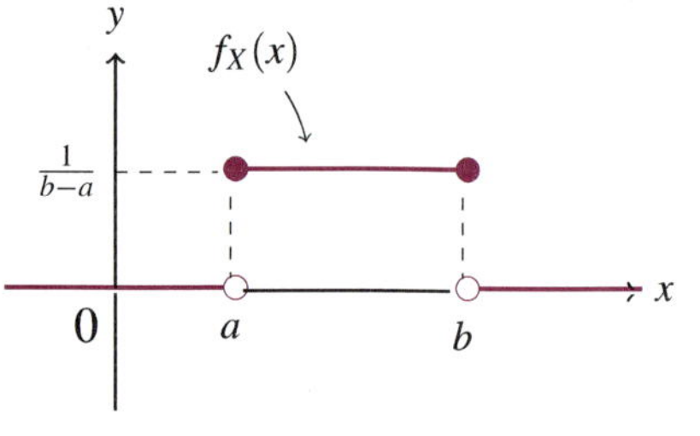

Figure 4.3: The p.d.f. of a uniform distribution on (a,b).

- Let the random variable X denote the outcome when a point is selected at random from an interval $[a, b]$. We want to find the probability of the event $X \leq x$, that is we would like to determine the probability that the point selected from $[a, b]$ would be less than or equal to x,

$$P(X \leq x) = \frac{\text{length of } [a,x]}{\text{length of } [a,b]} = \frac{x-a}{b-a}, \quad a \leq x \leq b.$$

Then, we get the p.d.f from the c.d.f, $f_X(x) = \dfrac{d}{dx} F_X(x) = \dfrac{1}{b-a}$ if $a \leq x \leq b$.

■

■ **Example 4.5** The p.d.f. is the derivative of the c.d.f. of a continuous distribution. The c.d.f of a continuous distribution is a definite integral of p.d.f :

- Consider the c.d.f of X, $F_X(x) = \frac{1}{1+e^{-x}}$, $-\infty < x < \infty$. Then, the p.d.f. is

$$f_X(x) = \frac{d}{dx}F_X(x) = \frac{d}{dx}\left(\frac{1}{1+e^{-x}}\right) = \frac{e^{-x}}{(1+e^{-x})^2}.$$

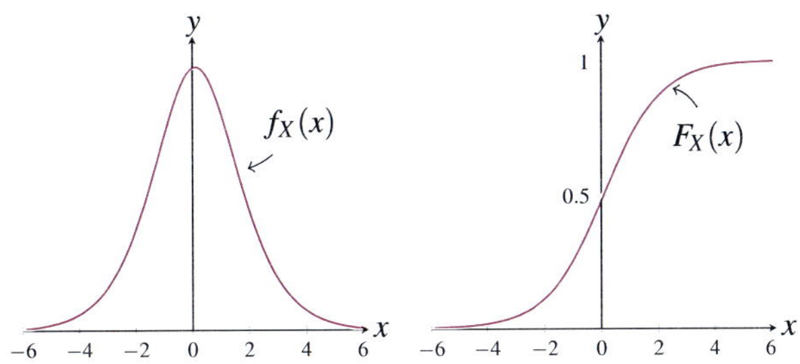

Figure 4.1: The shape of the p.d.f. and the c.d.f of X

- Let $f_X(x) = e^{-x}$ for $x > 0$ be the p.d.f. of X. Then the c.d.f. is

$$F_X(x) = \begin{cases} \int_0^x e^{-t}dt = 1 - \exp(-x), & x \geq 0, \\ 0, & x < 0. \end{cases}$$

- Let $f_X(x) = cx^{k-1}e^{-(\lambda x)^k}$ for $x > 0$, where $\lambda > 0$ and $k > 1$. Then the c.d.f. is

$$F_X(x) = c\int_0^x t^{k-1}e^{-(\lambda t)^k}dt = \frac{c}{k\lambda^k}\int_0^{h(x)} e^{-v}dv = \frac{c}{k\lambda^k} - e^{-v}\Big|_0^{h(x)} = \frac{1-e^{-h(x)}}{k\lambda^k}c,$$

where $h(x) = (\lambda x)^k$ and we use the transformation $v = (\lambda t)^k$, with $dv = k\lambda^k t^{k-1}dt$. Then, since $h(x) \to \infty$ as $x \to \infty$, we must take $c = k\lambda^k$, so that $F_X(x) = 1 - e^{-(\lambda x)^k}$ for $x > 0$.

■

Using Theorem 4.1.1, we can find the distribution of a new random variable $Y = r(X)$, a one-to-one transformation of a continuous random variable X. We can start with the c.d.f. of Y, $F_Y(y) = P(Y \leq y)$ and convert the event $\{Y \leq y\}$ into an event with X, $\{r(X) \leq y\}$. Then we can differentiate $F_Y(y)$ to get the p.d.f of Y, known as the *change of variable* formula as summarized in Theorem 4.1.2.

Theorem 4.1.2 — Change of Variable. Let r be a one-to-one increasing function with inverse r^{-1}. Let X have c.d.f. F_X, and let $Y = r(X)$.

- Then $Y \leq y$ if and only if $X \leq r^{-1}(y)$, and the c.d.f. of Y is

$$F_Y(y) = P(Y \leq y) = P(X \leq r^{-1}(y)) = F_X(r^{-1}(y)).$$

- The p.d.f. of Y is the derivative of $F_Y(y)$, so long as r^{-1} has a derivative. That is

$$f_Y(y) = \frac{d}{dy}F_X(r^{-1}(y)) = f_X(r^{-1}(y))\frac{d}{dy}r^{-1}(y).$$

- Similarly, if r is one-to-one and decreasing, $Y \le y$ if and only if $X \ge r^{-1}(y)$. In this case, the p.d.f. of Y is (if r^{-1} is differentiable) given by

$$f_Y(y) = \frac{d}{dy}1 - F_X(r^{-1}(y)) = -f_X(r^{-1}(y))\frac{d}{dy}r^{-1}(y).$$

- The two cases (increasing and decreasing r) can be combined to yield

$$f_Y(y) = f_X(r^{-1}(y))\left|\frac{d}{dy}r^{-1}(y)\right|.$$

Remark In the change of variables formula, the p.d.f. of Y can be written as $f_Y(y) = f_X(x)\frac{dx}{dy}$.

- It is convenient to write the change of variables formula as $f_Y(y)dy = f_X(x)dx$.
- To compute $\frac{dx}{dy}$, we can take the reciprocal $\frac{dy}{dx}$. In this regard, dx or dy can be viewed as *units* to measure the length of interval in terms of x or y.

Example 4.6 Suppose that U has a uniform distribution on $(0,1)$. Define $X = 1/U^{1/3}$. Then, consider $r(u) = 1/u^{1/3}$, which is monotone, and $r^{-1}(x) = 1/x^3$. Thus, the derivative is $dr^{-1}(x)/dy = -3/x^4$. Hence, the p.d.f. of Y is

$$f_X(x) = f_U(1/x^3)\left|-3/x^4\right| = 3/x^4, \text{ for } 0 < 1/x^3 < 1,$$

which simplifies to $3/x^4$ for $x > 1$. ■

Example 4.7 Suppose that X has p.d.f f_X and c.d.f. F_X. Let $Y = bX + a$.

- Depending on the sign of b, we have

$$\begin{aligned} F_Y(y) &= P(Y \le y) = P(bX + a \le y) \\ &= \begin{cases} P\left(X \le \frac{y-a}{b}\right) & = F_x\left(\frac{y-a}{b}\right), & b > 0, \\ P\left(X \ge \frac{y-a}{b}\right) & = 1 - F_x\left(\frac{y-a}{b}\right), & b < 0. \end{cases} \end{aligned}$$

By differentiating F_Y, we get

$$\begin{aligned} f_Y(y) &= F_Y'(y) = \begin{cases} \frac{d}{dy}F_x\left(\frac{y-a}{b}\right) & = \frac{1}{b}f_X\left(\frac{y-a}{b}\right), & b > 0 \\ \frac{d}{dy}\left(1 - F_x\left(\frac{y-a}{b}\right)\right) & = -\frac{1}{b}f_X\left(\frac{y-a}{b}\right), & b < 0. \end{cases} \\ &= \frac{1}{|b|}f_X\left(\frac{y-a}{b}\right). \end{aligned}$$

We say that the random variable X has been shifted(location change) by a and scaled(scale change) by b.

- Let $U \sim \text{Uniform}(0,1)$. Then, $X = (b-a)U + a \sim \text{Uniform}(a,b)$ because

$$f_X(x) = \frac{1}{b-a} f_U\left(\frac{x-a}{b-a}\right) = \frac{1}{b-a}, \text{ for } a < x < b.$$

In other words, if X is a uniform random variable on the interval $[a,b]$, then $Y = sX + t$ is also a uniform random variable on the interval $[sa+t, sb+t]$ if $s > 0$ since $Y = s((b-a)U + a) + t = s(b-a)U + sa + t$. In addition, $W = \frac{X-a}{b-a}$ has a uniform distribution on the interval $[0,1]$.

■

Theorem 4.1.3 — Probability Integral Transform. Any distribution can be represented as a function of a uniform random variable. The method is known as the probability integral transform or as inverse transform sampling.

- Let X have c.d.f. $F_X(x)$, where $F_X(x)$ is continuous and strictly increasing. Define $Y = F_X^{-1}(U)$. Then, the c.d.f. of Y is F_X. In other words, $F_X^{-1}(U)$ has the same distribution with X, $F_X^{-1}(U) \stackrel{d}{=} X$.
- Let X have c.d.f. $F_X(x)$, where $F_X(x)$ is continuous and strictly increasing. Define $Y = F_X(X)$. Then, Y has a uniform distribution on $(0,1)$.

Proof. Since $F_X(x)$ is strictly increasing and continuous distribution, its inverse function, $F_X^{-1}(x)$, exists with $F_X(F_X^{-1}(x)) = x$ and $F_X^{-1}(F_X(x)) = x$, and is strictly increasing.

- Define $Y = F_X^{-1}(U)$. Then, since $u \le F(x)$ is equivalent to $F^{-1}(u) \le x$, for all $u \in (0,1)$ and $x \in R$. we have

$$F_Y(y) = P(Y \le y) = P(F_X^{-1}(U) \le y) = P(U \le F_X(y)) = \int_0^{F_X(y)} 1du = F_X(y),$$

which implies that the c.d.f. of Y is F_X and thus Y has the same distribution with X. Alternatively, we can directly get the p.d.f of Y from Theorem 4.1.2 by taking $r(x) = F^{-1}(x)$, which follows that

$$f_Y(y) = f_U(F_X(X))\frac{d}{dx}F_X(x) = f_X(x).$$

- Define $Y = F_X(X)$. Since F_X is a c.d.f that is continuous and strictly increasing, we see that Y is a random variable between 0 and 1. In addition, for $0 < y < 1$, we have

$$F_Y(y) = P(Y \le y) = P(F_X(X) \le y) = P\left(X \le F_X^{-1}(y)\right) = F_X(F_X^{-1}(y)) = y,$$

which implies that Y has a uniform distribution on $(0,1)$. Alternatively, we can directly get the p.d.f of Y from Theorem 4.1.2 by taking $r(x) = F_X(x)$,

$$f_Y(y) = f_X(F_X^{-1}(y))\frac{d}{dy}F_X^{-1}(y) = f_X(F^{-1}(y))\frac{1}{f_X^{-1}(F_X^{-1}(y))} = 1, \ 0 < y < 1.$$

Indeed, we can take $X = F_X^{-1}(U)$, with $U \sim \text{Uniform}(0,1)$. Then, it follows that $Y = F_X(X) = F_X(F_X^{-1}(U)) = U$.

■

■ **Example 4.8** Suppose that U has a uniform distribution on the interval $[0,1]$.

- Let $X = -\log(1-U)$, where $U \sim \text{Uniform}(0,1)$. Then, the p.d.f. of X is

$$\begin{aligned} f_X(x) &= \frac{d}{dx}F_X(x) = \frac{d}{dx}P(-\log(1-U) \le x) \\ &= \frac{d}{dx}\int_0^{1-e^{-x}} 1du = e^{-x}, \text{ for } x > 0. \end{aligned}$$

Since $x = r(u) = -\log(1-u)$, we also get $f_X(x) = f_U\left(r^{-1}(x)\right)\left|\frac{d}{dx}r^{-1}(x)\right| = e^{-x}$.

- Note that the c.d.f. of X is $F_X(x) = 1 - e^{-x}$. Let $Y = F_X(X) = 1 - e^{-X}$. Then, the p.d.f. of Y is

$$\begin{aligned} f_Y(y) &= \frac{d}{dy}F_Y(y) = \frac{d}{dy}P\left(1 - e^{-X} \le y\right) = P(X \le -\log(1-y)) \\ &= \frac{d}{dy}F_X(-\log(1-y)) = \frac{d}{dy}y = 1, \text{ for } 0 < y < 1, \end{aligned}$$

which implies that Y has a uniform distribution on $(0,1)$. Also note that $1-Y$ has the same distribution with Y, $1 - Y \stackrel{d}{=} Y \sim \text{Uniform}(0,1)$.

■

■ **Example 4.9** Suppose that U has a uniform distribution on the interval $[0,1]$. Find a function $r(u)$ such that the random variable $X = r(U)$ has the following p.d.f.

$$f_X(x) = \begin{cases} \frac{5x^{3/2}}{2}, & \text{for } 0 < x < 1, \\ 0, & \text{otherwise.} \end{cases}$$

- We can use the reasoning of the probability integral transform. That is, the c.d.f. that we want for X is

$$F_X(x) = \int_{-\infty}^{x} f_X(t)dt = \int_0^x \frac{5t^{3/2}}{2}dt = x^{5/2}, \ \text{if } 0 < x < 1,$$

(the c.d.f. equals 1 for $x \ge 1$ and 0 for $x < 0$).

- We want $r(u) = F_X^{-1}(u)$, the inverse function of c.d.f. of X. Setting $F_X(x) = u$ we get $u = x^{5/2}, x = u^{2/5}$. Hence, the function we want is $r(u) = u^{2/5}$.

■

When $Y = r(X)$ is not a one-to-one(monotone) function, we can divide into sets where $r(x)$ is monotone and then apply Theorem 4.1.2 in each set. Then, the resulting p.d.f. of Y is written as the sum of p.d.f.'s of X under each set.

Theorem 4.1.4 Let X have p.d.f. $f_X(x)$, let $Y = r(X)$. Suppose there exists a partition, $A_0, A_1, \ldots, A_k$ of the sample space such that $P(X \in A_0) = 0$ and $f_X(x)$ is continuous on each A_i. Further, suppose there exist functions $r_1(x), \ldots, r_k(x)$, defined on $A_1, \ldots, A_k$, respectively, satisfying

- $r(x) = r_i(x)$ for $x \in A_i$, and $r_i(x)$ is monotone on A_i.
- The set $B_i = \{y : y = r_i(x) \text{ for some } x \in A_i\}$ is the same for each $i = 1, \ldots, k$, and $r_i^{-1}(y)$ has a continuous derivative on B_i, for each $i = 1, \ldots, k$.

Then, $f_Y(y) = \begin{cases} \sum_{i=1}^k f_X(r_i^{-1}(y)) \left| \frac{d}{dy} r_i^{-1}(y) \right|, & y \in B_i,\ i = 1, \ldots, k, \\ 0, & \text{otherwise.} \end{cases}$

■ **Example 4.10 — Squared transformation.** Suppose that X has a p.d.f. $f_X(x)$. Let $Y = X^2$. Then, for $y > 0$, the event $\{Y \le y\}$ corresponds to the set $\{-\sqrt{y} \le X \le \sqrt{y}\}$ and it follows that the c.d.f. $F_Y(y)$ of Y is given by

$$F_Y(y) = P\{Y \le y\} = P\{-\sqrt{y} \le X \le \sqrt{y}\} = \int_{-\sqrt{y}}^{\sqrt{y}} f_X(x)dx$$
$$= \int_{-\infty}^{\sqrt{y}} f_X(x)dx - \int_{-\infty}^{-\sqrt{y}} f_X(x)dx = F_X(\sqrt{y}) - F_X(-\sqrt{y}) \quad (y > 0).$$

Thus, by differentiating $F_Y(y)$, we get the p.d.f. of Y

$$f_Y(y) = \frac{1}{2\sqrt{y}}[f_X(\sqrt{y}) + f_X(-\sqrt{y})], \quad y > 0,$$

which corresponds to the result with $k = 2$ in Theorem 4.1.4. ■

■ **Example 4.11** Let $f_X(x) = \begin{cases} 1/2, & -1 < x < 1 \\ 0, & \text{otherwise.} \end{cases}$ be the p.d.f. of X. Define the random variable Y by $Y = X^2$. We wish to find the p.d.f. of Y.

- For $0 \le y \le 1$, we have

$$F_Y(y) = P(Y \le y) = P(X^2 \le y) = P(-\sqrt{y} \le X \le \sqrt{y}) = \int_{-\sqrt{y}}^{\sqrt{y}} \frac{1}{2} dx = \sqrt{y}.$$

- Clearly Y always takes on a value between 0 and 1, so the c.d.f. of Y is given by

$$F_Y(y) = \begin{cases} 0, & \text{if } y \le 0, \\ \sqrt{y}, & \text{if } 0 \le y \le 1, \\ 1, & \text{if } y \ge 1. \end{cases}$$

- From this we easily calculate that the density function of Y is

$$f_Y(y) = \begin{cases} 1/(2\sqrt{y}), & \text{if } 0 \le y \le 1, \\ 0, & \text{otherwise.} \end{cases}$$

- Alternatively, we can get $f_Y(y) = \frac{1}{2\sqrt{y}}[f_X(\sqrt{y}) + f_X(-\sqrt{y})] = \frac{1}{2\sqrt{y}}$ from the result in Theorem 4.1.4.

■

In addition, when the p.d.f. of X is not symmetrically defined and $Y = r(X)$ is not a one-to-one function, the change of variable formula in Theorems 4.1.2 and 4.1.4 may not be applicable, but we can still find the c.d.f of Y, $F_Y(y)$. Then, the p.d.f of Y, $f_Y(y)$ is obtained by differentiation. But, $f_Y(y)$ would have a case by case formula, depending on the possible range of y.

Example 4.12 Let X have a p.d.f. $f_X(x) = \begin{cases} \frac{1}{4}, & -1 < x < 3, \\ 0, & \text{otherwise.} \end{cases}$ Let $Y = X^2$. Find the p.d.f. of Y.

- We first compute the c.d.f. of Y. Since $-1 < X < 3$, we have $0 < Y < 9$ but note that the support of X(or, the p.d.f. of X) is not symmetric. Thus, it follows that $\{Y \le y\} = \{-\sqrt{y} < X < \sqrt{y}\}$ for $0 < y < 1$ and $\{Y \le y\} = \{-1 < x < \sqrt{y}\}$ for $1 < y < 9$. Hence, the c.d.f. of Y is given by

$$F_Y(y) = \begin{cases} 0, & y < 0, \\ \int_{-\sqrt{y}}^{\sqrt{y}} \frac{1}{4} dy = \frac{1}{2}\sqrt{y}, & 0 < y < 1, \\ \int_{-1}^{\sqrt{y}} \frac{1}{4} dy = \frac{1}{4}(\sqrt{y}+1), & 1 < y < 9, \\ 1, & y \ge 9. \end{cases}$$

- Differentiating $F_Y(y)$ we get the p.d.f. of Y

$$f_Y(y) = \begin{cases} \frac{1}{4\sqrt{y}}, & 0 < y < 1, \\ \frac{1}{8\sqrt{y}}, & 1 < y < 9, \\ 0, & \text{otherwise.} \end{cases}$$

■

Example 4.13 Let X have a p.d.f $f_X(x) = \begin{cases} 0.5e^{-x}, & x > 0, \\ 0.5, & -1 < x \le 0, \\ 0, & \text{otherwise.} \end{cases}$ Find the p.d.f. of $Y = X^2$.

- Since $X \ge -1$, we have $Y > 0$ but note that the support of X(or, the p.d.f. of X) is not symmetric. When $0 \le y \le 1$, $\{Y \le y\} = \{-\sqrt{y} \le X \le \sqrt{y}\}$ and $\{Y \le y\} = \{-1 \le X \le \sqrt{y}\}$ when $y > 1$.

- Thus, when $0 \le y \le 1$, $F_Y(y) = P(-\sqrt{y} \le X \le \sqrt{y})$ is the size of the shaded area in the left panel of Figure 4.5. Therefore, we have

$$F_Y(y) = \frac{1}{2}\sqrt{y} + \int_0^{\sqrt{y}} \frac{1}{2}e^{-x}dx = \frac{1}{2}\sqrt{y} + \frac{1}{2}\left(1 - e^{-\sqrt{y}}\right).$$

- On the other hand, when $y > 1$, $P(-\sqrt{y} \le X \le \sqrt{y})$ is the shaded area given in the right panel of Figure 4.5. Thus, the lower limit of the integral is set to -1 because of the range of x, and we have

$$F_Y(y) = \frac{1}{2} + \int_0^{\sqrt{y}} \frac{1}{2}e^{-x}dx = \frac{1}{2} + \frac{1}{2}\left(1 - e^{-\sqrt{y}}\right).$$

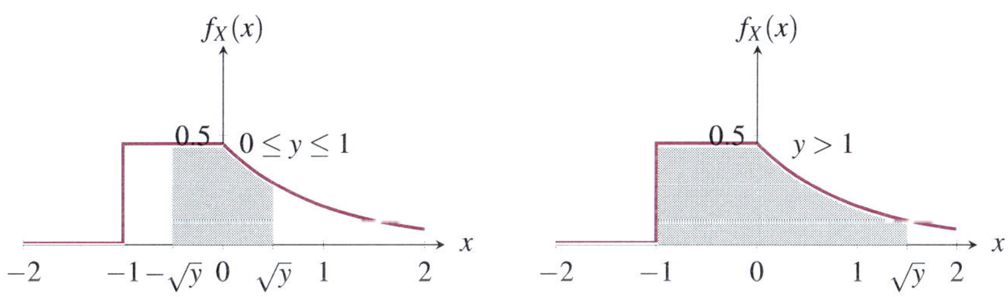

Figure 4.5: The shaded area of $\{(x,y) : x^2 \le y\}$ depending on the value of y

- Hence, the p.d.f of Y is obtained as

$$f_Y(y) = \frac{d}{dy}F_Y(y) = \begin{cases} \frac{1}{4\sqrt{y}}\left(1 + e^{-\sqrt{y}}\right), & 0 < y \le 1 \\ \frac{1}{4\sqrt{y}}e^{-\sqrt{y}}, & y > 1 \\ 0, & \text{otherwise.} \end{cases}$$

■

4.2 Moments

The definition of the expected value of a continuous random variable is the same as that of a discrete random variable by simply replacing the p.m.f. with the p.d.f., and summation with integration.

Definition 4.2.1 The expected value(mean) of a continuous random variable X with p.d.f. $f_X(x)$ is $\mathrm{E}(X) = \int x f_X(x)dx$, if the integral exists, i.e. $\int_{-\infty}^{\infty} |x| f_X(x)dx < \infty$.

Remark The expected value of a continuous random variable X can be understood as the weighted average of the values X can take, each value weighted by the probability that X lies in an infinitesimal neighborhood around that value.

- If we think of the integral as a "sum" of values x times probabilities $f_X(x)\,dx$, we can see the connection to the discrete case.

- We define the Riemann sum and substitute it into the definition of the integral

$$\lim_{\Delta x \to 0} \sum_{k=1}^{n} f_X(x_k)\Delta x = \int_a^b f_X(x)dx, \quad \Delta x = (b-a)/n.$$

- The mathematical properties with an integral are similar to the discrete case as a limiting form of a sum.

Example 4.14 Let U be uniformly distributed on the interval $[0,1]$. Then

$$\mathrm{E}(U) = \int_0^1 u du = \frac{1}{2}.$$

In general, if $X \sim \text{Uniform}(a,b)$, then $X = (b-a)U + a$. Thus,

$$\mathrm{E}(X) = \int_a^b x\frac{1}{b-a}dx = \frac{a+b}{2} = \mathrm{E}((b-a)U+a) = \frac{b-a}{2} + a.$$

■

Example 4.15 Let X have a p.d.f. $f_X(x) = \begin{cases} \frac{\alpha}{\beta}\left(\frac{\beta}{x}\right)^{\alpha+1}, & x > \beta, \\ 0, & \text{otherwise,} \end{cases}$ which is known as the p.d.f. of *Pareto distribution*. Assume that $\beta = 1$. Find the expected value of X.

- If $\alpha \leq 1$, the expected value of X is infinite because

$$\mathrm{E}(X) = \int_1^\infty x\frac{\alpha}{x^{\alpha+1}}dx = \lim_{N\to\infty}\int_1^N x\frac{\alpha}{x^{\alpha+1}}dx = \alpha \lim_{N\to\infty}\int_1^N \frac{1}{x^\alpha}dx,$$

which diverges to ∞ for $\alpha \leq 1$.

- However, for $\alpha > 1$

$$\mathrm{E}(X) = \int_1^\infty \frac{\alpha}{x^\alpha}dx = \frac{\alpha}{(\alpha-1)}.$$

■

Remark The improper integral $\int_1^\infty \frac{1}{x^p}\,dx$ converges to $\frac{1}{p-1}$ if $p > 1$ and diverges if $0 < p \leq 1$.

- Suppose that $p \neq 1$. Then,

$$\int_1^\infty \frac{1}{x^p}dx = \lim_{N\to\infty}\int_1^N \frac{1}{x^p}dx = \lim_{N\to\infty}\left[\frac{x^{-p+1}}{-p+1}\right]_1^N = \lim_{N\to\infty}\frac{1}{1-p}\left(N^{-p+1}-1\right)$$

$$= \lim_{N\to\infty}\frac{1}{1-p}\left(\frac{1}{N^{p-1}}-1\right) = \begin{cases} \text{converges to } \frac{1}{p-1}, & \text{if } p > 1, \\ \text{diverges,} & \text{if } p < 1. \end{cases}$$

- If $p=1$, the integral also diverges,

$$\int_1^\infty \frac{1}{x}dx = \lim_{N\to\infty}\int_1^N \frac{1}{x}dx = \lim_{N\to\infty}[\log x]_1^N = \lim_{N\to\infty}(\log N - \log 1) = \infty.$$

Example 4.16 Let X have the p.d.f. $f_X(x) = \frac{1}{\pi}\frac{1}{1+x^2}$, for $-\infty < x < \infty$. This distribution is called the *Cauchy distribution*. The integral $\int xf(x)dx$ is undefined, and *the mean does not exist.*

- $f(x)$ is nonnegative and its integral is equal to one.

$$\int_{-\infty}^{\infty} f_X(x)dx = \int_{-\infty}^{\infty}\frac{1}{\pi}\frac{1}{1+x^2}dx = \frac{1}{\pi}\left[\tan^{-1}x\right]_{-\infty}^{\infty} = \frac{1}{\pi}\left(\frac{\pi}{2}+\frac{\pi}{2}\right) = 1,$$

and the c.d.f. of X is $F_X(x) = \int_{-\infty}^{x} f(t)\,dt = \int_{-\infty}^{x}\frac{1}{\pi\left[1+t^2\right]}dt = \frac{1}{\pi}\tan^{-1}x + \frac{1}{2}.$

- We can find the expected value of X if the integral $\int xf(x)dx$ converges absolutely, that is $\int |x|f(x)\,dx < \infty$. If this integral diverges, then the expected value of X does not exist. Note that $\int_1^\infty \frac{x}{1+x^2}dx$ diverges from *Limit comparison test* for improper integrals, since $\int_1^\infty \frac{1}{x^p}dx$ diverges if $p \le 1$ and $\lim_{x\to\infty}\left(\frac{x}{1+x^2}\right)\Big/\left(\frac{1}{x}\right) = 1.$

- In fact, we can check $\int_1^\infty \frac{x}{1+x^2}dx = \lim_{N\to\infty}\int_1^N \frac{x}{1+x^2}dx = \lim_{N\to\infty}\frac{1}{2}\log(1+x^2)\Big|_1^N = \infty.$
  Thus $\int_0^\infty \frac{x}{1+x^2}dx = \int_0^1 \frac{x}{1+x^2}dx + \int_1^\infty \frac{x}{1+x^2} \ge \int_1^\infty \frac{x}{1+x^2}dx$ also diverges, which follows by *comparison test* for improper integrals.

- Therefore, although $\frac{x}{1+x^2}$ is an odd function, $\int_{-\infty}^{\infty}\frac{x}{1+x^2}dx$ is not defined. Indeed, it is not equal to $\lim_{b\to\infty}\int_{-b}^{b}\frac{x}{1+x^2}dx = 0$. Hence, the above integral does not exist, and thus the expected value does not exist.

■

Remark **(Comparison test for improper integrals)** Let $f\colon [a,\infty) \to R$ and $g\colon [a,\infty) \to R$ be functions that are Riemann integrable on $[a,b]$ for all $b > a$. Suppose that for all $x \ge a$ we have $|f(x)| \le g(x)$.

- If $\int_a^\infty g(x)dx$ converges, then $\int_a^\infty f(x)dx$ converges.
- If $\int_a^\infty f(x)dx$ diverges, then $\int_a^\infty g(x)dx$ diverges.

(Limit comparison Test for improper integrals) Let $f(x)$ and $g(x)$ be nonnegative functions that are Riemann integrable on $[a,b]$. If $\lim_{x\to\infty}\frac{f(x)}{g(x)} = k$, where k is a positive constant, then $\int_a^\infty f(x)dx$ and $\int_a^\infty g(x)dx$ are either both convergent or both divergent.

Theorem 4.2.1 Suppose that the random variable X has a continuous distribution with c.d.f $F_X(x)$ and p.d.f $f_X(x)$. Suppose also that $P(X \geq 0) = 1$ and that $\mathrm{E}(X)$ exists. Then

$$\mathrm{E}(X) = \int_0^\infty [1 - F_X(x)]dx.$$

Proof. We can use double integrals and the interchange in the order of integration:

$$\begin{aligned}\int_0^\infty P(X > x)dx &= \int_0^\infty \left[\int_x^\infty f_X(y)dy\right] dx = \int_0^\infty \int_0^\infty I[y \geq x] f_X(y)dydx \\ &= \int_0^\infty \int_0^y dx f_X(y)dy = \int_0^\infty y f_X(y)dy = \mathrm{E}(X).\end{aligned}$$

Alternatively, we can use integration by parts:

$$\begin{aligned}\int_0^y x f_X(x)dx &= -x(1 - F_X(x))\Big|_0^y + \int_0^y (1 - F_X(x))dx, \\ y(1 - F_X(y)) &= y\int_y^\infty f_X(x)dx \leq \int_y^\infty x f_X(x)dx.\end{aligned}$$

Then, because $\mathrm{E}(X) < \infty$, we have $\lim_{y\to\infty}\int_y^\infty x f_X(x)dx = 0$. Hence, we obtain

$$\mathrm{E}(X) = \int_0^\infty x f_X(x)dx = \lim_{y\to\infty}\int_0^y x f_X(x)dx = \int_0^\infty (1 - F_X(x))dx.$$

■

Remark For a continuous random variable X, $\mathrm{E}(X) = \int_0^\infty P(X \geq x)dx - \int_0^\infty P(X \leq -x)dx$.

- As in Theorem 4.2.1, $\int_0^\infty P(X \geq x)dx = \int_0^\infty y f(y)dy$. Similarly, we have

$$\int_0^\infty P(X \leq -x)dx = \int_0^\infty \int_{-\infty}^{-x} f_X(y)dydx = \int_{-\infty}^0 \int_0^{-y} f_X(y)dxdy = \int_{-\infty}^0 -y f_X(y)dy.$$

- Hence, it follows that

$$\int_0^\infty P(X \geq x)dx - \int_0^\infty P(X \leq -x)dx = \int_0^\infty y f(y)dy + \int_{-\infty}^0 y f_X(y)dy = \mathrm{E}(X).$$

- A simple version of the Fubini's theorem for interchanging the order of integration is stated as follows: for a continuous function $f(x,y) : [a,b] \times [c,d] \to R$,

$$\int_a^b \int_c^d f(x,y)dydx = \int_c^d \int_a^b f(x,y)dxdy.$$

Theorem 4.2.2 — Law of the unconscious statistician. Let X be a random variable, and let $Y = r(X)$, where r is a real valued function of a real variable. If X has a continuous distribution, then

$$\mathrm{E}(Y) = \int y f_Y(y)dy = \int r(x) f_X(x)dx.$$

In general $\mathrm{E}[r(X)] \neq r[\mathrm{E}(X)]$, but if $Y = r(X) = aX + b$, then $\mathrm{E}(Y) = a\mathrm{E}(X) + b$.

Proof. A general proof will not be given here, but we consider a specical case. Let X have p.d.f. f_X and let $Y = r(X)$ where r is strictly increasing. Let $f_Y(y)$ be the p.d.f. of Y. Then

$$\mathrm{E}(Y) = \int y f_Y(y)dy = \int y f_X(r^{-1}(y)) \frac{d}{dy} r^{-1}(y)dy = \int r(x) f(x)dx.$$

Alternatively, we can use Theorem 4.2.1 assuming that $r(X) \geq 0$. That is,

$$\begin{aligned} \mathrm{E}[r(X)] &= \int_0^\infty P(r(X) \geq x)dx = \int_0^\infty \int_{\{t:r(t)\geq x\}} f_X(t)dtdx \\ &= \int_0^\infty f_X(t) \int_0^{r(t)} dxdt = \int_0^\infty r(t) f_X(t)dt \end{aligned}$$

as required. The interchange in the order of integration is justified by Fubini's theorem on double integrals, similarly to Theorem 4.2.1.

■

Remark For a general random variable X, we define $X^+(\omega) \equiv \max(X(\omega), 0) \geq 0$, called the positive part of X, and $X^-(\omega) \equiv \max(-X(\omega), 0) \geq 0$, called the negative part of X, respectively.

- For each $\omega \in \Omega$, if $X(\omega) \geq 0$ then $X^+(\omega) \geq 0$ and $X^-(\omega) = 0$. Thus, $X(\omega) = X^+(\omega) - X^-(\omega)$.
- Hence, we define $\mathrm{E}(X) = \mathrm{E}(X^+) - \mathrm{E}(X^-)$, including ∞ or $-\infty$, unless $\mathrm{E}(X^+) = \mathrm{E}(X^-) = \infty$, in which case $\mathrm{E}(X)$ does not exist as in Cauchy random variable of Example 4.16.

■ **Example 4.17** Suppose that a point is chosen at random on a stick of unit length and that the stick is broken into two pieces at that point. Find the expected value of the length of the longer piece.

- Let X be the length of the longer piece, and let the break point be $U \sim \text{Uniform}(0,1)$. Then, $X = \max\{U, 1-U\}$ and $\{X \le x\}$ is equivalent to $\{U \le x,\ 1-U \le x\}$, which is $1-x \le U \le x$ for $1/2 < x < 1$. Thus,

$$F_X(x) = P(X \le x) = \begin{cases} P(1-x \le U \le x) = \int_{1-x}^{x} 1du = 2x-1, & 1/2 < x < 1, \\ 0, & \text{otherwise.} \end{cases}$$

  Hence, $f_X(x) = 2$, $1/2 < x < 1$, which means that $X \sim \text{Uniform}(1/2, 1)$ and that $\text{E}(X) = 3/4$.

- Alternatively, we can use the law of unconscious statistician in Theorem 4.2.2.

$$\begin{aligned} \text{E}(X) &= \text{E}(\max\{U, 1-U\}) = \int_0^1 \max\{u, 1-u\} \cdot 1du \\ &= \int_0^{1/2} (1-u)du + \int_{1/2}^1 udu = \frac{3}{4}. \end{aligned}$$

■

The properties of expectation and variance are the same for discrete and continuous random variables; We can just replace $\sum$ with $\int$.

Definition 4.2.2 The variance of a continuous random variable X is defined as,

$$\begin{aligned} \text{Var}(X) &= \text{E}[(X - \text{E}[X])^2] = \text{E}(X^2) - (\text{E}[X])^2 \\ &= \int_{-\infty}^{\infty} x^2 f(x)dx - \left(\int_{-\infty}^{\infty} xf(x)dx\right)^2. \end{aligned}$$

■ **Example 4.18** Suppose $X \sim U[a,b]$. Then

$$\text{E}(X) = \int_a^b x\frac{1}{b-a}dx = \frac{1}{2}(a+b), \quad \text{Var}(X) = \int_a^b x^2 \frac{1}{b-a}dx - (\text{E}[X])^2 = \frac{1}{12}(b-a)^2.$$

■

Theorem 4.2.3 For each random variable X, and every positive integer n, the expectation $\text{E}(X^n)$ is called the nth moment of X. If $\text{E}(|X|^n) < \infty$ for some positive integer n, then $\text{E}(|X|^k) < \infty$ for every positive integer k such that $k < n$.

Proof. Note that

$$\begin{aligned} \text{E}(|X|^k) &= \int_{-\infty}^{\infty} |x|^k f(x)dx = \int_{|x|\le 1} |x|^k f(x)dx + \int_{|x|>1} |x|^k f(x)dx \\ &\le \int_{|x|\le 1} 1 \cdot f(x)dx + \int_{|x|>1} |x|^k f(x)dx \le P(|X| \le 1) + \text{E}(|X|^n). \end{aligned}$$

Since $\text{E}(|X|^n) < \infty$, we have $\text{E}(|X|^k) < \infty$. ■

Definition 4.2.3 A function $g : (a,b) \to \mathbb{R}$ is **convex** if for all $x_1, x_2 \in (a,b)$ and $\lambda_1 \geq 0, \lambda_2 \geq 0$ with $\lambda_1 + \lambda_2 = 1$,

$$\lambda_1 g(x_1) + \lambda_2 g(x_2) \geq g(\lambda_1 x_1 + \lambda_2 x_2).$$

It is **strictly convex** if strict inequality holds when $x_1 \neq x_2$ and $0 < \lambda_1 < 1$. A function g is **concave** (strictly concave) if -g is convex (strictly convex).

Remark If g is a twice differentiable function and $g''(x) \geq 0$ for all $x \in (a,b)$ then g is convex. It is strictly convex if $g''(x) > 0$ for all $x \in (a,b)$.

Theorem 4.2.4 — Jensen's inequality. For a random variable X, if $g(x)$ is a convex function, then $\mathrm{E}[g(X)] \geq g(\mathrm{E}[X])$.If g is concave, then $\mathrm{E}[g(X)] \leq g(\mathrm{E}[X])$.

Proof. Let g be a real-valued convex function, and suppose that $\mu = \mathrm{E}(X)$ and $\mathrm{E}[g(X)]$ exist. We may assume that there exists a line $a + bx$ such that $g(\mu) = a + b\mu$ and $g(x) \geq a + bx$ for all x. For example, there exists a tangent line to g at the point μ such that the tangent lies below the function, $g(x) \geq g(\mu) + g'(\mu)(x - \mu)$. We can also consider $g(x) - g(\mu) = \int_\mu^x g'(x)dx \geq \int_\mu^x g'(\mu)dx = g'(\mu)(x - \mu)$, since $g''(x) \geq 0$ and $g'(x)$ is non-decreasing, assuming that $g(x)$ is twice differentiable. Taking expectations on both sides of $g(x) \geq a + bx$ gives $\mathrm{E}[g(X)] \geq a + b\mu = g(\mu) = g(\mathrm{E}[X])$, which establishes the inequality.

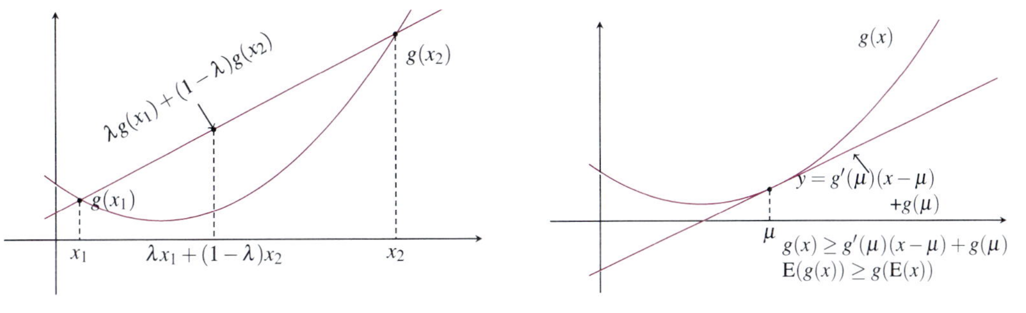

Figure 4.6: An illustration of Jensen's inequality with a convex function

■

■ **Example 4.19** Let $g(x) = x^2$. Then $g(x)$ is strictly convex on $\mathbb{R}$. Thus, Jensen's inequality in Theorem 4.2.4 says that $\mathrm{E}(X^2) \geq (\mathrm{E}[X])^2$ which is also true since variance is nonnegative. ■

Definition 4.2.4 Let X be a continuous random variable whose p.d.f. is $f_X(x)$. A real valued function $M_X(\cdot) : \mathbb{R} \to \mathbb{R}$ defined by

$$M_X(t) = \mathrm{E}\left(e^{tX}\right) = \int e^{tx} f_X(x)dx$$

is called the moment generating function(m.g.f.) of X if this expected value exists for all t in the interval $-h < t < h$ for some $h > 0$.

Remark In general, not every random variable has a moment generating function. But if the moment generating function of a random variable exists, then it is unique.

Theorem 4.2.5 Let $M_X(t)$ be the moment generating function of the random variable X. If

$$M_X(t) = a_0 + a_1 t + a_2 t^2 + \cdots + a_n t^n + \cdots$$

is the Taylor series expansion of $M_X(t)$, then

$$\mathrm{E}(X^n) = (n!)\, a_n,\ \mathrm{E}(X^n) = \frac{d^n}{dt^n} M_X(t)\Big|_{t=0},\ n = 1,2,3,\ldots$$

Proof. Let $M_X(t)$ be the moment generating function of the random variable X. The Taylor series expansion of $M_X(t)$ about 0 is given by

$$M_X(t) = M_X(0) + \frac{M_X'(0)}{1!} t + \frac{M_X''(0)}{2!} t^2 + \frac{M_X^{(3)}(0)}{3!} t^3 + \cdots + \frac{M_X^{(n)}(0)}{n!} t^n + \cdots$$

Note that by the Taylor series expansion of e^{tX} and the interchange in the order of infinite sum and integration, we have

$$\mathrm{E}(e^{tX}) = \mathrm{E}\left(\sum_{n=0}^{\infty} \frac{X^n t^n}{n!}\right) = \sum_{n=0}^{\infty} \frac{\mu_n t^n}{n!},\ \mu_n = \mathrm{E}(X^n).$$

Thus, by comparing $M_X(t)$ with $\mathrm{E}(e^{tX})$, term by term, it follows that $\mathrm{E}(X^n) = M_X^{(n)}(0)$ for $n \geq 1$, $M_X(0) = 1$, and that

$$M_X(t) = 1 + \frac{\mathrm{E}(X)}{1!} t + \frac{\mathrm{E}(X^2)}{2!} t^2 + \frac{\mathrm{E}(X^3)}{3!} t^3 + \cdots + \frac{\mathrm{E}(X^n)}{n!} t^n + \cdots$$

Hence, we obtain $a_n = \dfrac{\mathrm{E}(X^n)}{n!}$ which is $\mathrm{E}(X^n) = (n!)\, a_n$, which proves the theorem. ■

■ **Example 4.20** Suppose that $X \sim \text{Uniform}(0,1)$. Find $\mathrm{E}\left(e^{tX}\right)$. Then,

- $M_X(t) = \mathrm{E}\left(e^{tX}\right) = \int_0^1 e^{tx}dx = \left[\frac{1}{t}e^{tx}\right]_0^1 = \frac{e^t - 1}{t}$, which exists for all t, including $t = 0$, where it takes the value 1.

- Note that $\mathrm{E}(e^{tX}) = \mathrm{E}\left(\sum_{k=0}^{\infty} \frac{X^k t^k}{k!}\right) = \sum_{k=0}^{\infty} \frac{\mathrm{E}(X^k)t^k}{k!}$, provided the interchange of expectation and summation is justified. Thus, we have

$$M_X(t) = \mathrm{E}\left(e^{tX}\right) = \sum_{n=0}^{\infty} \frac{\mathrm{E}(X^n)}{n!}t^n = \frac{e^t - 1}{t} = \frac{\sum_{n=0}^{\infty} t^n/n! - 1}{t} = \sum_{n=0}^{\infty} \frac{t^n}{(n+1)!},$$

which implies that $\mathrm{E}(X^n) = 1/(n+1)$, $n = 0, 1, \ldots$. Hence, for example, variance is obtained as $\mathrm{Var}(X) = \mathrm{E}(X^2) - [\mathrm{E}(X)]^2 = 1/12$.

■

Definition 4.2.5 — Medians. The mean (expectation) of a distribution is its center of mass and is thought to be a measure of central location. Another measure of central location is the median. A *median* of a random variable X (or of its distribution) is any number m such that $P(X \leq m) \geq 1/2$ and $P(X \geq m) \geq 1/2$.

- Every distribution has at least one median. Some distributions have a unique median while some have a whole interval of medians.

- $F_X^{-1}(x)$ is called the *quantile function* of the distribution, and x is the p *quantile* or the $100 \times$ *pth percentile* of the distribution.

- The 0.5 quantile or the 50th percentile of a distribution is a median, m. If X is continuous random variable, there is a unique x such that $F_X(x) = p$, that is, $m = F_X^{-1}(0.5)$.

■ **Example 4.21** Let X have the p.d.f. $f_X(x) = \frac{1}{\pi(1+x^2)}$, for all x, which is the Cauchy distribution given in Example 4.16.

- As discussed in Example 4.16, the expected value of X does not exist, but it is symmetric around 0, so the median is 0, since $P(X \leq 0) \geq 1/2$ and $P(X \geq 0) \geq 1/2$.

- Alternatively, since $F_X(x) = \frac{1}{\pi}\tan^{-1}x + \frac{1}{2}$ as in Example 4.16, it follows that $F_X^{-1}(x) = \tan\pi(x - 0.5)$, which also implies that $m = F_X^{-1}(0.5) = 0$.

■

Both the mean and the median are used to locate the center of the distribution, but they are different in specifying some prediction problems. Here, we consider predicting a random variable X by choosing a constant c that will minimize some measure of how far X is from c.

Theorem 4.2.6 — Mean vs. median. Suppose that X is continuous random variable with p.d.f. $f_X(x)$. Then,

① The value of c that minimizes $\mathrm{E}(X-c)^2$ is the mean of X, that is, $c = \mathrm{E}(X)$.

② The value of c that minimizes $\mathrm{E}|X-c|$ is the median of X, that is, $c = m$, where m is the median.

Proof. Here, we assume X has finite mean and variance.

① Let $\mu = \mathrm{E}(X)$. Then, we write $(X-c)^2 = (X-\mu)^2 + (\mu-c)^2 + 2(X-\mu)(\mu-c)$. The mean of this expression is $\mathrm{Var}(X) + (\mu-c)^2$. Clearly $c = \mu$ provides the smallest value. Alternatively, differentiating $g(c) = \mathrm{E}(X-c)^2$ with respect to c, we get the solution of $g'(c) = 0$, the minimizer of $g(c)$, which is μ.

② Let m be the 0.5 quantile, $m = F_X^{-1}(0.5)$. Assume that $m \le a$. Then,

$$|X-a| - |X-m| = \begin{cases} a - X - (m-X) = a - m, & \text{if } X \le m, \\ |X-a| - (X-m) \ge X - a - (X-m) = m - a, & \text{if } X > m. \end{cases}$$

Thus, decomposing the expected value of $|X-a| - |X-m|$ into two parts, we have

$$\begin{aligned} &\mathrm{E}(|X-a| - |X-m|) = \int_{-\infty}^{\infty} (|x-a| - |x-m|) f_X(x)dx \\ &\ge \int_{-\infty}^{m} (a-m) f_X(x)dx + \int_{m}^{\infty} (m-a) f_X(x)dx \\ &= (a-m)\left[F_X(m) - (1 - F_X(m))\right] = (a-m)(2F_X(m) - 1) \ge 0, \end{aligned}$$

which shows that $\mathrm{E}|X-m| \le \mathrm{E}|X-a|$. Similarly if $a < m$, we can show that $\mathrm{E}(|X-a|) - \mathrm{E}(|X-m|) \ge 0$. So, $a = m$ provides the smallest value. Alternatively, we minimize $\mathrm{E}|X-c|$ with respect to c, which is written as

$$\begin{aligned} g(c) &= \mathrm{E}|X-c| = \int_{-\infty}^{\infty} |x-c| f_X(x)dx = \int_{-\infty}^{c} (c-x) f_X(x)dx + \int_{c}^{\infty} (x-c) f_X(x)dx \\ &= cF_X(c) - \int_{-\infty}^{c} x f_X(x)dx + \int_{c}^{\infty} x f_X(x)dx - c(1 - F_X(c)). \end{aligned}$$

By differentiating $g(c) = \mathrm{E}|X-c|$ with respect to c and using the fundamental theorem of calculus, we thus get the solution of $g'(c) = 0$, the minimizer of $g(c)$,

$$g'(c) = F_X(c) + cf_X(c) - cf_X(c) - cf_X(c) - (1 - F_X(c)) + cf_X(c) = 0,$$

which is $c = F_X^{-1}(0.5)$, the median.

■

4.3 Normal distribution

The normal distribution (also called a Gaussian distribution) is the most important continuous distribution because it arises in many different contexts. The normal distribution made its appearance as early as 1733 in the work of Abraham de Moivre as an approximation to probabilities involving the number of successes in a succession of tosses of a fair coin, which is known as de Moivre-Laplace theorem, and later generalized to the central limit theorem.

Definition 4.3.1 The normal distribution (or Gaussian distribution) with parameters μ and σ^2 has p.d.f.

$$f_X(x) = \frac{1}{\sqrt{2\pi\sigma^2}} e^{-\frac{(x-\mu)^2}{2\sigma^2}}, -\infty < x < \infty.$$

To indicate that X has this distribution we write $X \sim N(\mu, \sigma^2)$.

Example 4.22 The normal p.d.f.(distribution) with $\mu = 0$ and $\sigma = 1$ is called the standard normal p.d.f (distribution), usually denoted by $\phi(x)$,

$$\phi(x) = \frac{1}{\sqrt{2\pi}} \exp\left(-\frac{x^2}{2}\right), \text{ for } -\infty < x < \infty.$$

- $\phi(x)$ is symmetric, $\phi(x) = \phi(-x)$. In other words, $\phi(x)$ is an even function. $\phi(x)$ is strictly positive everywhere on the real line.

- Note that
$$\begin{aligned}\phi'(x) &= -\frac{1}{\sqrt{2\pi}} x e^{-\frac{x^2}{2}} = -x\phi(x),\\ \phi''(x) &= -\phi(x) - x\phi'(x) = (x^2-1)\phi(x) = \frac{1}{\sqrt{2\pi}}(x^2-1)e^{-\frac{x^2}{2}}.\end{aligned}$$

- $\phi(x)$ has one peak, a maximum at $x = 0$, and two inflection points at $x = -1$ and $x = 1$, being convex for $x < -1$ and $x > 1$ and concave for $-1 < x < 1$ and $\phi(x)$ decays exponentially for values far from 0 with $\lim_{x\to\pm\infty} \phi(x) = 0$.

x		-1		0		1	
$\phi'(x)$	$+$	$+$	$+$	0	$-$	$-$	$-$
$\phi''(x)$	$+$	0	$-$	$-$	$-$	0	$+$
$\phi(x)$	increases convex	inflection point	increases concave	maximum point	decreases concave	inflection point	decreases convex

- $\phi(x)$ is the classic bell-shaped curve as shown in Figure 4.7.

- The c.d.f. of standard normal distribution is denoted by $\Phi(x)$, $\Phi(x) = \int_{-\infty}^{x} \phi(t)dt$. The integral for $\Phi(x)$ cannot be given as an explicit form but $\Phi(0) = \frac{1}{2}$. In addition,

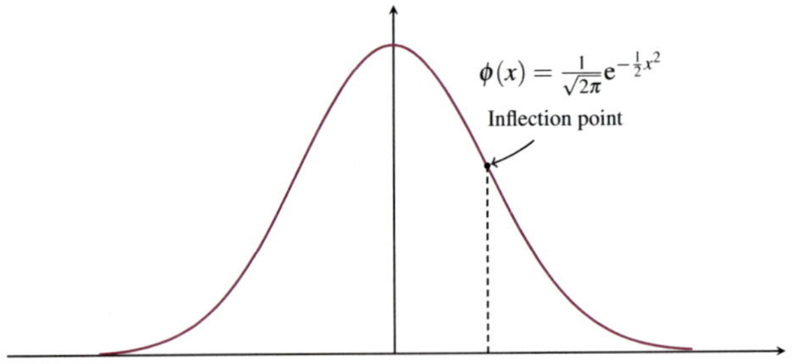

Figure 4.7: The p.d.f. of a standard normal distribution

since $\phi(x)$ is an even function

$$\Phi(-x) = \int_{-\infty}^{-x} \phi(t)dt = \int_{x}^{\infty} \phi(t)dt = 1 - \int_{-\infty}^{x} \phi(t)dt = 1 - \Phi(x).$$

■

Remark The function $\phi(z) = c\exp\left(-\frac{1}{2}z^2\right)$ can be a p.d.f. since $\int_{-\infty}^{\infty} \phi(z)dz < \infty$. Note that the improper integral $\int_0^{\infty} \exp\left(-\frac{1}{2}z^2\right) dz$ converges. Specifically, we have that

$$\int_{-N}^{N} \exp\left(-\frac{1}{2}z^2\right) dz < 2\left(\int_0^1 dz + \int_1^N e^{-z}dz\right) < 2\left(1 + e^{-1} - e^{-N}\right).$$

Thus, $\int_{-\infty}^{\infty} \exp\left(-\frac{1}{2}z^2\right) dz = \lim_{N\to\infty} \int_{-N}^{N} \exp\left(-\frac{1}{2}z^2\right) dz < 2(1 + e^{-1})$. Hence, $\phi(z)$ is a p.d.f. with the normalizing constant $c = \left(\int_{-\infty}^{\infty} \exp\left(-\frac{1}{2}z^2\right) dz\right)^{-1} = \frac{1}{\sqrt{2\pi}}$, which is analytically evaluated in Theorem 4.3.1.

Theorem 4.3.1 — Standard Normal. The standard normal distribution, $Z \sim N(0,1)$, is the distribution whose p.d.f. is

$$fz(z) = \frac{1}{\sqrt{2\pi}} \exp\left[-\frac{z^2}{2}\right], \quad \text{also denoted by } \phi(z),$$

and its c.d.f. is usually denoted by $\Phi(z) = \int_{-\infty}^{z} (1/\sqrt{2\pi})e^{-x^2/2}dx$. Then, we have

$$\mathrm{E(Z)} = 0, \ \mathrm{Var(Z)} = 1, \ \mathrm{E}(Z^n) = \begin{cases} 0 & n \text{ is odd,} \\ \frac{n!}{2^{n/2}(n/2)!} & n \text{ is even,} \end{cases} \quad M_Z(t) = \exp\left(\frac{1}{2}t^2\right).$$

Proof. It follows by the integral test that the area under the curve is positive and bounded. Since $\exp(-z^2/2) < z^{-4}$ for sufficiently large $|z|$, this distribution has finite mean and variance.

- The area under the curve of the standard normal density is unit. That is, we need to check that $I = \int_{-\infty}^{\infty} \phi(x)dx = 1$. As $\phi(z)$ is positive it is clear that I is positive. Thus, by considering its square, we get

$$\begin{aligned} I^2 &= \frac{1}{2\pi}\left[\int_{-\infty}^{\infty} e^{-\frac{1}{2}x^2}dx\right]\left[\int_{-\infty}^{\infty} e^{-\frac{1}{2}y^2}dy\right] = \frac{1}{2\pi}\int_{-\infty}^{\infty}\int_{-\infty}^{\infty} e^{-\frac{1}{2}(x^2+y^2)}dxdy \\ &= \frac{1}{2\pi}\int_0^{2\pi}\int_0^{\infty} re^{-\frac{1}{2}r^2}drd\theta = \frac{1}{2\pi}\int_0^{2\pi} d\theta = 1, \end{aligned}$$

where the polar transformation $x = r\cos\theta$ and $y = r\sin\theta$ is used. Hence, it follows that $I = \int_{-\infty}^{\infty} \phi(x)dx = 1$.

- The mean, $\mathrm{E}(X)$ is 0 because the p.d.f. is symmetric around 0. For $\mathrm{Var}(Z)$, integration by parts easily shows that

$$\mathrm{Var(Z)} = \frac{1}{\sqrt{2\pi}}\int_{-\infty}^{\infty} z^2 e^{-\frac{1}{2}z^2}dz = \left[-\frac{1}{\sqrt{2\pi}}ze^{-\frac{1}{2}z^2}\right]_{-\infty}^{\infty} + \frac{1}{\sqrt{2\pi}}\int_{-\infty}^{\infty} e^{-\frac{1}{2}z^2}dz = 1.$$

In general, for each integer $n \geq 0$, define the nth moment by $\mu_n = \int_{-\infty}^{\infty} x^n\phi(x)dx$. For each n, we can find a constant A_n such that $|x|^n \leq A_n e^{|x|}$. Thus, the comparison test shows that the integral is convergent for each n. If n is odd then $x^n\phi(x)$ is odd, hence $\mu_n = 0$. If $n = 2k$ is even then a simple integration by parts sets up an induction. For each $k \geq 1$, we have

$$\begin{aligned} \mu_{2k} &= \int_{-\infty}^{\infty} x^{2k}\phi(x)dx = \frac{1}{\sqrt{2\pi}}\int_{-\infty}^{\infty} x^{2k-1}d(-e^{-x^2/2}) \\ &= \frac{-x^{2k-1}e^{-x^2/2}}{\sqrt{2\pi}}\Bigg|_{-\infty}^{\infty} + \frac{2k-1}{\sqrt{2\pi}}\int_{-\infty}^{\infty} x^{2k-2}e^{-x^2/2}dx = (2k-1)\mu_{2k-2}. \end{aligned}$$

As $\mu_0 = 1$, it follows that $\mu_{2k} = (2k-1)(2k-3)\ldots 5\cdot 3\cdot 1\cdot \mu_0 = \dfrac{(2k)!}{2^k k!}$.

- Alternatively, we can compute the moment generating function of a standard normal distribution

$$\begin{aligned} M_Z(t) &= \mathrm{E}\left(e^{tZ}\right) = \int_{-\infty}^{\infty} e^{tz}\phi(z)dz = \int_{-\infty}^{\infty} e^{tz}\frac{1}{\sqrt{2\pi}}e^{-\frac{1}{2}z^2}dz \\ &= e^{\frac{1}{2}t^2}\int_{-\infty}^{\infty}\frac{1}{\sqrt{2\pi}}e^{-\frac{1}{2}(z-t)^2}dz = e^{\frac{1}{2}t^2}. \end{aligned}$$

Note that

$$\mathrm{E}(e^{tZ}) = \mathrm{E}\left(\sum_{n=0}^{\infty}\frac{Z^n t^n}{n!}\right) = \sum_{n=0}^{\infty}\frac{\mu_n t^n}{n!} = e^{\frac{1}{2}t^2} = \sum_{k=0}^{\infty}\frac{\left(\frac{1}{2}t^2\right)^k}{k!} = \sum_{k=0}^{\infty}\frac{(2k)!}{2^k k!}\frac{t^{2k}}{(2k)!}.$$

Hence $\mu_{2k+1} = 0$ and $\mu_{2k} = \dfrac{(2k)!}{2^k k!} = (2k-1)\times(2k-3)\times\cdots\times 5\times 3\times 1$.

■

Remark Let $(x,y)=\mathbf{g}(u,v):\begin{cases} x &= g_1(u,v), \\ y &= g_2(u,v). \end{cases}$ Then, the Jacobian matrix is

$$J(u,v)=\frac{\partial(x,y)}{\partial(u,v)}=\frac{\partial(g_1,g_2)}{\partial(u,v)}=\begin{pmatrix} \frac{\partial g_1}{\partial u} & \frac{\partial g_1}{\partial v} \\ \frac{\partial g_2}{\partial u} & \frac{\partial g_2}{\partial v} \end{pmatrix}.$$

and the Jacobian is its determinant, $\det[J(u,v)]=\frac{\partial g_1}{\partial u}\frac{\partial g_2}{\partial v}-\frac{\partial g_2}{\partial u}\frac{\partial g_1}{\partial v}$.

- $\int\int f(x,y)dxdy=\int\int f(g(u,v))\left|\det\left[g'(u,v)\right]\right|dudv$, where $g'(u,v)=\frac{\partial(g_1,g_2)}{\partial(u,v)}$.
- For the polar coordinate map, $(x,y)=g(r,\theta)=(r\cos\theta,r\sin\theta)$. We have

$$\int\int f(x,y)dxdy=\int\int f(r\cos\theta,r\sin\theta)\;r\,drd\theta.$$

■ **Example 4.23** Suppose that Z has the standard normal distribution, $Z\sim N(0,1)$. Find the expected value of $|Z|$.

- Define $Y=|Z|$. Then, differentiating the c.d.f. of Y, we get the p.d.f of Y,

$$f_Y(y)=\frac{d}{dy}F_Y(y)=\frac{d}{dy}P(-y\le Z\le y)=\frac{d}{dy}2\int_0^y\phi(z)dz=2\phi(y),\;y>0.$$

  Thus, the expected value of Y is $\mathrm{E}(Y)=\int_0^\infty 2y\phi(y)dy$. Since $\phi(y)'=-y\phi(y)$, we have $\mathrm{E}(Y)=-2\phi(y)\Big|_0^\infty=2\phi(0)=\sqrt{\frac{2}{\pi}}$.

- Alternatively, we can use the law of unconscious statistician

$$\mathrm{E}|Z|=\int_{-\infty}^{\infty}|z|\phi(z)dz=2\int_0^\infty z\phi(z)=\sqrt{\frac{2}{\pi}}.$$

■

■ **Example 4.24** It is an important and useful fact that if Z has a standard normal distribution and if $X=\mu+\sigma Z$, where $\sigma>0$, then X has a normal random distribution with mean μ and variance σ^2.

- Since $F_X(x)=P(\mu+\sigma Z\le x)=\Phi\left(\frac{x-\mu}{\sigma}\right)$, the p.d.f. of X is

$$f_X(x)=\phi\left(\frac{x-\mu}{\sigma}\right)\frac{1}{\sigma}=\frac{1}{\sqrt{2\pi}\sigma}\exp\left[-\frac{(x-\mu)^2}{2\sigma^2}\right],$$

  which implies that $X\sim N(\mu,\sigma^2)$.

- Note that as shown in Figure 4.8, the p.d.f. of X has a high peak with a small value of σ and is concentrated around the mean μ, whereas the p.d.f. with a large value of σ spreads around mean μ that determines the center of the distribution.

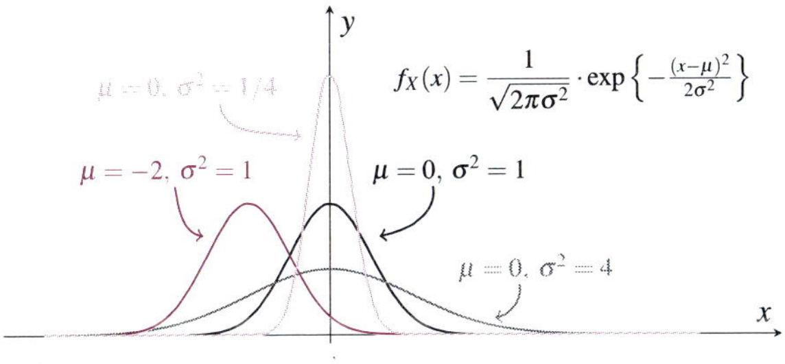

Figure 4.8: The p.d.f of a normal random variable X, $X \sim N(\mu, \sigma^2)$

- Since $X = \mu + \sigma Z$, we have

$$\mathrm{E}(X) = \mu, \quad \mathrm{Var}(X) = \sigma^2, \quad M_X(t) = \mathrm{E}(e^{tX}) = \exp\left(\mu t + \frac{1}{2}\sigma^2 t^2\right).$$

- If $X \sim N(\mu, \sigma^2)$, then the random variable $Z = \dfrac{X-\mu}{\sigma} \sim N(0,1)$. We compute the p.d.f. of Z by taking derivative of c.d.f of Z.

$$\begin{aligned} F_Z(z) &= P(Z \le z) = P\left(\frac{X-\mu}{\sigma} \le z\right) = \int_{-\infty}^{\sigma z+\mu} \frac{1}{\sigma\sqrt{2\pi}} e^{-\frac{1}{2}\left(\frac{x-\mu}{\sigma}\right)^2} dx, \\ f_Z(z) &= \frac{d}{dz} F_Z(z) = \frac{1}{\sqrt{2\pi}} e^{-\frac{1}{2}z^2}. \end{aligned}$$

- More generally, we see that if X has a normal distribution with mean μ and variance σ^2 then $Y = aX + b$ has a normal distribution with mean $a\mu + b$ and variance $a^2\sigma^2$. ■

Remark Note that $X - \mu$ has the same distribution as $\mu - X$ when $X \sim N(\mu, \sigma^2)$, for which we say that X has a symmetric distribution about μ or X is symmetric about μ.

- Let X be symmetric about its mean μ. Then for any odd number k, $\mathrm{E}(X-\mu)^k = \mathrm{E}(\mu - X)^k$ if it exists. Thus, $\mathrm{E}(X-\mu)^k = 0$.
- The skewness of a random variable X with mean μ and variance σ^2 defined by $\mathrm{E}\left(\frac{X-\mu}{\sigma}\right)^3$ measures how symmetric X is. For example, the skewness of X, $X \sim N(\mu, \sigma^2)$, is zero.
- The kurtosis of a random variable X with mean μ and variance σ^2 defined by $\mathrm{E}\left(\frac{X-\mu}{\sigma}\right)^4 - 3$ measures whether the tails of p.d.f. of X decays faster or slower than a normal distribution. Note that the fourth moment of the standard normal random variable Z, $Z \sim N(0,1)$ is 3. Thus, the kurtosis of X, $X \sim N(\mu, \sigma^2)$, is zero.

■ **Example 4.25** If $X \sim N(\mu, \sigma^2)$, then the random variable $W = \left(\frac{X-\mu}{\sigma}\right)^2$ has the p.d.f. $f_W(w) = \dfrac{1}{\sqrt{2\pi w}} e^{-\frac{1}{2}w}$, $w > 0$, which is known as the chi-squared distribution with 1 degree of freedom, $W \sim \chi^2(1)$.

- Let $F_W(w)$ be the c.d.f. of W, which is

$$\begin{aligned} F_W(w) &= P(W \le w) = P\left(\left(\frac{X-\mu}{\sigma}\right)^2 \le w\right) \\ &= P\left(-\sqrt{w} \le \frac{X-\mu}{\sigma} \le \sqrt{w}\right) = \int_{-\sqrt{w}}^{\sqrt{w}} \phi(z)\, dz. \end{aligned}$$

Thus, we have $f_W(w) = \dfrac{d}{dw} F_W(w) = \dfrac{d}{dw}\displaystyle\int_{-\sqrt{w}}^{\sqrt{w}} \phi(z) dz = \dfrac{1}{\sqrt{2\pi w}} e^{-\frac{1}{2}w}$.

■

■ **Example 4.26 — Tail bound (1).** Let $X \sim N(0,1)$ and denote $\phi(x)$ and $\Phi(x)$ are p.d.f. and c.d.f. of a standard normal random variable, respectively. Then, we have for $t > 0$

$$P(X > t) = 1 - \Phi(t) = \int_t^\infty \phi(x) dx \le \frac{1}{2} e^{-t^2/2}.$$

- Let $A = \{(x,y) : x > t,\ y > t\}$. By squaring the integral, we see that

$$[P(X > t)]^2 = \left(\int_t^\infty \frac{e^{-x^2/2}}{\sqrt{2\pi}} dx\right)^2 = \iint_A \frac{e^{-(x^2+y^2)/2}}{2\pi} dy dx.$$

- Note that $A \subset B = \{(x,y) : x^2 + y^2 > 2t^2,\ x > 0,\ y > 0\}$. By a polar coordinate change, we have that $B = \{(r, \theta) : r > \sqrt{2}t,\ 0 < \theta < \pi/2\}$ and that

$$\left(\int_t^\infty \frac{e^{-x^2/2}}{\sqrt{2\pi}} dx\right)^2 \le \iint_B \frac{e^{-(x^2+y^2)/2}}{2\pi} dy dx = \int_0^{\pi/2} \int_{\sqrt{2}t}^\infty \frac{e^{-r^2/2}}{2\pi} r dr d\theta = \frac{e^{-t^2}}{4}.$$

- Hence, we get $P(X > t) \le \frac{1}{2} e^{-t^2/2}$.

■

■ **Example 4.27 — Mill's ratio : Tail bound (2).** We define Mills' ratio $r(x)$ by $r(x)\phi(x) = 1 - \Phi(x)$, where $\phi(x)$ and $\Phi(x)$ are p.d.f. and c.d.f. of a standard normal random variable, respectively. Then for $x > 0$, it follows that $r(x) = \dfrac{1-\Phi(x)}{\phi(x)} < \dfrac{1}{x}$.

- Since $\phi'(x) = -x\phi(x)$, we have for $x > 0$,

$$1 - \Phi(x) = \int_x^\infty \phi(t) dt \le \int_x^\infty \frac{t}{x} \phi(t) dt = -\frac{1}{x} \int_x^\infty \phi'(t) dt = \frac{1}{x} \phi(x).$$

- Let $X \sim N(0,1)$. Then $P(|X| > x) = 2(1 - \Phi(x)) \le \dfrac{2}{x} \phi(x) = \sqrt{\dfrac{2}{\pi}} \dfrac{\exp(-x^2/2)}{x}$.

- Often, we can compute the tail probability using a Normal probability table or a computer using R. For example, we have

  $$P(|X| \le 1.96) = 2\Phi(1.96) - 1 \approx 95\%, \quad P(|X| \ge 3) = 2(1 - \Phi(3)) \approx 0.27\%.$$

  Indeed, integrating by parts we have an approximation of $P(X > x) = 1 - \Phi(x)$,

  $$1 - \Phi(x) = \frac{1}{\sqrt{2\pi}} \int_x^\infty e^{-\frac{t^2}{2}} dt = \frac{\phi(x)}{x} - \int_x^\infty \frac{\phi(t)dt}{t^2} \approx \frac{\phi(x)}{x},$$

  because the second integral is negligible as $x \to \infty$.

- More generally, for any $x > 0$, with a change of the variable of integration to $u = t^2/2$, we obtain $1 - \Phi(x) = \dfrac{1}{\sqrt{2\pi}} \displaystyle\int_x^\infty e^{-t^2/2} dt = \dfrac{1}{2\sqrt{\pi}} \displaystyle\int_{x^2/2}^\infty u^{-1/2} e^{-u} du$. Then, we consider a successive integration by parts,

  $$\begin{aligned}
  1 - \Phi(x) &= \left.\frac{-u^{-1/2}e^{-u}}{2\sqrt{\pi}}\right|_{x^2/2}^{\infty} - \frac{1}{4\sqrt{\pi}} \int_{x^2/2}^\infty u^{-3/2} e^{-u} du \\
  &= \frac{e^{-\frac{x^2}{2}}}{x\sqrt{2\pi}} - \frac{1}{4\sqrt{\pi}} \int_{\frac{x^2}{2}}^\infty u^{-\frac{3}{2}} e^{-u} du = \frac{e^{-x^2/2}}{x\sqrt{2\pi}} + \left.\frac{u^{-\frac{3}{2}}e^{-u}}{4\sqrt{\pi}}\right|_{\frac{x^2}{2}}^{\infty} + \frac{3}{8\sqrt{\pi}} \int_{\frac{x^2}{2}}^\infty u^{-\frac{5}{2}} e^{-u} du \\
  &= \frac{e^{-x^2/2}}{x\sqrt{2\pi}} - \frac{e^{-x^2/2}}{x^3\sqrt{2\pi}} + \frac{3}{8\sqrt{\pi}} \int_{x^2/2}^\infty u^{-5/2} e^{-u} du.
  \end{aligned}$$

  Note that we get $1 - \Phi(x) < \dfrac{\phi(x)}{x}$ from the second equality. In addition, since the integral on the right of the last equality is strictly positive, we obtain the lower bound $1 - \Phi(x) > \dfrac{\phi(x)}{x} - \dfrac{\phi(x)}{x^3}$. Therefore, combining these upper and the lower bounds, we obtain $\dfrac{\phi(x)}{x}\left(1 - \dfrac{1}{x^2}\right) < 1 - \Phi(x) < \dfrac{\phi(x)}{x}$, which holds for each $x > 0$. In particular, $1 - \Phi(x) \approx \dfrac{1}{x}\phi(x)$ as $x \to \infty$.

■

R has a lot of built-in commands for distributions. For each distribution, there are 4 commands to work with it, d (density), p (c.d.f), q (the inverse of the c.d.f.), and r (random number generation from the distribution).

- For the normal distribution, the four commands are dnorm, pnorm, qnorm, and rnorm. For example, to evaluate the p.d.f. of X, $X \sim N(\mu, \sigma^2)$, we use dnorm(x,mu,sigma), where mu is the mean μ and sigma is the standard deviation σ.

```
dnorm(x, mean = 0, sd = 1);  pnorm(q, mean = 0, sd = 1)
qnorm(p, mean = 0, sd = 1);  rnorm(n, mean = 0, sd = 1)
```

- To compute $P(|X| > x)$, when $X \sim N(0,1)$, we use 2*(1-pnorm(x,0,1)).

We can also check $1-\Phi(x) \approx \frac{1}{x}\phi(x)$ as x increases, using R.

```
2*(1-pnorm(3,0,1))
[1] 0.002699796

(1-pnorm(3,0,1))/(dnorm(3,0,1)/3)
[1] 0.9137709
(1-pnorm(4,0,1))/(dnorm(4,0,1)/4)
[1] 0.9466095
```

Definition 4.3.2 A random variable X is said to have a lognormal distribution if its probability density function is given by

$$f_X(x) = \begin{cases} \frac{1}{x\sigma\sqrt{2\pi}} e^{-\frac{1}{2}\left(\frac{\log x-\mu}{\sigma}\right)^2}, & \text{if } 0 < x < \infty, \\ 0, & \text{otherwise.} \end{cases}$$

where $-\infty < \mu < \infty$ and $0 < \sigma^2 < \infty$ are arbitrary parameters.

■ **Example 4.28** Y is said to have a lognormal distribution with parameters μ and σ^2 if $\log Y$ has a normal distribution with parameters μ and σ^2. Then,

$$\mathrm{E}(Y) = e^{\mu+\frac{1}{2}\sigma^2}, \quad \mathrm{Var}(Y) = \left[e^{\sigma^2} - 1\right] e^{2\mu+\sigma^2}.$$

- Let X have a normal distribution, $X \sim N(\mu, \sigma^2)$, and define $Y = e^X$. Since $X = \log Y$, we have $dx/dy = 1/y$, and hence

$$f_Y(y) = \frac{1}{y} f_X(\log y) = \frac{1}{y\sigma\sqrt{2\pi}} e^{-\frac{1}{2}\left(\frac{\log y-\mu}{\sigma}\right)^2}, \quad y > 0,$$

which is the p.d.f. of a lognormal distribution.

- Let n be a positive integer. We compute the nth moment of Y.

$$\begin{aligned} \mathrm{E}(Y^n) &= \int_0^\infty y^n f_Y(y)\,dy = \int_0^\infty y^n \frac{1}{y\sigma\sqrt{2\pi}} e^{-\frac{1}{2}\left(\frac{\log y-\mu}{\sigma}\right)^2} dy \\ &= \int_{-\infty}^\infty e^{nx} \frac{1}{\sigma\sqrt{2\pi}} e^{-\frac{1}{2}\left(\frac{x-\mu}{\sigma}\right)^2} dx = M_X(n) = e^{\mu n+\frac{1}{2}\sigma^2 n^2}, \end{aligned}$$

where we substitute $x = \log y$ in the last integral. Alternatively, we get $\mathrm{E}(Y^n) = \mathrm{E}\left([e^X]^n\right) = M_X(n)$. Therefore, letting $n = 1$, we get $\mathrm{E}(Y) = e^{\mu+\frac{1}{2}\sigma^2}$, and $\mathrm{E}(Y^2) = e^{2\mu+2\sigma^2}$ with $n = 2$. Hence, $\mathrm{Var}(Y) = \mathrm{E}(Y^2) - \mathrm{E}(Y)^2 = \left[e^{\sigma^2} - 1\right] e^{2\mu+\sigma^2}$.

■

4.4 Waiting time distributions

Consider a situation in some event, called an arrival, occurs over time. For example, the events could be customers arriving at a store during business hours, hits at a web site, traffic accidents, and others. The Poisson process is an important and often very effective model for such systems. The Poisson process can be viewed as a continuous-time analog of the Bernoulli process and applies to cases with dividing time into discrete but infinitesimal lengths. The Poisson process is usually defined in terms of "arrivals", a collection of random variables N_t where $t \geq 0$ is a real number. Here, N_t is the count of how many "arrivals" have occurred from time 0 to time t.

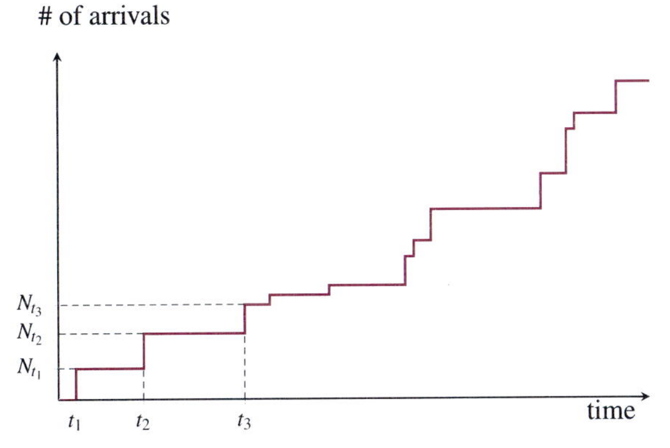

Figure 4.9: Poisson process with # of arrivals by time t

Definition 4.4.1 — Poisson process. The Poisson process is defined by assumptions about the arrivals. Heuristically, we assume the following:

- There are no arrivals before time 0.
- The number of arrivals in any time interval has a Poisson distribution with parameter proportional to the length of the time interval.
- The numbers of arrivals in two disjoint time intervals are independent.

These three assumptions completely determine the joint distribution of the random variables N_t for all $t \geq 0$. Mathematically, these assumptions are stated as follows:

① $N_0 = 0$,

② For $t > s$, $N_t - N_s$ has a Poisson$(\lambda(t-s))$ distribution,

③ For $t' > s' > t > s$, $N_{t'} - N_{s'}$ and $N_t - N_s$ are independent random variables.

The last condition is often called the property of "independent increments".

Remark The Poisson process has one parameter, the "rate" λ. This describes the average rate of arrivals, in arrivals per unit time. When specifying the parameter, we call this process the Poisson(λ) process. Notice that $\lambda \times$ a time interval is in the units of number of arrivals.

■ **Example 4.29** Suppose that customers arrive at a queue according to a Poisson process with rate λ. Find the distribution of the time that we have to wait until the first customer arrives.

- Let T be the time at which the first customer arrives. Let X be the number of customers who arrive by time t.
- Then $X \geq 1$ if and only if $T \leq t$. Also, X has a Poisson distribution with mean λt. Hence,
$$\begin{aligned} F_T(t) &= P(T \leq t) = P(X \geq 1) = 1 - P(X = 0) = 1 - \exp(-\lambda t), \\ \frac{d}{dt} F_T(t) &= f_T(t) = \lambda \exp(-\lambda t),\ t > 0. \end{aligned}$$
- T is a waiting time that can have any value $t > 0$ in the positive real line, which is an *exponential* random variable. When T represents a waiting time until some event occurs, then we can think of λ as the "rate" at which the event occurs.

■

Definition 4.4.2 A continuous random variable whose p.d.f is given by

$$f(x) = \begin{cases} \lambda e^{-\lambda x}, & \text{if } x \geq 0, \\ 0, & \text{otherwise} \end{cases} \quad \text{for some } \lambda > 0.$$

is said to have an *exponential distribution* with parameter λ, denoted by $X \sim \text{Exp}(\lambda)$. The corresponding c.d.f., is given by $F_X(x) = \begin{cases} 1 - e^{-\lambda x}, & \text{if } x \geq 0, \\ 0, & \text{if } x < 0. \end{cases}$

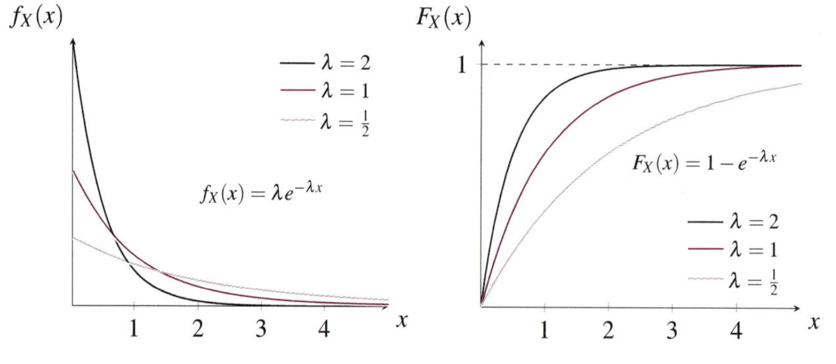

Figure 4.10: The p.d.f and c.d.f. of exponential random variable, $X \sim \text{Exp}(\lambda)$

Theorem 4.4.1 Suppose that X has an exponential distribution with parameter $\lambda = 1$, $X \sim \text{Exp}(1)$. Then,

- $\text{E}(X) = 1$, $\text{Var}(X) = 1$, and $M_X(t) = \dfrac{1}{1-t}$, $t < 1$.
- $Y = X/\lambda \sim \text{Exp}(\lambda)$. Conversely, if $Y \sim \text{Exp}(\lambda)$, then $\lambda Y \sim \text{Exp}(1)$. In addition, $\text{E}(Y) = 1/\lambda$, $\text{Var}(Y) = 1/\lambda^2$, and $M_Y(t) = \dfrac{\lambda}{\lambda - t}$, $t < \lambda$.
- Exponential random variables are memoryless (has no memory) :

$$P(X > s+t|X > t) = P(X > s)$$

Proof. We use standard integration by parts to calculate moments of exponential random variables and their probabilities.

- Note that $\text{E}(X) = \int_0^\infty xe^{-x}dx = 1$ and $\text{E}(X^2) = \int_0^\infty x^2e^{-x}dx = 2\int_0^\infty xe^{-x}dx = 2$. Thus, $\text{Var}(X) = \text{E}(X^2) - (\text{E}[X])^2 = 1$. The m.g.f is computed

$$M_X(t) = \int_0^\infty e^{tx}e^{-x}dx = \frac{1}{1-t}\int_0^\infty e^{-(1-t)x}dx = \frac{1}{1-t}, \quad t < 1.$$

  Alternatively, we can consider Taylor series expansion of $M_X(t)$ to find the moments,

$$M_X(t) = \sum_{n=0}^{\infty} \frac{M_X^{(n)}t^n}{n!} \sum_{n=0}^{\infty} \frac{\text{E}(X^n)t^n}{n!} = \frac{1}{1-t} = \sum_{n=0}^{\infty} t^n = 1 + t + t^2 + \cdots,$$

  which implies that $\text{E}(X^n) = n!$, $n = 0, 1, \ldots$.

- Since $X \sim \text{Exp}(1)$, we have

$$f_Y(y) = \frac{d}{dy}F_Y(y) = \frac{d}{dy}P(Y \le y) = \frac{d}{dy}P(X \le \lambda y) = \frac{d}{dy}\int_0^{\lambda y} e^{-x}dx = \lambda e^{-\lambda y}, \; y > 0.$$

  Hence, we have

$$\text{E}(Y) = \text{E}(X)/\lambda, \; \text{Var}(Y) = \text{Var}(X)/\lambda^2, \; M_Y(t) = \text{E}\left(e^{tX/\lambda}\right) = \frac{\lambda}{\lambda - t}, \; t < \lambda.$$

- Note that $P(X > x) = \int_x^\infty e^{-t}dt = e^{-x}$. Thus,

$$P(X > s+t|X > t) = \frac{P(X > s+t, \; X > t)}{P(X > t)} = \frac{P(X > s+t)}{P(X > t)} = \frac{e^{-(s+t)}}{e^{-t}} = P(X > s).$$

  Hence, we see that the function $L(s) = P(X > s) = 1 - F(s)$ satisfies $L(s+t) = L(s)L(t)$ for all $s, t > 0$.

■

Remark Exponential random variables are the continuous analogue of geometric random variables.

- If we consider repeated Bernoulli trials with success probability p, the number of trials up to and including the first success is a geometric random variable with parameter p, which has p.m.f and tail probabilities given by $p(1-p)^{x-1}$ and $(1-p)^x$, respectively, for $x = 1, 2, \ldots$.

- We can think of a geometric random variable as giving the time of the first success in a sequence of Bernoulli trials, where time is quantized into integer values.

- That is, we can think of the Poisson (arrival) process as a limit of a Bernoulli (trial) process where the number of trials per unit time goes to infinity and the probability of a success goes to 0. To see how this works, suppose we divide the interval $[0, 1)$ into n equal-sized pieces, where n is large.

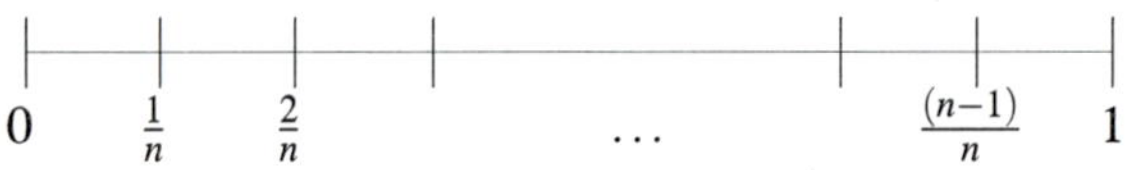

- Let $n \in N$ and suppose that Bernoulli trials with success probability p are performed at times $\frac{1}{n}, \frac{2}{n}, \ldots$; that is, time is quantized into units of duration $\frac{1}{n}$.

- Let X denote the time of the first success. For a positive real number x, we set $k = \lfloor nx \rfloor$. Note that $\frac{k}{n} \le x < \frac{k+1}{n}$ and thus, $P\{X > x\} = P\left\{X > \frac{k}{n}\right\} = (1-p)^k$. Assume that n is large and that p is small such that $np \approx \lambda$. Noting that $x \approx \frac{k}{n}$ and $e^{-t} \approx 1 - t$ for small t,
$$P\{X > x\} = (1-p)^k \approx e^{-pk} = e^{\left\{-np\frac{k}{n}\right\}} \approx e^{-npx} \approx e^{-\lambda x}.$$

- The exponential distribution can be regarded as the limit of the waiting time for the first success in a sequence of Bernoulli trials with probability $p \to 0$, with $n \to \infty$ trials in each unit time interval, and with $np \to \lambda$. Note that λ is the expected number of successes in a unit time interval. For example, in the survival time modeling, this is the expected number of fatalities, which explains why the expected lifetime is $1/\lambda$.

- In the discrete case, the epoch of first success is geometrically distributed with $P(X = k) = (1-p)^{k-1}p$. If we let $x = k/n$, then $\Delta x = 1/n = p/\lambda$, and we have
$$(1-p)^k p = (1-\lambda/n)^{nx}\lambda\Delta x \approx \lambda e^{-\lambda x}\Delta x, \quad \text{as } n \to \infty.$$

Remark Suppose that $X \sim \text{Exp}(\lambda)$. The memoryless property $P(X \ge s+t | X \ge s) = P(X \ge t)$ can be explained as follows:

- Imagine that the exponentially distributed random variable X represents the lifetime of an item, the residual life of an item has the same exponential distribution as the original lifetime, regardless of how long the item has already been in use.

- After you have waited s minutes, the probability you will have to wait another t minutes is exactly the same as the probability that you will have to wait t minutes with no previous waiting time. In other words, if $X \sim \text{Exp}(\lambda)$, then $Y = X - s$ is still distributed as $\text{Exp}(\lambda)$ conditional on $B = \{X \geq s\}$,

$$\begin{aligned} F_{Y|B}(y|B) &= P(Y \leq y|X \geq s) = 1 - P(X > y + s|X \geq s) \\ &= 1 - P(X > y) = 1 - e^{-\lambda y},\ y > 0. \end{aligned}$$

Thus, $Y|B \stackrel{d}{=} X$, which we will discuss in detail in Chapter 6 for conditional distributions.

- The following table summarizes distributions of $N(t)$, that is, the number of arrivals by time t, and waiting time T until to get one arrival :

Type of Process	Poisson process	Bernoulli process
Arrival time (t)	Continuous	Discrete
Arrival rate/probability	λ per unit time	p per each trial
Distribution of $N(t)$	Poisson (λt)	Binomial (np)
Distribution of T	Exponential (λ)	Geometric(p)

■ **Example 4.30** When you enter the bank, you find that all two tellers are busy serving two other customers, and there are no other customers in queue. You are told that your service will begin as soon as one of two customers leaves. Assume that the service times for you and for each of two customers being served by one of two tellers are independent identically distributed exponential random variables. What is the probability that you will be the last to leave the bank?

- When you start being served by one of two tellers, the remaining time of the other customer being served and your own remaining time are identically distributed as an exponential distribution. Therefore, you and the other customer have equal probability $1/2$ of being the last to leave the bank.

■

■ **Example 4.31** Airport limousines depart from the KU station to the airport from the early morning till late at night. The limousines leave from the KU station with independent interarrival times that are exponentially distributed with an expected value of 30 minutes. Suppose you arrived at the KU station at 2 p.m. on a given day, not knowing how long ago the previous limousine came. What is the expected waiting time for you to get the limousine leaving for the airport?

- By the memoryless property of the exponential distribution, the time from your arrival at 2 p.m. until the next departure of a limousine has an exponential distribution with an expected value of 30 minutes. In other words, regardless of when you arrive at the station, how much longer the next limousine will take to arrive is independent of how long ago the previous limousine arrived.

■

■ **Example 4.32 — Laplace distribution.** The Laplace distribution has a p.d.f. $f_X(x) = \frac{1}{2}e^{-|x|}$, $-\infty < x < \infty$, also called a symmetrized exponential (or double exponential, DE) distribution, $X \sim \text{Laplace(DE)}(0,1)$.

- The random variable X is equally likely to be either positive or negative, and its absolute value is exponentially distributed with parameter 1 as shown in Figure 4.11.

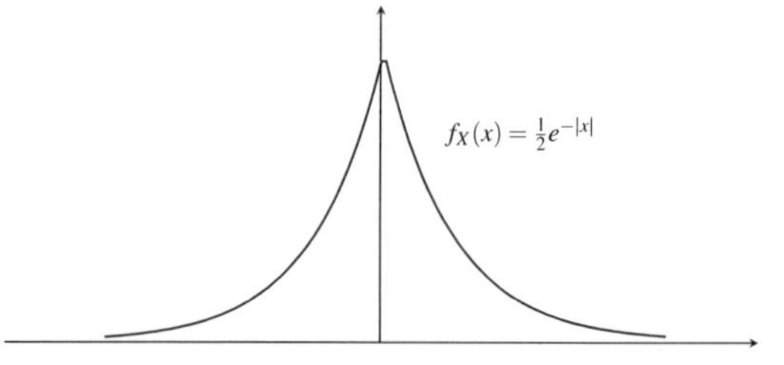

Figure 4.11: The p.d.f. of the Laplace distribution

- The c.d.f. of X is given by $F_X(x) = \begin{cases} \frac{1}{2}\int_{-\infty}^{x} e^t dt = \frac{1}{2}e^x, & x < 0, \\ \frac{1}{2}\int_{-\infty}^{0} e^t dt + \frac{1}{2}\int_0^x e^{-t} dt = 1 - \frac{1}{2}e^{-x}, & x > 0. \end{cases}$

  Let $Y = |X|$. Then, the c.d.f. of Y is easily obtained as

  $$F_Y(y) = F_X(y) - F_X(-y) = 1 - \frac{1}{2}e^{-y} - \frac{1}{2}e^{-y} = 1 - e^{-y},\ y > 0,$$

  which implies that $|X|$ has an exponential distribution with parameter 1.

- Note that $\mathrm{E}(X) = 0$ since it exists and the p.d.f. is symmetric about 0 and that $\mathrm{Var}(X) = \mathrm{E}(X^2) - [\mathrm{E}(X)]^2 = \mathrm{E}(|X|^2) = 2$. Also, the m.g.f. of X is given by

  $$\begin{aligned} M_X(t) &= \int_{-\infty}^{\infty} e^{tx}\frac{1}{2}e^{-|x|}\,dx = \frac{1}{2}\left[\int_{-\infty}^{0} e^{(1+t)x}\,dx + \int_0^{\infty} e^{(-1+t)x}\,dx\right] \\ &= \frac{1}{2}\left[\frac{1}{1+t} - \frac{1}{-1+t}\right] = \frac{1}{1-t^2} = \sum_{n=0}^{\infty} t^{2n}, \text{ for } |t| < 1, \end{aligned}$$

  which implies that $\mathrm{E}\left(X^{2n-1}\right) = 0$ and $\mathrm{E}\left(X^{2n}\right) = (2n)!$, $n = 1, 2, \ldots$.

■

Definition 4.4.3 — Gamma function. For each strictly positive real number $\alpha > 0$, the gamma function is defined by

$$\Gamma(\alpha) = \int_0^{\infty} x^{\alpha-1}\exp(-x)dx.$$

The condition $\alpha > 0$ is assumed for the convergence of the integral.

Theorem 4.4.2 The gamma function has the following recursion property :

$$\Gamma(\alpha+1)=\alpha\Gamma(\alpha), \text{ for all } \alpha>0.$$

In particular, when α is a positive integer, $\Gamma(\alpha+1)=\alpha!$. Indeed, the gamma function is a continuous extension of the factorial function.

Proof. We use integration by parts to get the recursion property $\Gamma(\alpha+1)=\alpha\Gamma(\alpha)$.

- We first consider α is a positive integer and show that the integral $\int_0^\infty x^{\alpha-1}e^{-\lambda x}dx$ exists (is finite). Because $x^{\alpha-1}e^{-\frac{\lambda}{2}x}\to 0$ as $x\to\infty$, there is some $M<\infty$ such that for $x>M$, $x^{\alpha-1}e^{-\frac{\lambda}{2}x}<1$. Hence, for $N>M$

$$\int_0^N x^{\alpha-1}e^{-\lambda x}dx<\int_0^M x^{\alpha-1}e^{-\lambda x}dx+\int_M^N e^{-\frac{\lambda}{2}x}dx<\int_0^M x^{\alpha-1}e^{-\lambda x}dx+2\lambda^{-1}e^{-\frac{\lambda}{2}M}<\infty.$$

- Using integration by parts, we have with $u=x^\alpha$ and $dv=\exp(-x)dx$. Then $du=\alpha x^{\alpha-1}$, and $v=-\exp(-x)$. Thus, we obtain

$$\Gamma(\alpha+1)=-x^\alpha\exp(-x)\Big|_0^\infty+\alpha\int_0^\infty x^{\alpha-1}\exp(-x)dx=\alpha\Gamma(\alpha), \text{ for all } \alpha>0.$$

- Also, $\Gamma(\alpha+2)=(\alpha+1)\Gamma(\alpha+1)=\alpha(\alpha+1)\Gamma(\alpha)$. Thus, if α is a positive integer n, then $\Gamma(n+1)=n\Gamma(n)=n(n-1)\Gamma(n-1)=\cdots=n!\Gamma(1)$, where $\Gamma(1)=1$.

■

Remark The value of the gamma function $\Gamma(x)=\displaystyle\int_0^\infty t^{x-1}e^{-t}dt$ increases very rapidly as the argument of the function increases.

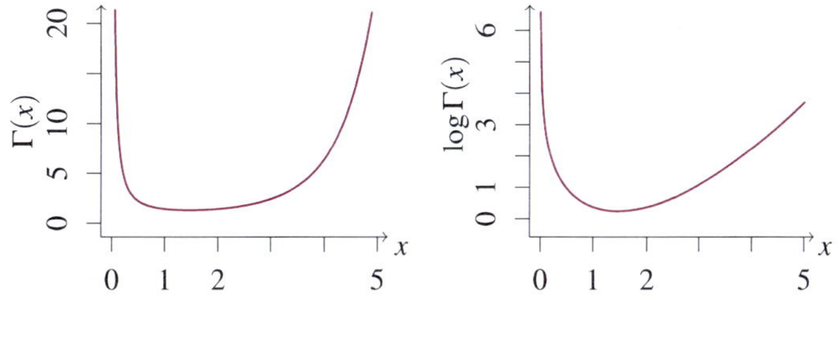

Figure 4.12: Graphical display of $\Gamma(x)$ and $\log\Gamma(x)$

- The gamma function has continuous derivatives of all orders, which are obtained by the interchange of differentiation under the integral sign

$$\Gamma'(x)=\int_0^\infty t^{x-1}e^{-t}\log t\,dt,\ \Gamma^{(n)}(x)=\int_0^\infty e^{-t}t^{x-1}(\log t)^n dt,\ x>0.$$

Thus, it follows that $\Gamma''(x)>0$ for $x>0$ and that $\Gamma(t)$ is a convex function as shown in Figure 4.12.

- It is often more convenient to work with the logarithm of the gamma function, $\log\Gamma(x)$, rather than with the gamma function itself. Indeed, $\log\Gamma(x+1) = \log x + \log\Gamma(x)$, for $x > 0$, and $\log\Gamma(x)$ is convex.

- The (lower) *incomplete gamma function* is defined to be the function obtained by restricting the region of integration to $(0,x)$, $\gamma(\alpha,x) = \int_0^x t^{\alpha-1}\exp(-t)dt$, $\alpha > 0$ and $x > 0$, whereas the upper incomplete gamma function is defined as $\Gamma(\alpha,x) = \int_x^\infty t^{\alpha-1}\exp(-t)dt$.

Example 4.33 There is one set of special values of the gamma function for non-integer arguments that we can compute $\Gamma\left(\frac{2k-1}{2}\right)$, $k = 1,2,\ldots$.

- Start with the normal p.d.f. with mean 0 and variance $1/2$.

$$\begin{aligned} 1 &= \int_{-\infty}^{\infty}\frac{1}{\sqrt{\pi}}\exp(-x^2)dx = \frac{2}{\sqrt{\pi}}\int_0^\infty \exp(-x^2)dx \\ &= \frac{1}{\sqrt{\pi}}\int_0^\infty y^{-1/2}\exp(-y)dy \equiv \Gamma\left(\frac{1}{2}\right)\frac{1}{\sqrt{\pi}}, \end{aligned}$$

where we did the transformation $y = x^2$. Hence $\Gamma\left(\frac{1}{2}\right) = \sqrt{\pi}$.

- Also, we can get $\Gamma(k/2)$ for a positive integer $k = 2n+1$, $n = 0,1,2,,\ldots$

$$\begin{aligned} \Gamma\left(\frac{3}{2}\right) &= \Gamma\left(\frac{1}{2}+1\right) = \frac{1}{2}\Gamma\left(\frac{1}{2}\right) = \frac{\sqrt{\pi}}{2}, \\ \Gamma\left(\frac{5}{2}\right) &= \Gamma\left(\frac{3}{2}+1\right) = \frac{3}{2}\Gamma\left(\frac{3}{2}\right) = \frac{3\sqrt{\pi}}{4}, \\ &\vdots \\ \Gamma\left(\frac{2n+1}{2}\right) &= \left(\frac{2n-1}{2}\right)\left(\frac{2n-3}{2}\right)\cdots\left(\frac{1}{2}\right)\Gamma\left(\frac{1}{2}\right) \\ &= \frac{1}{2^n}\sqrt{\pi}(2n-1)(2n-3)\cdots 1. \end{aligned}$$

- For example, when $Z \sim N(0,1)$, we can compute $\mu^{2n} = \mathrm{E}\left(Z^{2n}\right)$ using the gamma function and the change of variable,

$$\begin{aligned} \mathrm{E}\left(Z^{2n}\right) &= \int_{-\infty}^{\infty} z^{2n}\frac{1}{\sqrt{2\pi}}e^{-\frac{z^2}{2}}dz = \frac{2^n}{\sqrt{2\pi}}\int_0^\infty u^{\frac{2n-1}{2}}e^{-u}du,, \\ &= \frac{2^n}{\sqrt{\pi}}\Gamma\left(\frac{2n+1}{2}\right) = (2n-1)(2n-3)\cdots 1 \\ &= \frac{(2n)\times(2n-1)\times(2n-2)\times(2n-3)\times\cdots\times 2\times 1}{(2n)\times(2n-2)\times\cdots\times 4\times 2} = \frac{(2n)!}{2^n(n!)}. \end{aligned}$$

where we make the change of variables $u = \frac{z^2}{2}$ so that $du = zdz$ and $z = \sqrt{2u}$ for

$z > 0$. Thus, we have $\mu_2 = 1$, $\mu_4 = 3$, $\mu_6 = 15$, etc.

■

It is clear from the definition of the gamma function that $y^{\alpha-1}\exp(-y)/\Gamma(\alpha)$, for $y > 0$ is a p.d.f. Thus, we are ready to define the gamma distribution.

Definition 4.4.4 — Gamma distribution. A continuous random variable X is said to have a gamma distribution with parameters α and λ if its p.d.f is given by

$$X \sim \text{Gamma}(\alpha,\lambda), \quad f_X(x) = \begin{cases} \dfrac{\lambda^\alpha}{\Gamma(\alpha)} x^{\alpha-1}\exp(-\lambda x), & \text{for } x > 0, \\ 0, & \text{otherwise.} \end{cases}$$

where $\alpha > 0$ is called a shape parameter, and $\lambda > 0$ is called a rate parameter.

Remark The parameter α determines the shape of the gamma density function as shown in Figure 4.13. To focus on the role of the parameter α, we set $\lambda = 1$ and see the shape of the p.d.f. $f_X(x) = \frac{1}{\Gamma(\alpha)} x^{\alpha-1} e^{-x}$.

- If $\alpha = 1$ we obtain the (standard) exponential density, which decreases monotonically for $x > 0$.
- When $\alpha < 1$ the density also decreases monotonically for $x > 0$ but is unbounded near the origin.
- When $\alpha > 1$, we obtain a unimodal density that attains its global maximum at the point $x = \alpha - 1$. In addition, for small values, it is skewed, but gets more symmetric as α increases.

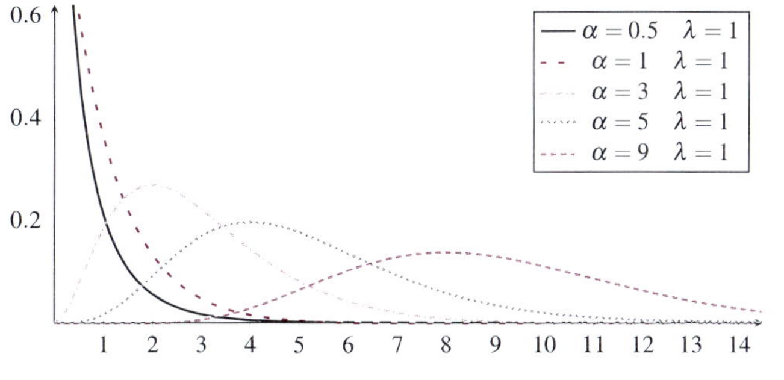

Figure 4.13: The shape of gamma p.d.f. depending on α

- When α is given by an integer n, the gamma distribution is often called the *Erlang distribution* with parameters n and λ.
- An alternative representation of a gamma random variable is based on the use of λ as a *scale* parameter :

$$X \sim \text{Gamma}(\alpha,\lambda), \quad f(x) = \begin{cases} \dfrac{1}{\lambda^\alpha\Gamma(\alpha)} x^{\alpha-1}\exp\left(-\dfrac{1}{\lambda}x\right), & \text{for } x > 0, \\ 0, & \text{otherwise.} \end{cases}$$

■ **Example 4.34 — Chi-squared distribution.** If X has a normal distribution with mean 0 and variance σ^2, then $Y = X^2$ has a gamma distribution with parameters $1/2$ and $1/[2\sigma^2]$, also known as the chi-squared(χ^2)-distribution. The p.d.f. of Y is

$$f_Y(y) = \frac{1}{\sqrt{2\pi}\sigma} y^{-1/2} \exp(-y/[2\sigma^2]),\ y > 0, \quad \sigma Y \sim \chi^2(1) \stackrel{d}{=} \text{Gamma}\left(\frac{1}{2}, \frac{1}{2}\right),$$

as also shown in Example 4.25. In particular, when $\alpha = \frac{n}{2}$ and $\lambda = \frac{1}{2}$, the p.d.f. $f_X(x) = \frac{1}{\Gamma\left(\frac{n}{2}\right)} \lambda^{\frac{n}{2}} x^{\frac{n}{2}-1} e^{-\frac{1}{2}x}$ is known as the chi-squared p.d.f., whose distribution is referred to as the χ^2-distribution with n degrees of freedom, $X \sim \chi^2(n)$.

■

Theorem 4.4.3 Suppose that $X \sim \text{Gamma}(\alpha, \lambda)$. Then,

- cX has a gamma distribution with parameters α and λ/c, $cX \sim \text{Gamma}(\alpha, \lambda/c)$.
- The exponential distribution with parameter λ is the gamma distribution with parameters 1 and λ.
- Let $k > -\alpha$. Then $\text{E}(X^k) = \frac{1}{\lambda^k} \frac{\Gamma(\alpha + k)}{\Gamma(\alpha)}$.

  Thus, we get

  $$\text{E}(X) = \frac{\alpha}{\lambda},\ \text{E}(X^2) = \frac{\alpha(\alpha+1)}{\lambda^2},\ \text{and } \text{Var}(X) = \frac{\alpha}{\lambda^2}.$$

- The moment generating function is $M_X(t) = \left(\frac{\lambda}{\lambda - t}\right)^{\alpha}$ for $t < \lambda$.

Proof. Note that the p.d.f. of X is $f_X(x) = \frac{\lambda^\alpha}{\Gamma(\alpha)} x^{\alpha-1} \exp(-\lambda x)$, for $x > 0$.

- Let $Y = cX$. Then, by the change of variables formula with $x = r^{-1}(y) = y/c$ and $dx/dy = 1/c$, the p.d.f of Y is

  $$\begin{aligned} f_Y(y) &= f_X(y/c)\left|\frac{dx}{dy}\right| = \frac{\lambda^\alpha}{\Gamma(\alpha)}(y/c)^{\alpha-1}\exp(-\lambda y/c)\frac{1}{c} \\ &= \frac{(\lambda/c)^\alpha}{\Gamma(\alpha)} y^{\alpha-1} \exp(-\lambda y/c),\ y > 0, \end{aligned}$$

  which is the p.d.f. of a gamma random variable with parameters α and λ/c.

- Taking $\alpha = 1$, the p.d.f. of $X \sim \text{Gamma}(1, \lambda)$ is $\exp(-\lambda x)$, for $x > 0$, which is a p.d.f. of the exponential random variable with a parameter λ.

- The kth moment is computed by recognizing the definition of the gamma function,

$$\int_0^\infty \frac{\lambda^\alpha}{\Gamma(\alpha)} x^k x^{\alpha-1} \exp(-\lambda x) dx = \frac{\lambda^\alpha}{\Gamma(\alpha)} \frac{\Gamma(\alpha+k)}{\lambda^{\alpha+k}} = \frac{1}{\lambda^k} \frac{\Gamma(\alpha+k)}{\Gamma(\alpha)}, \quad k > -\alpha.$$

For $k = 1, 2$, we get $\mathrm{E}(X)$ and $\mathrm{E}(X^2)$. Hence, combining these gives $\mathrm{Var}(X) = \dfrac{\alpha}{\lambda^2}$.

- The moment generating function of X is computed

$$\begin{aligned} M_X(t) &= \mathrm{E}\left(e^{tX}\right) = \int_0^\infty \frac{\lambda^\alpha}{\Gamma(\alpha)} x^{\alpha-1} e^{-\lambda x} e^{tx} dx \\ &= \frac{\lambda^\alpha}{(\lambda-t)^\alpha} \int_0^\infty \frac{(\lambda-t)^\alpha}{\Gamma(\alpha)} x^{\alpha-1} e^{-(\lambda-t)x} dx = \left(\frac{\lambda}{\lambda-t}\right)^\alpha, \quad t < \lambda, \end{aligned}$$

where the last equality holds since the expression inside the integral is the p.d.f of a gamma random variable with parameters α and $\lambda - t$.

■

■ **Example 4.35 — Waiting times in the Poisson process.** Suppose that customers arrive at a queue according to a Poisson process with rate λ. Find the distribution of the time that we have to wait until the third customer arrives.

- Let T_3 be the time at which the third customer arrives. Let Y_t be the number of customers who arrive by time t. Then $Y_t \geq 3$ if and only if $T_3 \leq t$. Also, Y_t has a Poisson distribution with mean λt. Hence,

$$\begin{aligned} F_{T_3}(t) &= P(T_3 \leq t) = P(Y_t \geq 3) = 1 - \sum_{y=0}^{2} P(Y_t = y), \\ \frac{d}{dt} F_{T_3}(t) &= f_{T_3}(t) = \frac{\lambda^3}{\Gamma(3)} t^2 \exp(-\lambda t), \quad t > 0. \end{aligned}$$

- Hence, T_3 has a Gamma distribution with parameters 3 and λ, $T_3 \sim \mathrm{Gamma}(3, \lambda)$. In other words, $T_3 = X_1 + X_2 + X_3$, where X_i is the waiting time of the ith arrival after the $i-1$st arrival that has an exponential distribution with parameter λ.

■

■ **Example 4.36 — Poisson-Gamma relationship.** There is a very important relationship between the exponential(Gamma) density and the Poisson distribution.

- Defining X_1, X_2, ... to be a sequence of independent exponentially distributed random variables with parameter λ. Let T_n denote the sum of $X_1, \ldots, X_n$, $T_n = \sum_{i=1}^n X_i$. Then, T_n has a Gamma distribution with parameters n and λ,

$$f_{T_n}(t) = \begin{cases} \frac{\lambda^n}{\Gamma(n)} t^{n-1} e^{-\lambda t}, & \text{if } t > 0, \\ 0, & \text{otherwise.} \end{cases}$$

- Consider a time interval of length t, and we let Y denote the random variable which counts the number of arrivals in the time interval. Then, we see that

$$P(Y=n)=P(T_n\le t \text{ and } T_{n+1}>t).$$

- Since the event $T_{n+1}\le t$ is a subset of the event $T_n\le t$, the above probability is seen to be equal to $P(Y=n)=P(T_n\le t)-P(T_{n+1}\le t)$. Hence $P(Y=n)=\dfrac{e^{-\lambda t}(\lambda t)^n}{n!}$, which implies $Y\sim \text{Poisson}(\lambda t)$.

- Alternatively, using the integration by parts, we have

$$\begin{aligned} P(T_n\le t) &= \int_0^t \frac{\lambda^n}{\Gamma(n)}x^{n-1}e^{-\lambda x}dx=\frac{\lambda^n}{n!}t^n e^{-\lambda t}+\int_0^t \frac{\lambda^{n+1}}{\Gamma(n+1)}x^{n+1-1}e^{-\lambda x}dx \\ &= P(Y=n)+P(T_{n+1}\le t),\ Y\sim \text{Poisson}(\lambda t). \end{aligned}$$

Hence, by induction on n or summing up $\sum_{k=n}^{\infty}P(Y=k)=P(T_n\le t)$, the c.d.f. of T_n is obtained as

$$P(T_n\le t)=\begin{cases} 1-e^{-\lambda t}\left(1+\frac{\lambda t}{1!}+\cdots+\frac{(\lambda t)^{n-1}}{(n-1)!}\right), & \text{if } t>0,\\ 0, & \text{otherwise.}\end{cases}$$

- In other words, the event that the waiting time of the nth arrival is less than or equal to x is equivalent to the event that the number of arrivals by time x is greater than or equal to n. That is, $P(Y\ge n)=P(X\le x)$, where $Y\sim\text{Poisson}(\lambda x)$ and $X\sim\text{Gamma}(n,\lambda)$, which is known as the Poisson-gamma relationship.

■

Remark The Poisson (arrival) process with a rate λ can be better understood by an analogy with Bernoulli (trial) process with a success probability p.

- The outcomes of different Bernoulli trials are independent.
- The number of Poisson arrivals in disjoint intervals are independent.
- The total waiting time until the k successes in the Bernoulli process has a negative Binomial distribution with parameters k and p.
- The total waiting time until the k arrival in the Poisson process has a Gamma distribution with parameters k and λ.
- The Gamma distribution is the continuous analog of the negative Binomial distribution.

Remark Let $\Gamma(x)$ be the gamma function defined by $\Gamma(x) = \int_0^\infty t^{x-1}e^{-t}dt$ and let $\Gamma(x,y)$ be the upper incomplete gamma function defined by $\Gamma(x,y) = \int_y^\infty t^{x-1}e^{-t}dt$.

- With the change of variable $s = t - y$, we have

$$\begin{aligned}
\Gamma(n,y) &= \int_y^\infty t^{n-1}e^{-t}dt = \exp(-y)\int_0^\infty (s+y)^{n-1}\exp(-s)ds \\
&= \exp(-y)\int_0^\infty \sum_{k=0}^{n-1}\binom{n-1}{k} s^{n-1-k}y^k \exp(-s)ds \\
&= \sum_{k=0}^{n-1}\exp(-y)y^k\binom{n-1}{k}\Gamma(n-k) = \sum_{k=0}^{n-1}\exp(-y)\frac{y^k}{k!}(n-1)!
\end{aligned}$$

  That is, for $n = 1,2,\ldots$, $\dfrac{\Gamma(n,y)}{\Gamma(n)} = \sum_{k=0}^{n-1}\dfrac{y^k}{k!}\exp(-y)$, $y > 0$, which can also follow from the Poisson-Gamma relationship in Example 4.36.

- Consider $\Gamma(n+1) = n!$. With the change of variable $t = n + y\sqrt{n}$, we have

$$\begin{aligned}
\Gamma(n+1) &= n! = \int_0^\infty t^{n+1-1}e^{-t}dt = \int_{-\sqrt{n}}^\infty (n+y\sqrt{n})^n e^{-(n+y\sqrt{n})}\sqrt{n}dy \\
&= \left(\frac{n}{e}\right)^n \sqrt{n}\int_{-\sqrt{n}}^\infty \left(1+\frac{y}{\sqrt{n}}\right)^n e^{-y\sqrt{n}}dy.
\end{aligned}$$

  Now, for any fixed y, we have, as $n \to \infty$,

$$\begin{aligned}
&\log\left\{\left(1+\frac{y}{\sqrt{n}}\right)^n e^{-y\sqrt{n}}\right\} = n\log\left(1+\frac{y}{\sqrt{n}}\right) - y\sqrt{n} \\
&= n\left\{\frac{y}{\sqrt{n}} - \frac{1}{2}\frac{y^2}{n} + \frac{y^3}{3n^{3/2}} + \cdots\right\} - y\sqrt{n} \to -\frac{1}{2}y^2,
\end{aligned}$$

  where the second equality holds from the Taylor series expansion of $\log(1+x)$ for $|x| < 1$. Hence, we have $\dfrac{n!}{(n/e)^n\sqrt{n}} \to \int_{-\infty}^\infty e^{-\frac{1}{2}y^2}dy = \sqrt{2\pi}$, which is the heuristic derivation of *Stirling's formula*, a famous and useful approximation for factorials,

$$n! \approx \sqrt{2\pi}e^{-n}n^{n+\frac{1}{2}}.$$

R Simulation is a technique that can be used to help us understand how a complicated system works if detailed analysis is unavailable. It is also very useful to figure out many probability problems which often do not have exact or explicit solutions. Because simulation introduces an element of randomness into an analysis, it is sometimes called *Monte Carlo* analysis(methods), named after the Monte Carlo Casino in Monaco. Often, a few lines of R codes are enough to conduct a simulation of various random variables and related probability problems.

- To generate an exponential random variable with parameter λ, $X \sim \text{Exp}(\lambda)$, we use `rexp(n=1,lambda)`.

- To generate a uniform random variable between a and b, $X \sim \text{Uniform}(a,b)$, we use `runif(n=1,a,b)`.

- From the probability integral transform in Theorem 4.1.3, we have for $U \sim \text{Uniform}(0,1)$, the random variable $F_X^{-1}(U)$ has a c.d.f $F_X(x)$. That is, the inverse probability transform is used to generate random variables, using uniform random numbers between 0 and 1 as inputs. For example, if $X \sim \text{Exp}(\lambda)$, then $-\log(1-U)/\lambda$ is an exponential random variable with parameter λ. We can simply generate a large number of Uniform(0,1) random numbers and transform them.

```
Nsim=10^4 #number of random variables(simulations)
U=runif(Nsim)
X=-log(1-U)/2  # transforms of Uniform(0,1) into Exp(2)
Y=rexp(Nsim,2) # direct generation of Exp(2)
par(mfrow=c(1,2)) # for comparison of two histograms
hist(X,freq=F,main="Exp from Uniform")
hist(Y,freq=F,main="Exp from R")
```

Suppose that two rookie barista, A and B, in a KU cafe start serving customers at the same time. They agree to meet for a discussion on how to serve a customer after each of them has served the first customers. Presumably, one of them will finish before the other and have to wait. How long, on average, will one of the servers have to wait for the other?

- Suppose that we model all service times, regardless of the barista, as i.i.d. random variables having the exponential distribution with parameter $\lambda = 0.2$ per customer. Let X be the time it takes A to serve one customer, and let Y be the time it takes B to serve one customer. We are asked to compute the mean of $|X-Y|$.

- We can find this mean analytically using a two-dimensional integral over the union of two non-rectangular regions. Indeed, $X-Y$ has a Double exponential distribution with parameter λ, $X-Y \sim \text{DE}(0,\lambda)$, which will be discussed further in Chapter 5, so $|X-Y|$ has an exponential distribution with parameter λ, as discussed in Example 4.32. Thus, $\text{E}(|X-Y|) = 1/\lambda = 5$.

- Suppose that we generate two independent exponential random variables as we desire. We can then obtain a pair (X,Y) and compute $|X-Y|$. We then repeat this process independently as many times as we want and average all the observed $|X-Y|$ values. The average value by simulation should be close to the exact value of $\text{E}(|X-Y|)$, which is 5.

```
nsim<-10^4 # number of simulation
x<-rexp(nsim,rate=0.2) # service time for barista A
y<-rexp(nsim,rate=0.2) # service time for barista B
mean(abs(x-y)) # average of |x-y| close to the exact value 5.
[1] 4.994769
```

To simulate n arrivals in a Poisson process with rate λ, we first generate the interarrival times, $X_1, X_2, \ldots$, a sequence of i.i.d. exponential random variables with parameter λ. Next, consider $S_n = X_1 + \cdots + X_n$, $n = 1, 2, \ldots$, the time of the nth arrival. Then, we define a Poisson process with rate λ, N_t :

$$N_t = \max\{n \, : \, X_1 + \cdots + X_n \leq t\} \text{ with } N_0 = 0.$$

```
lambda <- 5 #rate = 5
Xn <- rexp(100, lambda) #interarrival times
Sn <- cumsum(Xn) #arrival times
n <- max(which(Sn <= 3)) # number of arrivals by time t=3
```

4.5 Beta distributions

The beta distribution is one of the basic distributions in statistics that has many applications in frequentist as well as Bayesian statistics. It is a versatile distribution and as such it is used in modeling the behavior of random variables that are positive but bounded in possible values.

Definition 4.5.1 If X is a random variable whose p.d.f. is

$$f_X(x) = cx^{\alpha-1}(1-x)^{\beta-1}, \text{ for } 0 < x < 1,$$

where c is the necessary constant, we say that X has a *beta distribution with parameters* α *and* β, $X \sim \text{Beta}(\alpha, \beta)$. It is required that $\alpha > 0$ and $\beta > 0$.

- If $\alpha > 0$ and $\beta > 0$, $\int_0^1 x^{\alpha-1}(1-x)^{\beta-1}dx = \dfrac{\Gamma(\alpha)\Gamma(\beta)}{\Gamma(\alpha+\beta)}$, which is called the *beta function*, usually denoted by $B(\alpha, \beta)$.

- That is, if X has a beta distribution with parameters α and β, then

$$f_X(x) = \frac{\Gamma(\alpha+\beta)}{\Gamma(\alpha)\Gamma(\beta)} x^{\alpha-1}(1-x)^{\beta-1}, \text{ for } 0 < x < 1.$$

Remark Various choices of α and β provide different shapes of beta density functions. For example, if $\alpha < 1$ and $\beta < 1$, the density is U-shaped, whereas if $\alpha > 1$ and $\beta > 1$, it has a global maximum.

- When $\alpha = \beta = 1$, we obtain a beta distribution with parameters 1 and 1, $X \sim \text{Beta}(1, 1)$ with p.d.f $f_X(x) = 1$ in the unit interval, which is a uniform density.

- When $\alpha < 1$ and $\beta < 1$ the graph of the beta p.d.f. is concave upwards in the unit interval with a unique minimum and unbounded at the endpoints $x = 0$ and $x = 1$. When $\alpha > 1$ and $\beta > 1$, the graph of the beta p.d.f. is bell-shaped with a unique maximum while taking a value of zero at the endpoints $x = 0$ and $x = 1$ as we see in Figure 4.14.

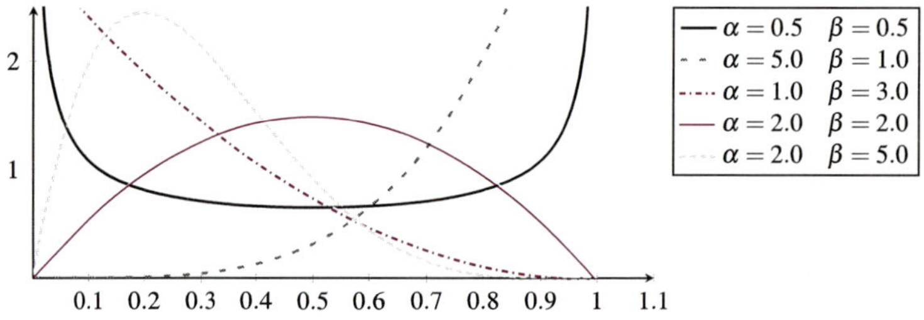

Figure 4.14: Beta p.d.f with various choices of α and β

Theorem 4.5.1 Let α and β be any two positive real numbers. Then

$$B(\alpha,\beta) = \frac{\Gamma(\alpha)\Gamma(\beta)}{\Gamma(\alpha+\beta)},$$

where $\Gamma(z) = \int_0^\infty x^{z-1}e^{-x}\,dx$ is the gamma function.

① $B(1,\beta) = \frac{1}{\beta}$, $B(\alpha,\beta) = B(\beta,\alpha)$, and $B(\alpha+1,\beta) = \frac{\alpha}{\alpha+\beta}B(\alpha,\beta)$.

② $B(\alpha,\beta) = 2\int_0^{\pi/2} (\sin\theta)^{2\alpha-1}(\cos\theta)^{2\beta-1}d\theta$.

Proof. Using integration by substitutions with the change of variables for double integrals, we evaluate $\Gamma(\alpha)\Gamma(\beta)$:

$$\begin{aligned}
\Gamma(\alpha)\Gamma(\beta) &= \int_0^\infty x^{\alpha-1}e^{-x}dx \int_0^\infty y^{\beta-1}e^{-y}dy \\
&= \int_0^\infty u^{2\alpha-2}e^{-u^2}2udu \int_0^\infty v^{2\beta-2}e^{-v^2}2vdv, \quad (x = u^2 \text{ and } y = v^2) \\
&= 4\int_0^\infty\int_0^\infty u^{2\alpha-1}v^{2\beta-1}e^{-(u^2+v^2)}dudv \\
&= 4\int_0^{\frac{\pi}{2}}\int_0^\infty r^{2\alpha+2\beta-2}(\cos\theta)^{2\alpha-1}(\sin\theta)^{2\beta-1}e^{-r^2}rdrd\theta, \quad (u = r\cos\theta \text{ and } v = r\sin\theta) \\
&= \left[\int_0^\infty (r^2)^{\alpha+\beta-1}e^{-r^2}dr^2\right]\left[2\int_0^{\frac{\pi}{2}}(\cos\theta)^{2\alpha-1}(\sin\theta)^{2\beta-1}d\theta\right] \\
&= \Gamma(\alpha+\beta)\left[2\int_0^{\frac{\pi}{2}}(\cos\theta)^{2\alpha-1}(\sin\theta)^{2\beta-1}d\theta\right] \\
&= \Gamma(\alpha+\beta)\int_0^1 t^{\alpha-1}(1-t)^{\beta-1}dt = \Gamma(\alpha+\beta)B(\alpha,\beta), \quad (t = \cos^2\theta).
\end{aligned}$$

proving the result. Thus, it is easy to see that ① and ② hold.

Alternatively, using the change of variables $u = x+y$ and $v = x/(x+y)$ for double integrals, we rewrite

$$\Gamma(\alpha)\Gamma(\beta) = \int_0^\infty \int_0^\infty x^{\alpha-1}y^{\beta-1}e^{-(x+y)}dxdy$$
$$= \int_0^1 \int_0^\infty u^{\alpha+\beta-1}e^{-u}v^{\alpha-1}(1-v)^{\beta-1}dudv$$
$$= \int_0^\infty u^{\alpha+\beta-1}e^{-u}du \int_0^1 v^{\alpha-1}(1-v)^{\beta-1}dv = \Gamma(\alpha+\beta)B(\alpha,\beta).$$

■

■ **Example 4.37 — Incomplete beta function.** Consider the following integral

$$\int_0^x t^{\alpha-1}(1-t)^{\beta-1}dt, \text{ where } \alpha > 0 \text{ and } \beta > 0,$$

which is known as the *incomplete beta function* denoted by $B_x(\alpha,\beta)$.

- Using integration by parts, we have

$$\int_0^x t^{\alpha-1}(1-t)^{\beta-1}dt = \frac{1}{\alpha}t^\alpha(1-t)^{\beta-1}\Big|_0^x + \frac{\beta-1}{\alpha}\int_0^x t^\alpha(1-t)^{\beta-2}dt$$
$$= \frac{1}{\alpha}x^\alpha(1-x)^{\beta-1} + \frac{\beta-1}{\alpha}\int_0^x t^\alpha(1-t)^{\beta-2}dt.$$

- The c.d.f of the binomial distribution, $X \sim \text{Binomial}(n,p)$ can be represented using the incomplete beta function. That is,

$$\int_0^{1-p} t^{n-k-1}(1-t)^k dt = \frac{1}{n-k}p^k(1-p)^{n-k} + \frac{k}{n-k}\int_0^{1-p} t^{n-k}(1-t)^{k-1}dt.$$

  Then, multiplying both sides of the equation by $(n-k)\binom{n}{k}$, we have

$$(n-k)\binom{n}{k}\int_0^{1-p} t^{n-k-1}(1-t)^k dt = P(X=k) + k\binom{n}{k}\int_0^{1-p} t^{n-k}(1-t)^{k-1}dt,$$
$$n\binom{n-1}{k}\int_0^{1-p} t^{n-k-1}(1-t)^k dt = P(X=k) + n\binom{n-1}{k-1}\int_0^{1-p} t^{n-k}(1-t)^{k-1}dt.$$

  Now, let $a_k = n\binom{n-1}{k}\int_0^{1-p} t^{n-k-1}(1-t)^k dt$ with $a_0 = n\int_0^{1-p} t^{n-1}dt = (1-p)^n = P(X=0)$. Then, we have $a_k = P(X=k) + a_{k-1}$ and thus

$$a_1 = P(X=1) + a_0, \quad a_2 = P(X=2) + a_1,$$
$$\vdots$$
$$a_{k-1} = P(X=k-1) + a_{k-2}, \quad a_k = P(X=k) + a_{k-1}.$$

Adding both sides of these equations, we have $a_k = \sum_{x=0}^{k} P(X=x) = P(X \le k)$. Hence, we have

$$P(X \le k) = \sum_{x=0}^{k} \binom{n}{x} p^x (1-p)^{n-x} = n\binom{n-1}{k} \int_0^{1-p} t^{n-k-1}(1-t)^k dt$$
$$= n\binom{n-1}{k} B_{1-p}(n-k,k+1) = (n-k)\binom{n}{k} B_{1-p}(n-k,k+1).$$

■

Theorem 4.5.2 Suppose that X has a beta distribution with parameters α and β. Then,

- For $k > -\alpha$, $\mathrm{E}(X^k) = \dfrac{\Gamma(\alpha+\beta)}{\Gamma(\alpha+\beta+k)} \dfrac{\Gamma(\alpha+k)}{\Gamma(\alpha)}$.
- In particular, $\mathrm{E}(X) = \dfrac{\alpha}{\alpha+\beta}$, $\mathrm{Var}(X) = \dfrac{\alpha\beta}{(\alpha+\beta)^2(\alpha+\beta+1)}$.

Proof. We can use the generalization of the beta integral formula to compute moments of beta distributions.

$$\begin{aligned}
\mathrm{E}(X^k) &= \int_0^1 \frac{\Gamma(\alpha+\beta)}{\Gamma(\alpha)\Gamma(\beta)} x^{k+\alpha-1}(1-x)^{\beta-1}dx \\
&= \frac{\Gamma(\alpha+\beta)}{\Gamma(\alpha)\Gamma(\beta)} \frac{\Gamma(\alpha+k)\Gamma(\beta)}{\Gamma(\alpha+\beta+k)} \int_0^1 \frac{\Gamma(\alpha+\beta+k)}{\Gamma(\alpha+k)\Gamma(\beta)} x^{k+\alpha-1}(1-x)^{\beta-1}dx \\
&= \frac{\Gamma(\alpha+\beta)}{\Gamma(\alpha+\beta+k)} \frac{\Gamma(\alpha+k)}{\Gamma(\alpha)}, \quad k > -\alpha.
\end{aligned}$$

- The expected value of X is given by

$$\mathrm{E}(X) = \int_0^1 x f(x)\, dx = \frac{1}{B(\alpha,\beta)} \int_0^1 x^{\alpha}(1-x)^{\beta-1}\, dx = \frac{B(\alpha+1,\beta)}{B(\alpha,\beta)} = \frac{\alpha}{\alpha+\beta}.$$

- Similarly, we have

$$\mathrm{E}\left(X^2\right) = \frac{\alpha(\alpha+1)}{(\alpha+\beta+1)(\alpha+\beta)}, \ \mathrm{Var}(X) = \mathrm{E}\left(X^2\right) - (\mathrm{E}[X])^2 = \frac{\alpha\beta}{(\alpha+\beta)^2(\alpha+\beta+1)}.$$

■

Remark The m.g.f. of X does not have a closed form but can be written as a infinite series

$$M_X(t) = \sum_{k=0}^{\infty} \frac{t^k}{k!} \mathrm{E}(X^k) = \sum_{k=0}^{\infty} \frac{t^k}{k!} \cdot \frac{\Gamma(\alpha+\beta)}{\Gamma(\alpha+\beta+k)} \cdot \frac{\Gamma(\alpha+k)}{\Gamma(\alpha)} = 1 + \sum_{k=1}^{\infty} \left(\prod_{j=0}^{k-1} \frac{\alpha+j}{\alpha+\beta+j} \right) \frac{t^k}{k!},$$

where the second equality holds from the property $\Gamma(\alpha+1) = \alpha\Gamma(\alpha)$.

Example 4.38 Suppose that a continuous random variable X has the following p.d.f.

$$f_X(x) = \begin{cases} \theta x^{\theta-1}, & \text{if } 0 < x < 1,\ \theta > 0, \\ 0, & \text{otherwise.} \end{cases}$$

- X has a beta distribution with parameters θ and 1, $X \sim \text{Beta}(\theta, 1)$.

- When $\theta = 1$, X has a uniform distribution between 0 and 1. In other words, the $\text{Beta}(1,1)$ is just the $\text{Uniform}(0,1)$ distribution.

- Let $Y = -\log X$. Then, Y has an exponential distribution with parameter θ since

$$f_Y(y) = f_X(e^{-y}) \left| \frac{d}{dy} e^{-y} \right| = \theta e^{-y(\theta-1)} e^{-y} = \theta e^{-\theta y},\ y > 0.$$

■

Example 4.39 — Arcsine distribution. The beta distribution with parameters $\alpha = \beta = 1/2$, $X \sim \text{Beta}\left(\frac{1}{2}, \frac{1}{2}\right)$ is often called the *arcsine distribution* with p.d.f.

$$f_X(x) = \frac{\Gamma\left(\frac{1}{2}+\frac{1}{2}\right)}{\Gamma\left(\frac{1}{2}\right)\Gamma\left(\frac{1}{2}\right)} x^{\frac{1}{2}-1}(1-x)^{\frac{1}{2}-1} = \frac{1}{\pi\sqrt{x(1-x)}}, \text{ for } 0 < x < 1.$$

- The mean is $\text{E}(X) = \frac{\alpha}{\alpha+\beta} = \frac{1}{2}$ and variance is $\text{Var}(X) = \frac{\alpha\beta}{(\alpha+\beta)^2(\alpha+\beta+1)} = \frac{1}{8}$. Note that the shape of $f_X(x)$ is U-shaped as shown in Figure 4.14.

- Using the change of variables, the c.d.f. of X is given by

$$F_X(x) = P(X \le x) = \int_0^x \frac{1}{\pi\sqrt{t(1-t)}} dt = \frac{2}{\pi} \int_0^{\sqrt{x}} \frac{1}{\sqrt{1-u^2}} du,\ \ (t = u^2,\ dt = 2udu)$$
$$= \frac{2}{\pi} \int_0^{\sin^{-1}(\sqrt{x})} 1 d\theta = \frac{2}{\pi} \arcsin\sqrt{x},\ \ (u = \sin\theta,\ du = \cos\theta d\theta).$$

■

Remark The beta distribution can be used as a prior distribution for the probability of success in Bernoulli trials (Bernoulli/binomial model). In fact, the Jeffreys prior distribution for the binomial (Bernoulli) model is the beta distribution with parameters $1/2$ and $1/2$, the arcsine distribution.

Example 4.40 — Pareto distribution. Suppose X has a beta distribution with parameters α and β. Define $Y = 1/X$.

- Then, Y has a support $\{y : 1 < y < \infty\}$, and its c.d.f. is given by

$$F_Y(y) = P\{Y \le y\} = P\{X \ge 1/y\} = \int_{1/y}^1 \frac{\Gamma(\alpha+\beta)}{\Gamma(\alpha)\Gamma(\beta)} x^{\alpha-1}(1-x)^{\beta-1} dx, \text{ for } y > 1.$$

Thus, we have the p.d.f. of Y by differentiation of $F_Y(y)$ or the change of variable formula

$$f_Y(y) = f_X(1/y)\left|\frac{d}{dy}\frac{1}{y}\right| = \frac{\Gamma(\alpha+\beta)}{\Gamma(\alpha)\Gamma(\beta)}\frac{(y-1)^{\beta-1}}{y^{\alpha+\beta}}, \text{ for } y > 1.$$

- When $\beta = 1$, $f_Y(y) = \alpha y^{-\alpha-1}$ for $y > 1$, and the distribution of Y is called the *Pareto distribution*, which was also considered in Example 4.15 for the existence of the expected value.

- We can also check the existence of the m.g.f. of Y, $M_Y(t) = \mathrm{E}(e^{tY})$ for Pareto distribution. Let k be the largest integer smaller than or equal to α, $\lfloor\alpha\rfloor = k$, so that $k \le \alpha < k+1$. Then, using the fact that $e^{tx} \ge (tx)^j/j!$ for $j = 0, 1, \ldots$ and $t \ge 0$, we have

$$\mathrm{E}(e^{tY}) = \alpha\int_1^\infty \frac{e^{ty}}{y^{\alpha+1}}dy \ge \frac{\alpha}{(k+1)!}\int_1^\infty \frac{(ty)^{k+1}}{y^{\alpha+1}}dy = \frac{\alpha t^{k+1}}{(k+1)!}\int_1^\infty \frac{1}{y^{\alpha-k}}dy,$$

which diverges since $0 < \alpha - k < 1$.

- More generally, a random variable Y has a Pareto distribution with parameters α and β, $Y \sim \text{Pareto}(\alpha, \beta)$ if the p.d.f. of Y is $f_Y(y) = \alpha\frac{\beta^\alpha}{y^{\alpha+1}}$ for $x > \beta$. Then, the c.d.f is $F_Y(y) = \int_\beta^y \alpha\frac{\beta^\alpha}{t^{\alpha+1}}dt = -\frac{\beta^\alpha}{t^\alpha}\Big|_\beta^y = 1 - \left(\frac{\beta}{y}\right)^\alpha$.

- Suppose that Z has an exponential distribution with parameter α, $Z \sim \text{Exp}(\alpha)$. Let $X = e^{-Z}$. Then, the p.d.f. of X is given by

$$f_X(x) = f_Z(-\log x)\left|\frac{d}{dx} - \log x\right| = \alpha e^{\alpha\log x}\frac{1}{x} = \alpha x^{\alpha-1},$$

which implies that $X \sim \text{Beta}(\alpha, 1)$. In addition, let $Y = \beta e^Z = \beta/X$. Then, the p.d.f. of Y is given by

$$f_Y(y) = f_Z\left(\log\frac{y}{\beta}\right)\left|\frac{d}{dy}\log\frac{y}{\beta}\right| = \alpha e^{-\alpha\log\frac{y}{\beta}}\frac{1}{y} = \alpha\frac{\beta^\alpha}{y^{\alpha+1}},$$

which implies that Y has a Pareto distribution with parameters α and β.

■

Remark The Pareto distribution is named after the Italian economist Vilfredo Pareto, who used it to model income distribution in the late nineteenth century to account for the observed fact that a relatively small proportion of the population controlled a significant portion of the resources.

- The Pareto distribution is an example of a "power-law" distribution whose p.d.f. is proportional to the power function x^{-a}. Such distributions have "heavy tails," which decays slower than those distributions such as exponential, gamma or normal, which go to zero exponentially fast. Thus, power law distributions have non-negligible probability to large values and thus have *heavy tails*.

- The Pareto distribution has been used to model population, magnitude of earthquakes, price returns of stocks, internet traffic, and others.

R To generate beta random variables, we can use the probability integral transform. However, the c.d.f. of beta is not analytically evaluated in general as shown in Example 4.37 with the incomplete beta function. Here, we consider two special cases of beta distributions to apply the probability integral transform and continue discussing more general way in Chapters 5 and 6.

- When X has the arcsine distribution, $X \sim \text{Beta}(1/2, 1/2)$, the c.d.f of X is $F_X(x) = \frac{2}{\pi}\arcsin\sqrt{x}$, and its inverse function $F_X^{-1}(x) = \left(\sin\frac{\pi}{2}x\right)^2$. Thus, we can generate the beta random variable with parameters $1/2$ and $1/2$ with $\left(\sin\frac{\pi}{2}U\right)^2$, $U \sim \text{Uniform}(0,1)$.

```
Nsim <- 10000; u <- runif(Nsim)
x <- sin(u*pi/2)^2
hist(x,freq=F); x=seq(0.001,1,0.01); fx=1/(pi*sqrt(x*(1-x)))
lines(x, fx, lty=1, lwd=2)
```

- When $X \sim \text{Beta}(\alpha, 1)$ with p.d.f. $f_X(x) = \frac{\Gamma(\alpha+1)}{\Gamma(\alpha)\Gamma(1)}x^{\alpha-1}(1-x)^{1-1}$, the c.d.f of X is $F_X(x) = \int_0^x \alpha t^{\alpha-1}dt = x^{\alpha}$. Thus, its inverse function is easily identified as $F_X^{-1}(x) = x^{1/\alpha}$. Thus, we can generate the beta random variable with parameters α and 1 with $F_X^{-1}(U) = U^{1/\alpha}$, $U \sim \text{Uniform}(0,1)$.

```
Nsim <- 10000; alpha<-3; u <- runif(Nsim);
x <- u^(1/alpha)
hist(x,freq=F); x=seq(0.001,1,0.01); fx=alpha*x^(alpha-1)
lines(x, fx, lty=1, lwd=2)
```

To generate the Pareto random variables, $Y \sim \text{Pareto}(\alpha, \beta)$, we can transform the beta random variable $X \sim \text{Beta}(\alpha, 1)$ into $Y = \beta/X$ or the exponential random variable Z into $Y = \beta e^Z$.

- Alternatively, we can use the probability integral transform since the inverse of c.d.f. is easily identified as $F_Y^{-1}(y) = \frac{\beta}{(1-y)^{1/\alpha}}$. Thus,

$$Y \stackrel{d}{=} \beta(1-U)^{-1/\alpha} \stackrel{d}{=} \beta U^{-1/\alpha}, \text{ where } U \sim \text{Uniform}(0,1).$$

```
Nsim <- 1000; alpha <-3; beta <- 1
pareto2 <- beta/runif(Nsim)^(1/alpha)
hist(pareto2, freq=F); y=seq(beta,10,0.01)
fy=alpha*beta^alpha/x^(alpha+1); lines(y, fy, lty=1, lwd=2)
```

4.6 Exercises

Problem 4.1 Let X be uniformly distributed on $[-1,1]$. Find the density of $Y = X^n$ for positive integers n.

Problem 4.2 Suppose X has a p.d.f. $f_X(x) = \begin{cases} e^{-x}, & \text{for } x > 0, \\ 0, & \text{otherwise.} \end{cases}$ Let $Y = (X-1)^2$. Find the p.d.f of Y.

Problem 4.3 Let X have the p.d.f. $f_X(x) = \dfrac{1}{x^2}$ for $x > 1$. Find the mean and median of X.

Problem 4.4 Let X have a p.d.f. $f_X(x) = \begin{cases} \frac{1}{5}, & -2 < x < 3, \\ 0, & \text{otherwise.} \end{cases}$ Find the p.d.f. of $Y = X^2$.

Problem 4.5 Suppose that the random variable U is uniformly distributed on $(0,1)$, $U \sim \text{Uniform}(0,1)$. Let the random variable X be defined by

$$X = \begin{cases} (2U)^{1/2}, & \text{if } U < 1/2, \\ 2-(2-2U)^{1/2}, & \text{if } U \geq 1/2. \end{cases}$$ Find the p.d.f. of X.

Problem 4.6 Let X have a p.d.f. given by $f_X(x) = \dfrac{e^{-x}}{(1+e^{-x})^2}$, $-\infty < x < \infty$.

① Find the expected value of X.

② Let U be a uniform random variable on $(0,1)$, $U \sim \text{Uniform}(0,1)$. Find a function $r(u)$ such that $r(U)$ has the same distribution with X.

Problem 4.7 Let X have a standard normal distribution, $X \sim N(0,1)$. Define $Y = \sqrt{|X|}$. Find the p.d.f of Y and expected value of Y. Also, define $T = X^{-2}$. Find the p.d.f. of T.

Problem 4.8 Let X have a normal distribution with mean μ and variance σ^2, $X \sim N(\mu, \sigma^2)$. Define $Y = X^2/\sigma^2$. Find the p.d.f. of Y and m.g.f.of Y.

Problem 4.9 Suppose that KU hospital keeps records of its emergency-room traffic and the number of patients arriving at the emergency-room. Those records indicate that, beginning at 6:00 pm on any given day, the number of patients arriving at the emergency-room has a Poisson process with parameter λ and that the expected time between patients is $1/\lambda$, where time is measured in hours.

① Given that the first patient doesn't arrive by 6:15 p.m., determine the probability that he or she arrives by 6:45 p.m.

② Let X be the time until the third patient will arrive at the emergency-room. Determine the variance of $1/X$.

Problem 4.10 Suppose that a point is chosen at random on a stick of unit length and that the stick is broken into two pieces at that point. Then, we form a right angle with two pieces of stick, forming the two shorter sides of a right-angled triangle. Let Θ be the smallest angle in this triangle. Define $Y = \tan\Theta$. Find the c.d.f. of Y and compute $\mathrm{E}(Y)$.

Problem 4.11 Let the random variable Z have the standard normal distribution, $Z \sim N(0,1)$.

① Compute the variance of $|Z|$.

② Let $Y = \max(Z-1, 0)$. Find the expected value of Y.

③ Compute $\mathrm{Var}[\Phi(Z)]$, where $\Phi(x)$ denotes the c.d.f. of Z, $\Phi(x) = P(Z \le x)$.

Problem 4.12 Suppose that the amount of service (ordering a coffee and getting it done) time at a KU driving-through coffee shop is exponentially distributed with an expected value of 10 minutes. You arrive at the driving-through line while one customer is being served and one other customer is waiting in the line. The staff of the coffee shop informs you that the customer has already ordered a Cafe Latte and waited for 5 minutes. What is the probability that the customer will need no more than 5 minutes extra? What is the probability that you have to wait more than 15 minutes before you get your coffee?

Problem 4.13 Let $Y = e^{X^2}$, where $X \sim N(0,1)$. Evaluate $\displaystyle\int_0^\infty P(Y^2 > t)dt$.

Problem 4.14 Suppose that X is a continuous random variable with p.d.f.

$$f_X(x) = \begin{cases} |x| & \text{for } -1 \le x \le 1 \\ 0 & \text{otherwise.} \end{cases}$$ Define $Y = X^2$. Find the distribution of Y.

Problem 4.15 Suppose that X is a continuous random variable with p.d.f.

$f_X(x) = c\exp\left(-|x|^3\right)$, $-\infty < x < \infty$. Find the expected value of $\mathrm{E}(|X|^4)$.

Problem 4.16 Buses arrive at KU bus stop according to a Poisson process with parameter λ. Let X be the waiting time until the first bus arrives at KU bus stop. Define $\lfloor X \rfloor$ be the integer part of X. Find the expected value of $X^2 - \lfloor X \rfloor$.

Problem 4.17 Suppose that X has a beta distribution with parameters m and n, both positive integers, $X \sim \mathrm{Beta}(m,n)$. Show that $P(X \le x) = P(Y \ge m)$ for $0 < x < 1$, where Y has a binomial distribution with parameters $m+n-1$ and x, $Y \sim \mathrm{Binomial}(m+n-1, x)$.

5. Joint distributions

5.1 Bivariate random variables

In this chapter, we generalize the concept of distribution of a continuous random variable to the joint distribution of two (bivariate) continuous random variables. A random variable that is a point in $\mathbb{R}^2$ is called bivariate. The idea discussed in this chapter can be easily extended to more than two (multivariate) random variables.

Two random variables X and Y have a continuous joint distribution with *joint p.d.f.* $f_{X,Y}$ if, for each subset A of $\mathbb{R}^2$,

$$P[(X,Y) \in A] = \int_A \int f_{X,Y}(x,y)dxdy.$$

Definition 5.1.1 — Joint p.d.f. The joint probability density function (joint p.d.f.) of the bivariate random variables X and Y is an integrable function $f_{X,Y}(x,y)$ such that

① $f_{X,Y}(x,y) \geq 0$ for $-\infty < x < \infty$ and $-\infty < y < \infty$

② $\int_{-\infty}^{\infty}\int_{-\infty}^{\infty} f_{X,Y}(x,y)\,dx\,dy = 1.$

Remark Any function that satisfies the two conditions in Definition 5.1.1 is the joint p.d.f. for some probability distribution.

- The set $\{(x,y) \, : \, f_{X,Y}(x,y) > 0\}$ in the condition ① is called the *support* of the joint distribution of (X,Y).

- The condition ② in Definition 5.1.1 means that the total volume below the surface $z = f_{X,Y}(x,y)$ and above the xy-plane in R^2 defined as a double integral need to be 1.

- An example of a joint p.d.f is $f_{X,Y}(x,y) = \frac{1}{2\pi}e^{-\frac{x^2+y^2}{2}}$ for $-\infty < x,y < \infty$. This joint p.d.f. is known as a *bivariate normal p.d.f.*, a product of two independent standard normal p.d.f's X and Y. Figure 5.1 provides the 3-dimensional plot for $z = f_{X,Y}(x,y)$.

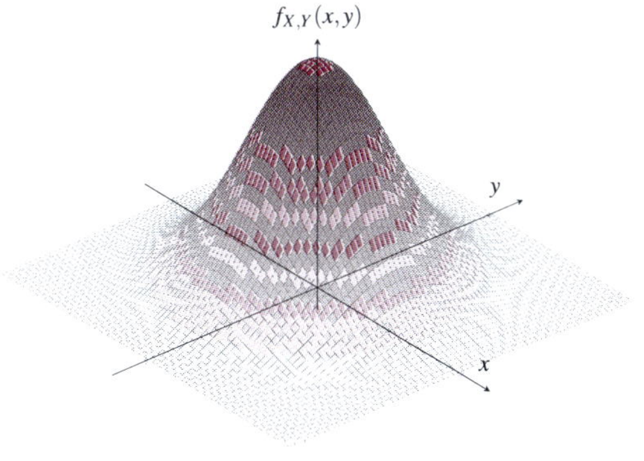

Figure 5.1: An example of a joint p.d.f.

- Each individual point, and every infinite sequence of points, in the xy-plane has probability mass 0 since the probability is computed by integrating the joint p.d.f. over the region under consideration.

Definition 5.1.2 — Joint c.d.f. If X and Y are continuous random variables, then the joint cumulative distribution function (joint c.d.f) of X and Y is

$$F_{X,Y}(x,y) = P(X \le x,\ Y \le y) = \int_{-\infty}^{x}\int_{-\infty}^{y} f_{X,Y}(s,t)\,dt\,ds.$$

Hence, it follows, upon differentiation, that

$$f_{X,Y}(x,y) = \frac{\partial^2}{\partial x \partial y}F_{X,Y}(x,y) = \frac{\partial^2}{\partial y \partial x}F_{X,Y}(x,y)$$

at every point (x,y) at which two second-order partial derivatives exist.

Remark The probability that X and Y will be in a specified rectangle can be obtained from the joint c.d.f. of X and Y.

$$\begin{aligned}
F_X(a) &= P\{X \le a\} = P\{X \le a,\ Y < \infty\} = P\left(\lim_{b\to\infty}\{X \le a,\ Y \le b\}\right) \\
&= \lim_{b\to\infty} P\{X \le a,\ Y \le b\} = \lim_{b\to\infty} F_{X,Y}(a,b) = F_{X,Y}(a,\infty), \\
F_Y(b) &= F_{X,Y}(\infty,b).
\end{aligned}$$

- $P(X > a,\ Y > b) = 1 - F_X(a) - F_Y(b) + F_{X,Y}(a,b)$,
- $P(a_1 < X \le a_2,\ b_1 < Y \le b_2) = F_{X,Y}(a_2,b_2) - F_{X,Y}(a_1,b_2)$
$$-F_{X,Y}(a_2,b_1) + F_{X,Y}(a_1,b_1).$$

The joint c.d.f. of X and Y is $P(X \le x,\ Y \le y)$ defined by the double sum, the double integral, or the sum and the integral, depending on the types of two random variables.

- Suppose that X is the discrete random variable and Y is continuous. Then the joint pmf/pdf is a nonnegative function f such that, for each pair of sets A and B of real numbers
$$P(X \in A, Y \in B) = \sum_{x \in A} \int_B f_{X,Y}(x,y)dy.$$
- Let X have a continuous distribution and let Y have a discrete distribution. If there is a function f such that
$$P(a < X \le b,\ Y = y) = \int_a^b f_{X,Y}(x,y)dx, \text{ for all } a < b, y.$$
Then X and Y have a mixed joint distribution with joint p.m.f./p.d.f. $f_{X,Y}(x,y)$.

Example 5.1 Suppose that X and Y have a continuous joint distribution for which the joint p.d.f. is defined as follows:

$$f_{X,Y}(x,y) = \begin{cases} cy^2 & \text{for } 0 \le x \le 2 \text{ and } 0 \le y \le 1, \\ 0 & \text{otherwise.} \end{cases}$$

- To determine the value of the constant c, we compute the integral

$$\int_0^1 \int_0^2 cy^2 dxdy = \int_0^1 2cy^2 dy = \frac{2c}{3} y^3 \bigg|_{y=0}^1 = \frac{2c}{3} = 1. \text{ Thus, we have } c = \frac{3}{2}.$$

- Find ① $P(X+Y > 2)$; ② $P(Y < 1/2)$; ③ $P(X \le 1)$; ④ $P(X = 3Y)$.

① $\int_1^2 \int_{2-x}^1 \frac{3}{2} y^2 dydx = \int_1^2 \frac{1}{2}[1-(2-x)^3]dx = \frac{1}{2} + \frac{1}{8}(2-x)^4\big|_{x=1}^2 = \frac{3}{8}$,

② $P(Y < 1/2) = \int_0^{1/2} \int_0^2 \frac{3}{2} y^2 dxdy = \int_0^{1/2} 3y^2 dy = \frac{1}{8}$,

③ $P(X \le 1) = \int_0^1 \int_0^1 \frac{3}{2} y^2 dxdy = \int_0^1 \frac{3}{2} y^2 dy = \frac{1}{2}$,

④ $P(X = 3Y) = 0$.

For ④, since the set of (x,y) pairs such that $x = 3y$ has 0 area, the integral of the p.d.f. over that set will also be 0, which implies that $P(X = 3Y) = 0$. ■

Remark Consider a nonnegative real-valued function (a joint p.d.f.) $z = f_{X,Y}(x,y) \geq 0$ defined on a rectangular region $A = [a,c] \times [b,d]$ as in Figure 5.2.

- The double integral of $f_{X,Y}(x,y)$ over a region A represents *the volume under the surface* $z = f_{X,Y}(x,y)$,

$$\iint_A f_{X,Y}(x,y)dydx = \int_a^c \int_b^d f_{X,Y}(x,y)dydx.$$

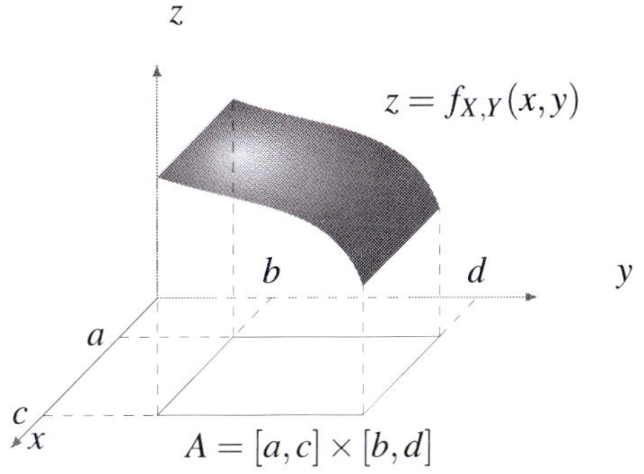

Figure 5.2: Surface $z = f_{X,Y}(x,y)$ over a region A

- The double integral over a general (nonrectangular) region A in Figure 5.3 can be defined in a similar way.

$$\iint_A f(x,y)\,dydx = \int_a^b \left[\int_{g_1(x)}^{g_2(x)} f(x,y)\,dy\right] dx = \int_c^d \left[\int_{h_1(y)}^{h_2(y)} f(x,y)\,dx\right] dy.$$

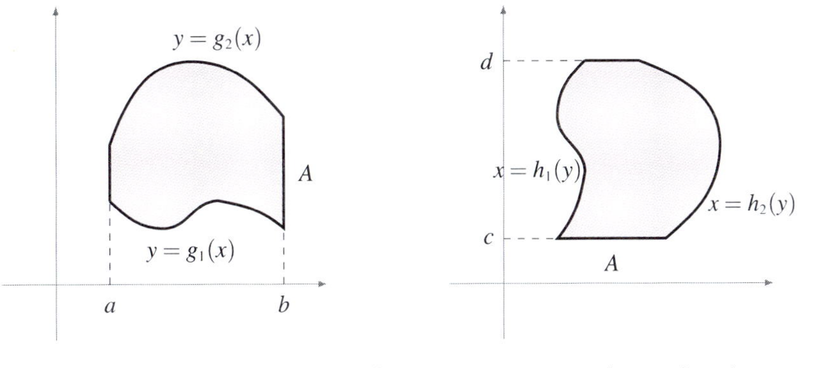

Figure 5.3: Double integral over a nonrectangular region A

■ **Example 5.2** Suppose that the joint p.d.f. of X and Y is specified as follows.

$$f(x,y) = \begin{cases} cx^2y & \text{for } x^2 \leq y \leq 1, \\ 0 & \text{otherwise} \end{cases}$$

Determine the value of the constant c and the value of $P(X \geq Y)$.

- Since it is a valid joint p.d.f., the value of the integral in the support of X and Y in the left panel of Figure 5.4 must be 1. Thus, we have

$$\int_{-1}^{1}\int_{x^2}^{1} cx^2 y\,dy\,dx = c\cdot\frac{4}{21} = 1, \text{ which implies that } c = \frac{21}{4}.$$

- Let A be the region $\{(x,y)\ :\ x \geq y,\ x^2 \leq y \leq 1\}$ as shown in the right panel of Figure 5.4. Then,

$$P(X \geq Y) = P((X,Y) \in A) = \int_A\int f_{X,Y}(x,y)\,dx\,dy = \int_0^1\int_{x^2}^{x} \frac{21}{4}\cdot x^2 y\,dy\,dx = \frac{3}{20}.$$

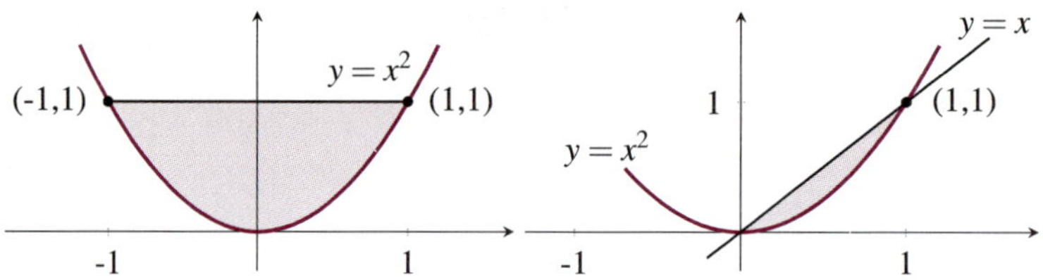

Figure 5.4: The support of the joint distribution and its subset A where $x \geq y$.

■

■ **Example 5.3 — Uniform distribution on a rectangle.** Suppose that the joint p.d.f. of X and Y is given by

$$f_{X,Y}(x,y) = \begin{cases} c, & \text{for } a_1 < x < a_2 \text{ and } b_1 < y < b_2, \\ 0, & \text{otherwise.} \end{cases}$$

That is, you pick a random point in the rectangle $\{(x,y)\ :\ a_1 < x < a_2,\ b_1 < y < b_2\}$.

- Since $\displaystyle\int_{a_1}^{a_2}\int_{b_1}^{b_2} c\,dy\,dx = 1$, $c = \dfrac{1}{(a_2-a_1)(b_2-b_1)}$, and it is uniformly distributed over the rectangle.

- Furthermore, if A is a subset the rectangle with area $|A|$, then we have

$$P\{(X,Y) \in A\} = \iint_{(x,y)\in A} f_{X,Y}(x,y)\,dx\,dy = \frac{\text{size (area) of } A}{(a_2-a_1)(b_2-b_1)}.$$

■

■ **Example 5.4 — Uniform triangle.** A point is chosen at random uniformly in the triangular area defined by the points $(1,0)$, $(2,0)$ and $(0,1)$. Let X be the x-coordinate of the chosen point and Y be the y-coordinate of the chosen point.

- Since the point (X,Y) is chosen uniformly in the triangle area $S = \{(x,y)\ :\ 1-y < x < 2-2y,\ 0 < y < 1\}$ in Figure 5.5, we know that the joint p.d.f. is flat (a constant)

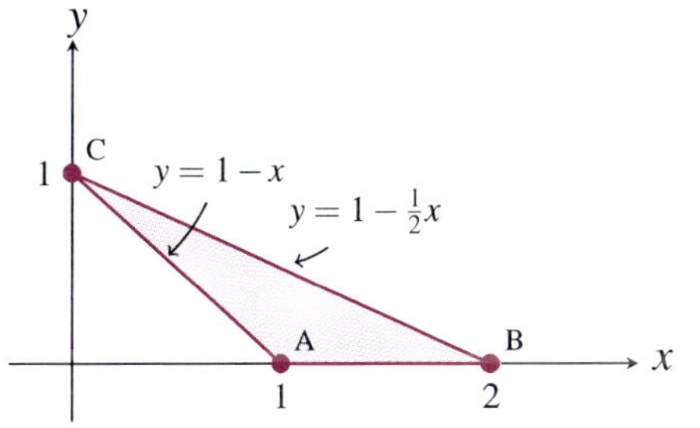

Figure 5.5: Triangular area defined by the points $(1,0)$, $(2,0)$ and $(0,1)$.

in the triangle and 0 outside of it. That is, $f_{X,Y}(x,y) = c$ for (x,y) in the triangle, and $f_{X,Y}(x,y) = 0$ for (x,y) outside of the triangle. Because

$$\begin{aligned}\int_{-\infty}^{\infty}\int_{-\infty}^{\infty} f_{X,Y}(x,y)dxdy &= \int_0^1\int_{1-y}^{2-2y} cdxdy = c\cdot\int_0^1(1-y)dy \\ &= \frac{1}{2}(\text{area of triangle})\cdot c = 1.\end{aligned}$$

Therefore, $c = 2$ and we have $f_{X,Y}(x,y) = \begin{cases} 2, & \text{for } (x,y)\in S, \\ 0, & \text{otherwise.}\end{cases}$

■

■ **Example 5.5 — Uniform circle.** A dart game involves throwing a dart at a circular target of radius r. Suppose we throw a dart once so that it hits the target, and we observe where it lands.

- The sample space Ω consists of all the points in the target
  $$\Omega = \{(x,y) \ : \ x^2+y^2 \le r^2\}.$$
- Consider two random variables X and Y for the point in the target, selected at random from inside the circle $x^2+y^2 \le r^2$. Then, the joint p.d.f. is a constant
  $$f_{X,Y}(x,y) = \begin{cases} c, & x^2+y^2\le r^2 \\ 0, & \text{otherwise.}\end{cases}$$
  Since it is a joint p.d.f.
  $$\int\int_{\Omega} cdxdy = c\cdot\text{area of target(circle with radius } r) = c\pi r^2 = 1,$$
  which implies that $c = \dfrac{1}{\pi r^2}$.
- For simplicity, we assume $r = 1$. Then, we can evaluate the integral using integration by substitutions with $x = \sin\theta$
  $$\begin{aligned}&\int\int_{\Omega} cdxdy = \int_{-1}^{1}\int_{-\sqrt{1-x^2}}^{\sqrt{1-x^2}} cdxdy = \int_0^1 4c\sqrt{1-x^2}dx \\ &= \int_0^{\pi/2} 4c\sqrt{1-\sin^2\theta}\cos\theta d\theta = \int_0^{\pi/2} 2c(\cos 2\theta+1)d\theta = c\pi.\end{aligned}$$

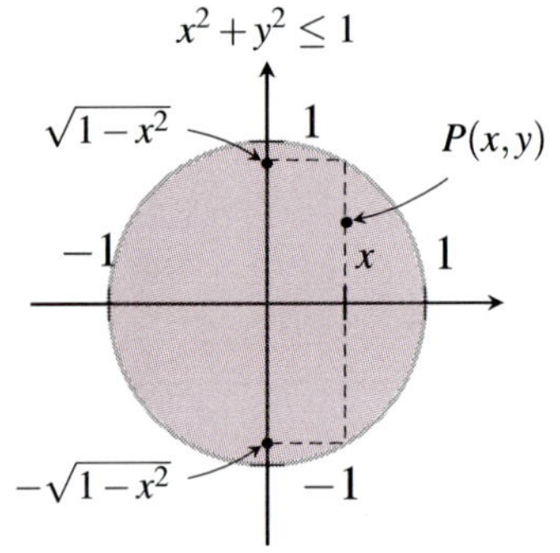

Figure 5.6: A circular target of radius $r=1$

- Alternatively, we can use a polar coordinate change with $x = r\cos\theta$ and $y = r\sin\theta$, $0<r<1, 0<\theta<2\pi$,

$$\int\int_{\Omega} cdxdy = \int_0^{2\pi}\int_0^1 crdrd\theta = \int_0^{2\pi}\frac{c}{2}d\theta = c\pi.$$

■

Theorem 5.1.1 — Marginal distribution. From the joint p.d.f. of (X,Y), we can find the p.d.f. of each of X and Y, which is called the *marginal* p.d.f. of X and Y, respectively.

$$f_X(x) = \int_{-\infty}^{\infty} f_{X,Y}(x,y)dy, \quad f_Y(y) = \int_{-\infty}^{\infty} f_{X,Y}(x,y)dx.$$

Proof. Since the marginal c.d.f. of X is obtained as $F_X(x) = P(X\le x) = P(X\le x, -\infty < Y<\infty) = \int_{-\infty}^{\infty}\int_{-\infty}^{x} f_{X,Y}(s,t)dsdt$, we have, $f_X(x) = \frac{d}{dx}F_X(x) = \int_{-\infty}^{\infty} f_{X,Y}(x,t)dt$ by the Fundamental Theorem of Calculus. Similarly, we have $f_Y(y) = \int_{-\infty}^{\infty} f_{X,Y}(x,y)dx$. ■

Remark *The other direction is not possible in general.* That is, the marginal distribution of X and Y can be obtained from their joint distribution, it is not possible to construct the joint distribution of X and Y from their marginal distributions without additional information.

■ **Example 5.6** Suppose that the joint p.d.f. of X and Y is given by

$$f_{X,Y}(x,y) = \begin{cases} cye^{-y^2} & 0<x<y<\infty, \\ 0 & \text{otherwise.}\end{cases}$$

- We compute the normalizing constant c :

$$\int_0^{\infty}\int_x^{\infty} ye^{-y^2}dydx = -\frac{1}{2}\int_0^{\infty}\left[e^{-y^2}\right]_x^{\infty}dx = \frac{1}{2}\int_0^{\infty} e^{-x^2}dx = \frac{1}{2}\times\frac{\sqrt{2\pi}\sqrt{\frac{1}{2}}}{2} = \frac{\sqrt{\pi}}{4},$$

using the fact that the normal density is symmetric around mean($= 0$) and integrates to 1. Hence, $c = 4/\sqrt{\pi}$.

- To derive $f_X(x)$ and $f_Y(y)$, we integrate the joint p.d.f with respect to y and x respectively :

$$\begin{aligned} f_X(x) &= \frac{4}{\sqrt{\pi}}\int_x^{\infty} ye^{-y^2}dy = \frac{2}{\sqrt{\pi}}e^{-x^2},\ x>0, \\ f_Y(y) &= \frac{4}{\sqrt{\pi}}\int_0^{y} ye^{-y^2}dx = \frac{4}{\sqrt{\pi}}y^2e^{-y^2},\ y>0. \end{aligned}$$

■

■ **Example 5.7 — Uniform triangle (revisited).** A point is chosen at random uniformly in the triangular area defined by the points $(1,0),(2,0)$ and $(0,1)$. Let X be the x-coordinate of the chosen point and Y be the y-coordinate of the chosen point, which has the joint p.d.f.

$$f_{X,Y}(x,y) = \begin{cases} 2, & \text{for } 1-y<x<2-2y,\ 0<y<1, \\ 0, & \text{otherwise.} \end{cases}$$

Find the marginal p.d.f.'s of X and Y.

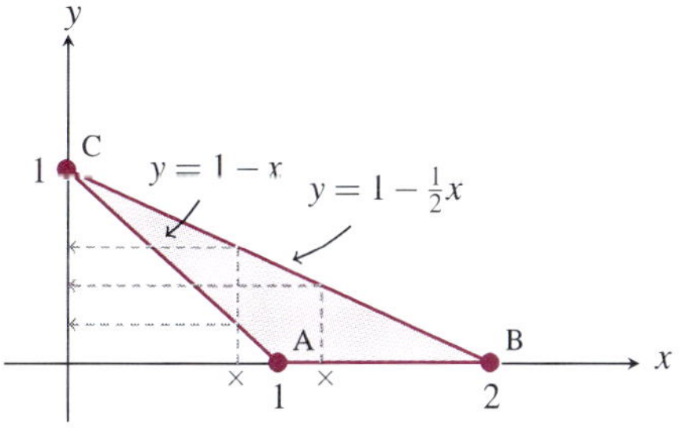

Figure 5.7: Triangle with vertices $(1,0)$, $(2,0)$ and $(0,1)$

- The marginal support of X is $\{x : 0 < x < 2\}$, and the evaluation of the joint p.d.f. for the marginal p.d.f. of X needs to be made carefully in two cases as in Figure 5.7

$$f_X(x) = \begin{cases} \int_{1-x}^{1-\frac{x}{2}} 2dy = x, & \text{for } 0<x<1, \\ \int_0^{1-\frac{x}{2}} 2dy = 2-x, & \text{for } 1\le x<2, \\ 0, & \text{otherwise.} \end{cases}$$

- The marginal support of Y is $\{y : 0 < y < 1\}$, and the p.d.f. of Y is obtained

$$f_Y(y) = \int_{1-y}^{2-2y} 2dx = 2-2y, \quad 0<y<1.$$

■

■ **Example 5.8 — Uniform circle (revisited).** Let (X,Y) be distributed uniformly on the circular disk centered at $(0,0)$ with radius r, which has the joint p.d.f.

$$f_{X,Y}(x,y) = \begin{cases} \frac{1}{\pi r^2}, & \text{for } x^2+y^2 \le r^2, \\ 0, & \text{otherwise.} \end{cases}$$

- From the equation of a circle with radius r and center at the origin is $x^2+y^2=r^2$, we get $y = \pm\sqrt{r^2-x^2}$.
- Thus, the marginal p.d.f. of X is given by

$$f_X(x) = \int_{-\sqrt{r^2-x^2}}^{\sqrt{r^2-x^2}} f_{X,Y}(x,y)dy = \frac{2}{\pi r^2}\sqrt{r^2-x^2}, \text{ for } -r < x < r.$$

- By symmetry, we have $f_Y(y) = \frac{2}{\pi r^2}\sqrt{r^2-y^2}$, for $-r < y < r$.
- Let $W = \sqrt{X^2+Y^2}$ be the distance of the chosen point from the origin. Then, the c.d.f of W is given by

$$F_W(w) = P(\sqrt{X^2+Y^2} \le w) = \int\int_{x^2+y^2 \le w^2} \frac{1}{\pi r^2} dxdy = \frac{\pi w^2}{\pi r^2} = \frac{w^2}{r^2} \text{ for } 0 < w < r.$$

Thus, the p.d.f. of W is given by $f_W(w) = \frac{d}{dw}F_W(w) = \frac{2w}{r^2}$ for $0 < w < r$. ■

Remark It is common to have to deal with both discrete and continuous random variables at the same time. In this case, there is a hybrid of the p.f. and p.d.f. that we call the joint p.f./p.d.f. for such a pair.

- Suppose that X is the discrete random variable and Y is continuous.
- Then the joint p.f./p.d.f. is a nonnegative function f such that, for each pair of sets A and B of real numbers

$$P(X \in A, Y \in B) = \sum_{x \in A} \int_B f_{X,Y}(x,y)dy.$$

■ **Example 5.9 — Mixed case.** Consider a random variable Y that tells us how likely patients are to respond positively. We can think of this as the proportion of all patients who would respond positively if everyone in a very large population were in the clinical trial.

- Let X be the actual number of patients who respond positively in a sample of size n. The following function is a joint p.f./p.d.f.:

$$f_{X,Y}(x,y) = \begin{cases} \binom{n}{x} y^x(1-y)^{n-x}, & \text{if } x \in \{0,\dots,n\} \text{ and } 0 < y < 1, \\ 0 & \text{otherwise.} \end{cases}$$

- The marginal p.d.f. of Y is obtained by summing over x for each fixed y. Since $\sum_{x=0}^{n}\binom{n}{x}y^x(1-y)^{n-x}=1$ by the binomial theorem, we have that $f_Y(y)=1$ for $0<y<1$ is the marginal p.d.f. of Y. The marginal distribution of Y is uniform on the interval $[0,1]$.

- To compute the marginal p.f. of X, we use the Beta integral formula. It follows that
$$f_X(x)=P(X=x)=\int_0^1\binom{n}{x}y^x(1-y)^{n-x}dy=\binom{n}{x}\frac{x!(n-x)!}{(n+1)!}=\frac{1}{n+1},$$
for $x=0,1,\ldots,n$, which is a discrete uniform on the integers $0,1,\ldots,n$.

■

5.2 Independent random variables

Two random variables X and Y are *independent* if $P(X\in A, Y\in B)=P(X\in A)P(Y\in B)$ for all sets A and B. If X and Y are independent and we let $A=(-\infty,x]$ and $B=(-\infty,y]$, we have $P(X\le x,\ Y\le y)=F_X(x)F_Y(y)$, for all x and y, where F_X and F_Y are the marginal c.d.f.'s of X and Y.

Definition 5.2.1 Two random variables X and Y are independent if, for all x and y,
$$P(X\le x,\ Y\le y)=P(X\le x)P(Y\le y).$$
In other words, $F_{X,Y}(x,y)=F_X(x)F_Y(y)$.

Theorem 5.2.1 For two continuous random variables X and Y with a joint p.d.f. $f_{X,Y}(x,y)$, if $f_{X,Y}(x,y)=f_X(x)f_Y(y)$ for all x and y, then X and Y are independent.

Proof. It follows from the relationship between a joint c.d.f. and a joint p.d.f. with partial derivatives $f_{X,Y}(x,y)=\frac{\partial^2}{\partial y\partial x}F_{X,Y}(x,y)=\frac{\partial^2}{\partial y\partial x}F_X(x)\cdot F_Y(y)=f_X(x)f_Y(y)$. ■

■ **Example 5.10** Let X and Y be distributed uniformly on the unit disk centered at $(0,0)$ with radius 1. That is, suppose that X and Y have the following joint p.d.f.
$$f_{X,Y}(x,y)=\begin{cases}\frac{1}{\pi}, & \text{for } x^2+y^2\le 1,\\ 0, & \text{otherwise.}\end{cases}$$
Are X and Y independent?

- To check, we could compute the marginal p.d.f.'s of X and Y. But it is clear that the marginal p.d.f. of X $f_X(x)$ is strictly positive for $-1<x<1$ and the marginal p.d.f. of Y $f_Y(y)$ is strictly positive for $-1<y<1$. The product of two such functions will be strictly positive on a square, not just the unit disk.

- Specifically, we can compute the marginal p.d.f of X and check if $f_{X,Y}(x,y) = f_X(x)f_Y(y)$. As shown in Example 5.8, the marginal p.d.f.'s of X is given by $f_X(x) = \frac{2}{\pi}\sqrt{1-x^2}$, for $-1 \le x \le 1$, which implies that X and Y are not independent. ■

Example 5.11 Suppose that the joint p.d.f. of X and Y is as follows:

$$f_{X,Y}(x,y) = \begin{cases} 2xe^{-y}, & \text{for } 0 \le x \le 1 \text{ and } 0 < y < \infty, \\ 0, & \text{otherwise.} \end{cases}$$ Are X and Y independent?

- The region where the p.d.f. is strictly positive is a rectangle and the p.d.f. factors into $2x$ times e^{-y} on that rectangle. This is sufficient to conclude that X and Y are independent.
- The two marginals are multiples of $2x$ for $0 \le x \le 1$ and e^{-y} for $0 < y < \infty$. ■

Theorem 5.2.2 — How to check for independence. Two random variables X and Y are independent if and only if

① the formula for the joint p.f., p.d.f., or p.f./p.d.f. factors into functions of x and y, and

② the joint p.f., p.d.f., or p.f./p.d.f. is strictly positive on a rectangular region with sides parallel to the coordinate axes.

In other words, suppose that the range of X and Y is a (possibly infinite) rectangle. If $f(x,y) = g(x)h(y)$ for some functions g and h (not necessarily probability density functions) then X and Y are independent.

Example 5.12 Suppose that the joint p.d.f. of X and Y is as follows:

$$f_{X,Y}(x,y) = \begin{cases} 24xy, & \text{for } x,y \ge 0 \text{ and } x+y \le 1, \\ 0, & \text{otherwise.} \end{cases}$$ Are X and Y independent?

- The formula $24xy$ factors into $24x$ and y on the region where $f_{X,Y} > 0$. But the region where $f_{X,Y} > 0$ is not a rectangle. Thus, X and Y are not independent. ■

Example 5.13 Suppose that X and Y are independent and both have the same density

$$f(x) = \begin{cases} 2x, & \text{if } 0 \le x \le 1, \\ 0, & \text{otherwise.} \end{cases}$$ Find $P(X+Y \le 1)$.

- Since X and Y are independent with the same density, the joint p.d.f. of X and Y is

$$f_{X,Y}(x,y) = f_X(x)f_Y(y) = \begin{cases} 4xy, & \text{for } 0 \le x \le 1 \text{ and } 0 \le y \le 1, \\ 0, & \text{otherwise.} \end{cases}$$

- Hence, the desired probability is $P(X+Y \leq 1) = \int_0^1 \int_0^{1-x} 4xydydx = \frac{1}{6}$. ■

■ **Example 5.14** Suppose that Jaden and Joanne decide to meet at the classroom of P&E 506. If they independently arrive at a time uniformly distributed between 12 noon and 1 pm, find the probability that the first to arrive has to wait no longer than 10 minutes.

- Let X and Y be the arrival time uniformly distributed in an hour between 12 noon and 1 pm. Then, we can think of two points X and Y chosen at random and independently along a line segment of length L $(L = 60)$. The question is to find the probability that $|X-Y| \leq \ell$, $(\ell = 10)$.

- Since two points X and Y are chosen at random, they are uniformly distributed on the square with length L,

$$f_{X,Y}(x,y) = \begin{cases} \frac{1}{L^2}, & (x,y) \in [0,L] \times [0,L], \\ 0, & \text{otherwise.} \end{cases}$$

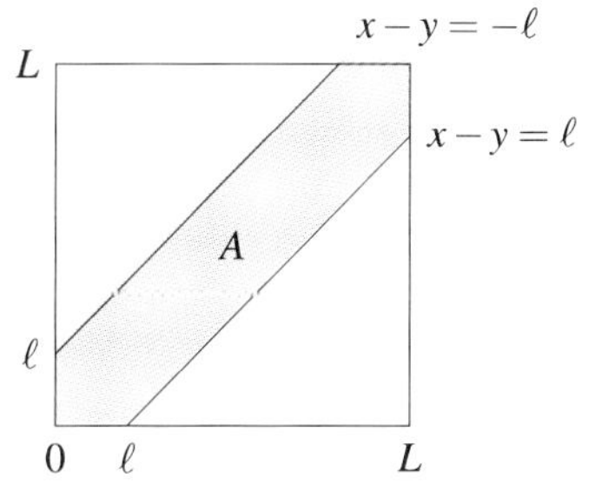

Figure 5.8: Line segment of length L on the square

- Let $A = \{(x,y) \,:\, |x-y| \leq \ell\}$. Then, the desired probability is

$$P(|X-Y| \leq \ell) = \iint_A f_{X,Y}(x,y)dxdy = \frac{\text{area of } A}{L^2} = \frac{L^2 - (L-\ell)^2}{L^2} = \frac{2L\ell - \ell^2}{L^2}.$$

■

■ **Example 5.15** Let X be the time (in minutes) that Joanne waits for a traffic light to turn green, and let Y be the time (in minutes, at a different intersection) that Jaden waits for a traffic light to turn green. Suppose that X and Y have joint p.d.f.

$$f_{X,Y}(x,y) = \begin{cases} ke^{-x-2y}, & \text{for } x > 0 \text{ and } y > 0 \\ 0, & \text{otherwise.} \end{cases}$$

- X and Y are defined inside of an unbounded rectangular region, and $f_{X,Y}(x,y)$ can be factored as $h_1(x) = e^{-x}$ and $h_2(y) = ke^{-2y}$. Therefore X and Y are independent, and $k = 2$ by choosing constants that make $h_1(x)$ and $h_2(y)$ integrate to one. Note that $X \sim \text{Exp}(1)$ and $Y \sim \text{Exp}(2)$.

- Find the probability that Joanne waits longer at her light than Jaden, that is, $P(X>Y)$,

$$P(X>Y)=\int_0^\infty\int_0^x 2e^{-x-2y}dydx=\int_0^\infty(1-e^{-2x})e^{-x}dx=1-\frac{1}{3}=\frac{2}{3}.$$

- In addition, let $Z=\min\{X,Y\}$. Find the c.d.f of Z.

$$\begin{aligned}F_Z(z) &= P(Z\le z)=P(\min\{X,Y\}\le z)\\ &= 1-P(\min\{X,Y\}>z)=1-P(X>z\;\&\;Y>z)\\ &= 1-P(X>z)P(Y>z)=1-e^{-z}e^{-2z}=1-e^{-3z}.\end{aligned}$$

- Suppose that X and Y are independent random variables, where X is exponentially distributed with expected value $1/\lambda_1$ and Y is exponentially distributed with expected value $1/\lambda_2$. Then, $P(\min\{X,Y\}\le z)=1-e^{-(\lambda_1+\lambda_2)z}$ for $z\ge 0$, which implies that $\min\{X,Y\}\sim\text{Exp}(\lambda_1+\lambda_2)$. In addition,

$$P(X>Y)=\frac{\lambda_2}{\lambda_1+\lambda_2},\quad P(X<Y)=\frac{\lambda_1}{\lambda_1+\lambda_2}.$$

■

■ **Example 5.16** Suppose that two points X and Y are chosen at random inside a square with side 1 and whose center is the origin, which has a joint p.d.f.,

$$f_{X,Y}(x,y)=\begin{cases}1, & (x,y)\in\left[-\frac{1}{2},\frac{1}{2}\right]\times\left[-\frac{1}{2},\frac{1}{2}\right],\\ 0, & \text{otherwise.}\end{cases}$$

Define $Z=XY$. Find the p.d.f. of Z.

- Note that the support of Z is $\left\{z:-\frac{1}{4}\le z\le\frac{1}{4}\right\}$. Thus, we consider separately two cases for $0<z<1/4$, and $-1/4<z<0$ to evaluate the c.d.f. of Z.

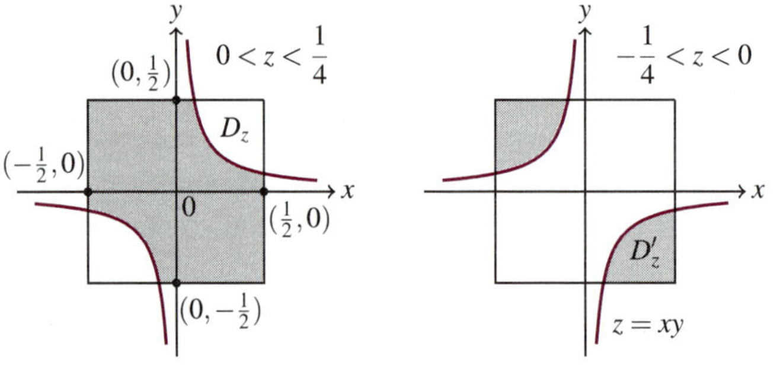

Figure 5.9: The shaded area of $\{(x,y):xy<z,\;-1/4<z<1/4\}$

- For $0<z<1/4$, note that $F_Z(z)=P(Z\le z)=P(XY\le z)=1-P(XY>z)$. Then, $P(XY>z)=2P(D_z)$, where $D_z=\{(x,y):xy>z,0<x,y<1/2\}$ since X and Y are defined in a symmetric area about the origin as in the left panel of Figure 5.9.

In addition, since the joint p.d.f. of X and Y is a constant, $P(D_z)$ corresponds to the area of D_z, which is computed as

$$P(D_z) = \int\int_{D_z} 1 dxdy = \text{the area of } D_z = \int_{2z}^{1/2} \frac{1}{2} - \frac{z}{x} dx = \frac{1}{4} - z + z\log 4z.$$

Thus, $F_Z(z) = 1 - 2\left(\frac{1}{4} - z + z\log 4z\right) = \frac{1}{2} + 2z - 2z\log 4z.$

- Likewise, for $-1/4 < z < 0$, $F_Z(z) = 2P(D_z')$, where $D_z' = \{(x,y) : xy < z, -1/2 < y < 0, 0 < x < 1/2\}$ as in the right panel of Figure 5.9. Thus, we compute

$$\begin{aligned} F_Z(z) &= 2P(D_z') = 2\int\int_{D_z'} 1 dxdy = 2 \cdot \text{area of } D_z' = 2\int_{-2z}^{1/2} \frac{z}{x} + \frac{1}{2}\, dx \\ &= 2\left(\frac{1}{4} + z - z\log(-4z)\right) = \frac{1}{2} + 2z - 2z\log(-4z). \end{aligned}$$

- Hence, differentiating $F_Z(z)$, we have the p.d.f. of Z,

$$f_Z(z) = \frac{d}{dz}F_Z(z) = \begin{cases} -2\log 4|z|, & \text{for } |z| < 1/4, \\ 0, & \text{otherwise.} \end{cases}$$

■

We now discuss the distribution of sum of two independent continuous random variables with thc convolution formula.

Theorem 5.2.3 — Sum of two independent random variables. Let X_1 and X_2 be independent random variables with p.d.f.'s f_1 and f_2 respectively. Then, the c.d.f. and p.d.f. h of $Y = X_1 + X_2$ are given by

$$\begin{aligned} F_Y(y) &= P(Y \le y) = P(X_1 + X_2 \le y) = \int_{-\infty}^{\infty}\int_{-\infty}^{y-x_2} f_1(x_1)f_2(x_2)dx_1dx_2, \\ f_Y(y) &= \frac{d}{dy}F_Y(y) = \int_{-\infty}^{\infty} f_1(y-x_2)f_2(x_2)dx_2. \end{aligned}$$

The p.d.f. f_Y above is called the *convolution* of f_1 and f_2.

Remark Convolution is a binary operation on the set of all density functions. Let f and g be two real valued (density) functions. The convolution of f and g, denoted by $f * g$, is defined as

$$(f * g)(z) = \int_{-\infty}^{\infty} f(z-y)g(y)\,dy = \int_{-\infty}^{\infty} g(z-x)f(x)\,dx = (g * f)(z).$$

■ **Example 5.17** The convolution of two independent normal p.d.f.'s with means μ_1 and μ_2 and variances σ_1^2 and σ_2^2 is again a normal p.d.f., with mean $\mu_1+\mu_2$ and variance $\sigma_1^2+\sigma_2^2$.

- Suppose that X_1 and X_2 are two independent random variables $X_i \sim N(0,1)$, $i=1,2$.
- Let $Y = X_1+X_2$. The p.d.f. of Y is obtained from the convolution of f_{X_1} and f_{X_2},

$$\begin{aligned} f_Y(y) &= f_{X_1} * f_{X_2}(y) = \int_{-\infty}^{\infty} \phi(y-x)\phi(x)dx = \frac{1}{2\pi}\int_{-\infty}^{\infty} e^{-(y-x)^2/2}e^{-x^2/2}dx \\ &= \frac{1}{2\pi}e^{-y^2/4}\sqrt{\pi}\left[\frac{1}{\sqrt{\pi}}\int_{-\infty}^{\infty} e^{-\left(x-\frac{y}{2}\right)^2}dx\right] = \frac{1}{\sqrt{4\pi}}e^{-y^2/4}, \end{aligned}$$

which implies that $Y = X_1+X_2 \sim N(0,2)$.

■

■ **Example 5.18** Suppose that X_1 and X_2 are two independent random variables $X_i \sim \text{Exp}(1)$, $i=1,2$. Find the p.d.f. of $Y = X_1+X_2$.

- Let $f_1(x) = f_2(x) = \exp(-x)$ for $x > 0$. Then

$$f_1(y-x)f_2(x) = \exp(x-y)\exp(-x), \text{ for } y-x>0 \text{ and } x>0.$$

- For a fixed y value, the values of x for which we must integrate are those such that both $x > 0$ and $y-x > 0$. That is, $0 < x < y$. The convolution is then

$$f_Y(y) = \int_0^y \exp(x-y)\exp(-x)dx = y\exp(-y), \text{ for } y>0,$$

which implies that $Y = X_1+X_2 \sim \text{Gamma}(2,1)$.

- Suppose we have two independent random variables that have exponential density with parameter λ, $X_i \sim \text{Exp}(\lambda)$, $i=1,2$. Then, $Y = X_1+X_2 \sim \text{Gamma}(2,\lambda)$.

$$\begin{aligned} f_Y(y) &= \int_{-\infty}^{\infty} f_{X_1}(y-x)f_{X_2}(x)\,dx = \int_0^y \lambda e^{-\lambda(y-x)}\lambda e^{-\lambda x}dx \\ &= \lambda^2 y e^{-\lambda y},\ y \ge 0. \end{aligned}$$

■

■ **Example 5.19** Suppose that X_1 and X_2 are i.i.d. random variables and the each of them has a uniform distribution on the interval $[0,1]$. Find the p.d.f. of $Y = X_1+X_2$.

- The p.d.f. of each random variable is 1 for $0 < x < 1$. Thus, we have

$$\begin{aligned} f_1(y-x_2)f_2(x_2) &= 1, \text{ for } 0<y-x_2<1 \text{ and } 0<x_2<1, \\ f_Y(y) &= \int_0^1 f_1(y-x_2)f_2(x_2)dx_2. \end{aligned}$$

- Now, if $0 \le y \le 1$, we have $f_Y(y) = \int_0^y dx_2 = y$, while if $1 < y \le 2$, we have $f_Y(y) = \int_{y-1}^1 dx_2 = 2 - y$, whose upper and lower limits for two integrals are easily identified from Figure 5.10.

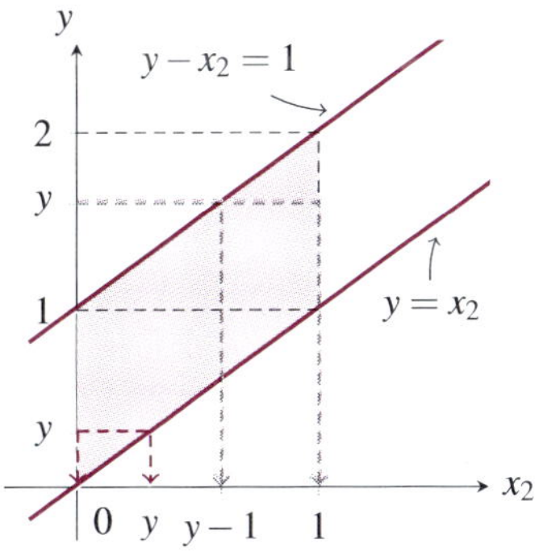

Figure 5.10: The set where the convolution of two uniforms is evaluated.

Hence, $f_Y(y) = \min\{y, 2-y\}$, $0 \le y \le 2$, that is,

$$f_Y(y) = \begin{cases} y & \text{if } 0 < y < 1, \\ 2-y & \text{if } 1 \le y < 2, \\ 0 & \text{otherwise,} \end{cases}$$

which is called the *triangular* distribution.

- Also note that $F_Y(y)$ is the shaded area of the intersection of the unit square $[0,1] \times [0,1]$ with the region $x_1 + x_2 \le y$ in Figure 5.11.

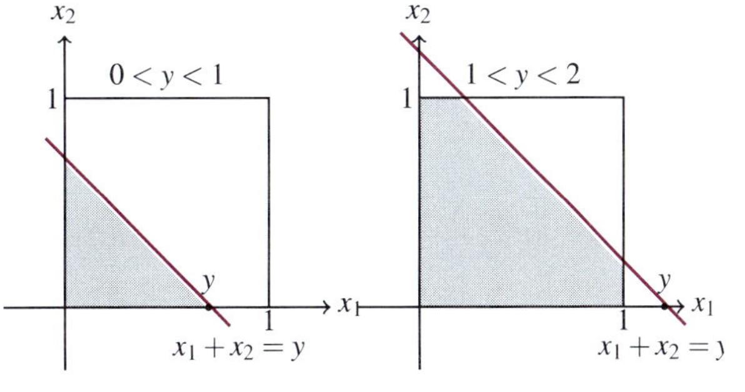

Figure 5.11: The shaded area of $\{(x_1, x_2) : 0 \le x_1, x_2 \le 1,\ x_1 + x_2 \le y,\ 0 < y < 2\}$

Thus, $F_Y(y) = P(Y < y) = y^2/2$ for $y \in [0,1]$, and $1 - (2-y)^2/2$ for $y \in [1,2]$.

- Suppose that we consider adding another independent uniform random variable U_3 to Y, $U_3 \sim \text{Uniform}(0,1)$ independent of U_1 and U_2. That is, define $W = Y + U_3$. Using the convolution formula again, the p.d.f. of W is given by

$$f_W(w) = \int_{-\infty}^{\infty} f_Y(y) f_{U_3}(w-y)dy = \int_{w-1}^{w} f_Y(y)dy,\ 0 < w < 3.$$

Since $0 < y < 2$, we need to consider three cases $0 \le w \le 1$, $1 < w \le 2$, and $2 < w \le 3$ separately. Hence, we obtain the p.d.f.

$$f_W(w) = \begin{cases} \frac{w^2}{2}, & \text{for } 0 < w \le 1, \\ -w^2 + 3w - \frac{3}{2}, & \text{for } 1 < w \le 2, \\ \frac{(w-3)^2}{2}, & \text{for } 2 < w \le 3, \\ 0, & \text{otherwise.} \end{cases}$$

■

■ **Example 5.20** Suppose X and Y are two independent standard normal random variables. Now suppose we locate a point P in the xy-plane with coordinates (X,Y) and ask: What is the p.d.f. of the distance of P from the origin?

- Note that $f_X(x) = f_Y(x) = \dfrac{1}{\sqrt{2\pi}} e^{-x^2/2}$ and that X^2 has a $\chi^2(1)$ distribution,

$$f_{X^2}(s) = \begin{cases} \frac{1}{2\sqrt{s}}(f_X(\sqrt{s}) + f_X(-\sqrt{s})) = \frac{1}{\sqrt{2\pi s}} e^{-s/2}, & \text{if } s > 0, \\ 0, & \text{otherwise.} \end{cases}$$

- Let $T = X^2 + Y^2$. Then, from the convolution formula, we have

$$f_T(t) = \int_{-\infty}^{\infty} f_{X^2}(t-s) f_{Y^2}(s)\, ds = \begin{cases} \frac{1}{2} e^{-t/2}, & \text{if } t \ge 0, \\ 0, & \text{otherwise.} \end{cases}$$

Hence, T has a gamma distribution with $\alpha = \frac{1}{2}$, $\beta = 1$, that is, $\chi^2(2)$ distribution.

- Now, the p.d.f. of $R = \sqrt{T}$ is obtained

$$f_R(r) = \begin{cases} \frac{1}{2} e^{-r^2/2} \cdot 2r = r e^{-r^2/2}, & \text{if } r \ge 0, \\ 0, & \text{otherwise,} \end{cases}$$

which is known as a *Rayleigh* p.d.f.

■

■ **Example 5.21** Suppose that $X_1 \sim \text{Exp}(1)$ and $X_2 \sim \text{Exp}(2)$ are independent. Find the distribution of $Y = X_1 + X_2$.

- We get the p.d.f. of Y using convolution :

$$\begin{aligned} f_Y(y) &= \int_{-\infty}^{\infty} f_{X_1}(y - x_2) f_{X_2}(x_2) dx_2 \\ &= \int_{-\infty}^{\infty} e^{-(y-x_2)} \cdot I(y - x_2 > 0) 2e^{-2x_2} I(x_2 > 0) dx_2 \\ &= \int_0^y 2e^{-y-x_2} dx_2 = 2e^{-y}\left(1 - e^{-y}\right) = -2e^{-2y} + 2e^{-y},\ y > 0, \end{aligned}$$

which is not a Gamma p.d.f.

■

Remark Note that Y does not have a Gamma distribution since two exponential random variables X_1 and X_2 are not identically distributed. More generally, when X_1 and X_2 are independent gamma random variables that has the identical rate (or scale) parameter, the sum of two independent gamma random variables has a gamma distribution as shown in Example 5.22.

- Example 5.18 presents the sum of two independent gamma random variables with the shape parameter 1 and the rate (scale) parameter 1, $Y = X_1 + X_2$, $X_i \sim \text{Gamma}(1,1) \stackrel{d}{=} \text{Exp}(1)$, $i = 1,2$, has a gamma distribution with the shape parameter 2 and the rate (scale) parameter 1, $Y \sim \text{Gamma}(2,1)$.

- Alternatively, it can be explained in terms of the Poisson process argument that the distribution of the sum of two independent exponential random variables is a gamma random variable as discussed Chapter 4. That is, since the waiting time until one arrival in the Poisson process with rate $\lambda = 1$ has an exponential distribution with parameter λ, the waiting time until two arrivals in the Poisson process with rate $\lambda = 1$, has a gamma distribution with parameters 2 and λ.

- Thus, the sum of n independent exponential random variables with parameter λ has a gamma distribution with parameters n and λ. More generally, the sum of two independent gamma random variables, $Y = X_1 + X_2$, $X_1 \sim \text{Gamma}(n,\lambda)$, $X_2 \sim \text{Gamma}(m,\lambda)$, where n and m are positive integers, has a gamma distribution with parameters $n+m$ and λ since by the Poisson process argument, it follows that $X_1 = \sum_{i=1}^n Z_i$ and $X_2 = \sum_{j=1}^m Z_i'$ where $Z_1,\dots,Z_n,Z_1',\dots,Z_m'$ are independent exponential random variables with the identical parameter λ.

- However, when we have two independent gamma random variables with the general positive (non-integer valued) shape parameters, $X_1 \sim \text{Gamma}(\alpha,\lambda)$ and $X_2 \sim \text{Gamma}(\beta,\lambda)$, where α and β are not integers but positive, the Poisson process argument would not work, and we can use the convolution approach as in Example 5.22.

■ **Example 5.22** Suppose that X and Y are independent random variables with $X \sim \text{Gamma}(\alpha,\lambda)$ and $Y \sim \text{Gamma}(\beta,\lambda)$. Let $Z = X + Y$. Find the distribution of Z.

- By the convolution, we have $f_Z(z) = \int_{-\infty}^{\infty} f_X(x) f_Y(z-x)dx$. Thus,

$$
\begin{aligned}
f_Z(z) &= \int_0^z \frac{1}{\Gamma(\alpha)}\lambda^\alpha x^{\alpha-1}e^{-\lambda x}\frac{1}{\Gamma(\beta)}\lambda^\beta (z-x)^{\beta-1}e^{-\lambda(z-x)}dx \\
&= \int_0^1 \frac{1}{\Gamma(\alpha)\Gamma(\beta)}\lambda^{\alpha+\beta}(zu)^{\alpha-1}(z-zu)^{\beta-1}e^{-\lambda z}du, \ (x = zu,\ dx = zdu) \\
&= \frac{z^{\alpha+\beta-1}}{\Gamma(\alpha+\beta)}e^{-\lambda z}\int_0^1 \frac{\Gamma(\alpha+\beta)}{\Gamma(\alpha)\Gamma(\beta)}u^{\alpha-1}(1-u)^{\beta-1}du = \frac{1}{\Gamma(\alpha+\beta)}z^{\alpha+\beta-1}e^{-\lambda z},
\end{aligned}
$$

which is the p.d.f. of a gamma random variable with parameters $\alpha+\beta$ and λ. Note that the last equality holds since the integral is for the beta p.d.f. with parameters α and β, which equals 1. Hence, $Z \sim \text{Gamma}(\alpha+\beta,\lambda)$.

■

Remark More generally, if X_1 and X_2 have joint p.d.f $f_{X_1,X_2}(x_1,x_2)$, then

$$f_{X_1+X_2}(y)=\int_{-\infty}^{\infty} f_{X_1,X_2}(y-x_2,x_2)dx_2=\int_{-\infty}^{\infty} f_{X_1,X_2}(x_1,y-x_1)dx_1.$$

Example 5.23 — Bivariate beta. Suppose that X_1 and X_2 have the joint p.d.f.

$$f_{X_1,X_2}(x_1,x_2)=\begin{cases}\frac{\Gamma(\alpha_1+\alpha_2+\alpha_3)}{\Gamma(\alpha_1)\Gamma(\alpha_2)\Gamma(\alpha_3)}x_1^{\alpha_1-1}x_2^{\alpha_2-1}(1-x_1-x_2)^{\alpha_3-1}, & \text{for } x_1>0,\ x_2>0,\\ & x_1+x_2<1,\\ 0, & \text{otherwise.}\end{cases}$$

- The joint p.d.f. is also known as Dirichlet density with parameters α_1, α_2 and α_3. Obviously, X_1 and X_2 are not independent.

- The p.d.f. of $Y=X_1+X_2$ is still obtained using the *general* convolution formula, for $0<y<1$,

$$\begin{aligned}f_Y(y) &= \int_{-\infty}^{\infty} f_{X_1,X_2}(y-x_2,x_2)dx_2\\ &= \int_0^y \frac{\Gamma(\alpha_1+\alpha_2+\alpha_3)}{\Gamma(\alpha_1)\Gamma(\alpha_2)\Gamma(\alpha_3)}(y-x_2)^{\alpha_1-1}x_2^{\alpha_2-1}(1-y)^{\alpha_3-1}dx_2\\ &= \frac{\Gamma(\alpha_1+\alpha_2+\alpha_3)}{\Gamma(\alpha_1)\Gamma(\alpha_2)\Gamma(\alpha_3)}y^{\alpha_1+\alpha_2-1}(1-y)^{\alpha_3-1}\int_0^1(1-u)^{\alpha_1-1}u^{\alpha_2-1}du\\ &= \frac{\Gamma(\alpha_1+\alpha_2+\alpha_3)}{\Gamma(\alpha_1+\alpha_2)\Gamma(\alpha_3)}y^{\alpha_1+\alpha_2-1}(1-y)^{\alpha_3-1},\end{aligned}$$

  which is a p.d.f. of a beta distribution with parameters $\alpha_1+\alpha_2$ and α_3.

■

Example 5.24 Suppose that X_1 and X_2 have the joint p.d.f.

$$f_{X_1,X_2}(x_1,x_2)=\begin{cases}cx_1^2x_2^2, & \text{for } x_1>0,\ x_2>0,\ x_1+x_2<1,\\ 0, & \text{otherwise.}\end{cases}\quad \text{Find } P(X_1+X_2\le y).$$

- The joint p.d.f. is identified as a Dirichlet density in Example 5.23 with parameters $\alpha_1=3$, $\alpha_2=3$ and $\alpha_3=1$. Thus $c=\dfrac{\Gamma(7)}{\Gamma(3)\Gamma(3)\Gamma(1)}$.

- For $0<y<1$, we have

$$\begin{aligned}P(X_1+X_2\le y) &= c\int_0^y x_1^2\int_0^{y-x_1} x_2^2dx_2dx_1=\frac{c}{3}\int_0^y x_1^2(y-x_1)^3dx_1\\ &= \frac{c}{3}y^6\int_0^1 t^2(1-t)^3dt=y^6.\end{aligned}$$

  The probability is zero for $y\le 0$ and is 1 for $y\ge 1$. Note that the last equality follows from the integral of the beta p.d.f. with parameters 3 and 4.

- By differentiation, we can also obtain the p.d.f. $Y = X_1 + X_2$ is $f_Y(y) = 6y^5, 0 < y < 1$. Alternatively, $Y = X_1 + X_2$ has a beta distribution with parameters 6 and 1 from the result in Example 5.23.

■

The Buffon's Needle Problem is considered to be the first and the most famous problem in geometric probability, problem, introduced in 1733 by Georges-Louis Leclerc, Comte de Buffon (1707– 1788), French naturalist and mathematician. Buffon estimated the number π by throwing needles on a ruled surface and recording how many times the needles crossed a line.

■ **Example 5.25 — Buffon's needle.** A needle of length L is dropped at random on a floor ruled with equally spaced parallel lines a distance D apart, where $L \le D$. What is the probability that the needle will intersect one of the parallel lines? This problem is known as Buffon's needle problem.

- Let X be the distance from the center of the needle to the closest line and Θ be the angle at which the needle falls, the angle between the needle and the parallel line, that is measured against a line parallel to the lines on the floor. It is reasonable to assume that independently $X \sim \text{Uniform}\left[0, \frac{1}{2}D\right]$ and $\Theta \in \text{Uniform}[0, \pi]$. Thus,

$$f_{X,\Theta}(x, \theta) = \begin{cases} \frac{2}{D}\frac{1}{\pi}, & \text{for } 0 \le x \le \frac{1}{2}D \text{ and } 0 \le \theta \le \pi, \\ 0, & \text{otherwise.} \end{cases}$$

- The needle intersects the nearest line if and only if the hypotenuse of the right-angled triangle is less than half the length L of the needle as shown in Figure 5.12, that is, $\dfrac{x}{\sin\theta} < \dfrac{1}{2}L$.

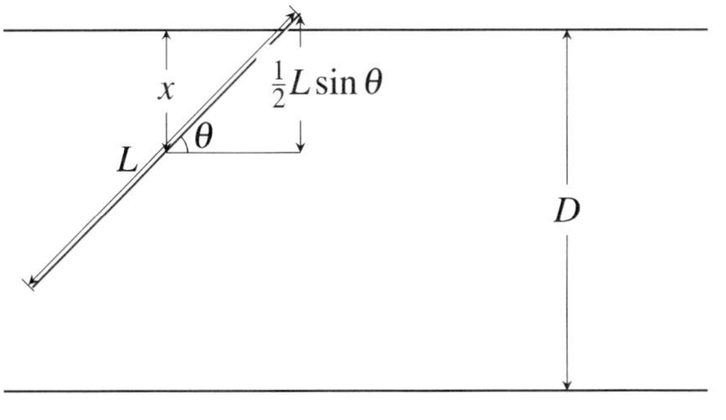

Figure 5.12: Buffon's needle

- Therefore, the desired probability is given by

$$P(\text{a needle intersects the nearest line}) = P\left(X \le \frac{1}{2}L\sin\Theta\right)$$

$$= \int_0^{\pi}\int_0^{\frac{1}{2}L\sin\theta} \frac{1}{\frac{1}{2}D\pi}\,dx\,d\theta = \frac{\int_0^{\pi}\frac{1}{2}L\sin\theta\,d\theta}{\frac{1}{2}D\pi} = \frac{-L\cos\theta\Big|_0^{\pi}}{D\pi} = \frac{2L}{D\pi}.$$

Alternatively, Θ can be assumed to be an acute angle, $\Theta \in \text{Uniform}[0, \pi/2]$, and the required probability would still be the same as $\dfrac{2L}{D\pi}$.

■

R Monte Carlo integration refers to statistical estimation of the value of an integral using evaluations of an integrand at a set of points drawn randomly from a distribution with support over the range of integration. Famous examples of Monte Carlo integration are about the Monte Carlo estimate of π.

- (Random points in the unit square) The unit disk $A = \{(x,y) : x^2 + y^2 \le 1\}$ is inscribed in the square $\Omega = [-1,1] \times [-1,1]$. Suppose that (X,Y) is a random point in the square Ω with the joint p.d.f, $f_{X,Y}(x,y) = \frac{1}{4} I_\Omega(x,y)$, and we would like to evaluate $P(A)$. If we generate a large number of points that are uniform on the square, the proportion of points falling inside the disk is approximately equal to the ratio of the disk's area to the square's area,

$$P(A) = \int_A \int f_{X,Y}(x,y)dxdy = \frac{\pi}{4} \approx \frac{1}{n} \sum_{i=1}^{n} I_\Omega(X_i, Y_i).$$

```
nsim<-10^6
x<-runif(nsim,-1,1); y<-runif(nsim,-1,1)
4*sum(x^2+y^2<1)/nsim
[1] 3.1409
```

- (Buffon's needle) Suppose that we take a card table and draw across the top surface a set of parallel lines a unit distance apart($D = 1$). We then drop a common needle of unit length($L = 1$) at random on this surface and observe whether or not the needle lies across one of the lines. The probability that the needle intersects a line is $\frac{2}{\pi}$. Suppose we drop a needle n times and it hits the line n times. Then, we can estimate π by

$$\pi \approx \frac{2n}{\sum_{i=1}^{n} I(X_i \le \frac{1}{2}\sin\Theta_i)}, \quad X_i \sim \text{Uniform}(0, 1/2), \ \Theta_i \sim \text{Uniform}(0, \pi).$$

```
buffon <- function(D, L) {
 x <- runif(1,0,D/2); theta <- runif(1, 0, pi/2)
 count <- (x <= L/2 * sin(theta))
}
L=1; D=1; N=10^5; times=replicate(N,buffon(L,D))
(2*L)/(D*mean(times))
[1] 3.148069
```

5.3 Transformation of two random variables

We have learned how to find the distribution of a transformation of a single random variable. These ideas extend fairly easily to transforming two random variables that have a joint continuous distribution. Suppose that the random vector (X_1,X_2) is interpreted as a random point picked in $\mathbb{R}^2$ according to joint p.d.f. $f_{X_1,X_2}(x_1,x_2)$, where (x_1,x_2) denotes the Cartesian coordinate. Instead of (x_1,x_2), another system of coordinates can be chosen with (y_1,y_2), where (x_1,x_2) and (y_1,y_2) are related by $y_1 = r_1(x_1,x_2)$ and $y_2 = r_2(x_1,x_2)$. Then, we get the joint p.d.f. of $Y_1 = r_1(X_1,X_2)$ and $Y_2 = r_2(X_1,X_2)$. In principle, the joint c.d.f. is obtained in a straightforward way. For Y_1 and Y_2 above, and $A = \{(x_1,x_2) : r_1(x_1,x_2) \le y_1,\ r_2(x_1,x_2) \le y_2\}$ then, $F_{Y_1,Y_2}(y_1,y_2) = \iint_A f_{X_1,X_2}(x_1,x_2)dx_1dx_2$. There are well-known results about changing variables within a multiple integral that provide the joint p.d.f. of Y_1 and Y_2. We state without proof a theorem for a transformation of two random variables, and additional details will be given in Theorem 5.3.3.

Theorem 5.3.1 Let X_1 and X_2 be jointly continuous random variable with a joint p.d.f. $f_{X_1,X_2}(x_1,x_2)$. Suppose that $Y_1 = r_1(X_1,X_2)$ and $Y_2 = r_2(X_1,X_2)$ for some functions r_1 and r_2. Assume that the functions r_1 and r_2 satisfy the following conditions:

① The equations $y_1 = r_1(x_1,x_2)$ and $y_2 = r_2(x_1,x_2)$ can be uniquely solved (one-to-one) for x_1 and x_2 in terms of y_1 and y_2 with solutions given by, say, $x_1 = s_1(y_1,y_2)$, $x_2 = s_2(y_1,y_2)$.

② The functions s_1 and s_2 have continuous partial derivatives at all points (y_1,y_2) and are such that the following 2×2 determinant, the so-called *Jacobian* J,

$$J(y_1,y_2) = \det\begin{pmatrix} \frac{\partial s_1}{\partial y_1} & \frac{\partial s_1}{\partial y_2} \\ \frac{\partial s_2}{\partial y_1} & \frac{\partial s_2}{\partial y_2} \end{pmatrix} \equiv \frac{\partial s_1}{\partial y_1}\frac{\partial s_2}{\partial y_2} - \frac{\partial s_1}{\partial y_2}\frac{\partial s_2}{\partial y_1} \neq 0, \text{ at all points } (y_1,y_2).$$

Under these two conditions, the random variables Y_1 and Y_2 have a joint p.d.f. given by

$$f_{Y_1,Y_2}(y_1,y_2) = f_{X_1,X_2}(s_1(y_1,y_2), s_2(y_1,y_2))\,|J(y_1,y_2)|.$$

■ **Example 5.26** A random point (X,Y) is uniformly distributed on the unit disk. Then the joint p.d.f. of X and Y is $f_{X,Y}(x,y) = \begin{cases} \frac{1}{\pi}, & \text{for } \{(x,y) : x^2+y^2 \le 1\}, \\ 0, & \text{otherwise.} \end{cases}$

- Let (R,Θ) denote the polar coordinates of the point (X,Y) with $X = R\cos\Theta$ and $Y = R\sin\Theta$, which map $\{(x,y) : x^2+y^2 \le 1\}$ one-to-one onto $\{(r,\theta) : 0 \le r \le 1,\ 0 < \theta \le 2\pi\}$. Then,

$$J(r,\theta) = \det\begin{pmatrix} \frac{\partial x}{\partial r} & \frac{\partial x}{\partial \theta} \\ \frac{\partial y}{\partial r} & \frac{\partial y}{\partial \theta} \end{pmatrix} \equiv \frac{\partial x}{\partial r}\frac{\partial y}{\partial \theta} - \frac{\partial x}{\partial \theta}\frac{\partial y}{\partial r} = r\cos^2\theta + r\sin^2\theta = r.$$

- Thus, the joint p.d.f. of R and Θ is given by

$$f_{R,\Theta}(r,\theta) = f_{X,Y}(r\cos\theta, r\sin\theta)|J(r,\theta)| = \frac{1}{\pi}r,\ 0 \le r \le 1,\ 0 < \theta \le 2\pi.$$

- Note that R and θ are not uniformly distributed but independent.

■

Example 5.27 Suppose that X and Y have a joint p.d.f

$$f_{X,Y}(x,y) = \begin{cases} 4xy, & \text{for } 0 \le x \le 1, 0 < y \le 1, \\ 0, & \text{otherwise.} \end{cases}$$

Let $U = X/Y$ and $V = XY$. Find the joint p.d.f. of U and V.

- Since $U = X/Y$ and $V = XY$, the transformation is one-to-one, and its inverse transform is given $X = \sqrt{UV}$ and $Y = \sqrt{V/U}$.
- Thus, the Jacobian is computed as

$$J = \det \begin{pmatrix} \frac{\partial x}{\partial u} & \frac{\partial x}{\partial v} \\ \frac{\partial y}{\partial u} & \frac{\partial y}{\partial v} \end{pmatrix} = \det \begin{pmatrix} \frac{1}{2}\sqrt{v/u} & \frac{1}{2}\sqrt{u/v} \\ -\frac{1}{2}\sqrt{v/u^3} & \frac{1}{2}\sqrt{1/(uv)} \end{pmatrix} = \frac{1}{2u}.$$

- Hence, for $(u,v) \in \{(u,v) \ : \ 0 \le uv \le 1, \ 0 \le v/u \le 1\}$, the joint p.d.f. of U and V is obtained as $f_{U,V}(u,v) = 4\sqrt{uv}\sqrt{\frac{v}{u}}|J| = \frac{2v}{u}$, and $f_{U,V}(u,v) = 0$ otherwise.
- Note that U and V are not independent, whereas X and Y are independent.

■

Theorem 5.3.2 — Linear Transformation. Let $\mathbf{X} = (X_1, X_2)^T$ be a two-dimensional random vector with a joint p.d.f. $f_{X_1,X_2}(x_1,x_2)$. Define a new two-dimensional random vector $\mathbf{Y} = (Y_1, Y_2)^T$ by

$$\mathbf{Y} = \begin{pmatrix} Y_1 \\ Y_2 \end{pmatrix} = \mathbf{AX} = \begin{pmatrix} a & b \\ c & d \end{pmatrix} \begin{pmatrix} X_1 \\ X_2 \end{pmatrix},$$

where $\mathbf{A}$ is a nonsingular 2×2 matrix. Then $\mathbf{Y}$ has a joint p.d.f.

$$f_{Y_1,Y_2}(y_1,y_2) = \frac{1}{|\det \mathbf{A}|} f_{X_1,X_2}\left(\mathbf{A}^{-1}\mathbf{y}\right),$$

where $\mathbf{A}^{-1}$ is the inverse of $\mathbf{A}$.

Proof. Each Y_i is a linear combination of X_1 and X_2, $Y_1 = aX_1 + bX_2$ and $Y_2 = cX_1 + dX_2$. Because $\mathbf{A}$ is nonsingular(invertible), this transformation is a one-to-one transformation. Thus, the inverse transformation can be represented by the equation $\mathbf{x} = \mathbf{A}^{-1}\mathbf{y}$. The Jacobian J of the transformation $J = \det\mathbf{A}^{-1}$. Hence, the joint p.d.f. of Y_1 and Y_2 is obtained as $f_{Y_1,Y_2}(y_1,y_2) = f_{X_1,X_2}\left(\mathbf{A}^{-1}\mathbf{y}\right)\left|\det\mathbf{A}^{-1}\right|$. Also, it is known from the theory of determinants that $\det\mathbf{A}^{-1} = \frac{1}{\det\mathbf{A}}$, which completes the proof. ■

Example 5.28 Let X_1 and X_2 be jointly continuous random variables with p.d.f. $f_{X_1,X_2}(x_1,x_2)$. Let $Y_1 = X_1 + X_2$, $Y_2 = X_1 - X_2$. Find the joint p.d.f. of Y_1 and Y_2.

- The inverse transformation is $x_1 = \dfrac{y_1+y_2}{2}$ and $x_2 = \dfrac{y_1-y_2}{2}$. Note that $\mathbf{y} = \mathbf{A}\mathbf{x}$, $\mathbf{A} = \begin{pmatrix} 1 & 1 \\ 1 & -1 \end{pmatrix}$ and that $\det(\mathbf{A}) = -2$. Thus $|J(y_1,y_2)| = 1/|\det(\mathbf{A})| = \dfrac{1}{2}$. Hence, we have $f_{Y_1,Y_2}(y_1,y_2) = f_{X_1,X_2}\left(\dfrac{y_1+y_2}{2}, \dfrac{y_1-y_2}{2}\right)\dfrac{1}{2}$.

- Let X_1 and X_2 be independent standard normal random variables. Then,

$$\begin{aligned} f_{Y_1,Y_2}(y_1,y_2) &= \frac{1}{2}\frac{1}{\sqrt{2\pi}}e^{-\frac{1}{2}\left(\frac{y_1+y_2}{2}\right)^2}\frac{1}{\sqrt{2\pi}}e^{-\frac{1}{2}\left(\frac{y_1-y_2}{2}\right)^2} \\ &= \frac{1}{\sqrt{2}\sqrt{2\pi}}e^{-\frac{1}{2}y_1^2/2} \times \frac{1}{\sqrt{2}\sqrt{2\pi}}e^{-\frac{1}{2}y_2^2/2}, \quad -\infty < y_1, y_2 < \infty, \end{aligned}$$

  which implies that the marginal p.d.f.'s $f_{Y_1}(y_1)$ and $f_{Y_2}(y_2)$ of Y_1 and Y_2 are p.d.f's of $N(0,2)$ and that Y_1 and Y_2 are independent. That is, $X_1+X_2 \sim N(0,2)$ and $X_1 - X_2 \sim N(0,2)$ are independent.

- Let X_1 and X_2 be independent uniform $(0,1)$ random variables. Then,

$$f_{Y_1,Y_2}(y_1,y_2) = \begin{cases} \frac{1}{2}, & 0 \le y_1+y_2 \le 2,\ 0 \le y_1-y_2 \le 2 \\ 0, & \text{otherwise} \end{cases}$$

■

Remark Let $\mathbf{u} = (a,b) = a\mathbf{e}_1 + b\mathbf{e}_2$ and $\mathbf{v} = (c,d)$ be two-dimensional vectors in the plane, where $\mathbf{e}_1 = (1,0)$ and $\mathbf{e}_2 = (0,1)$ are unit vectors along the x-axis and y-axis. Then, the sum of the two vectors, $\mathbf{u}+\mathbf{v} = (a+c, b+d)$ is constructed as shown in Figure 5.13.

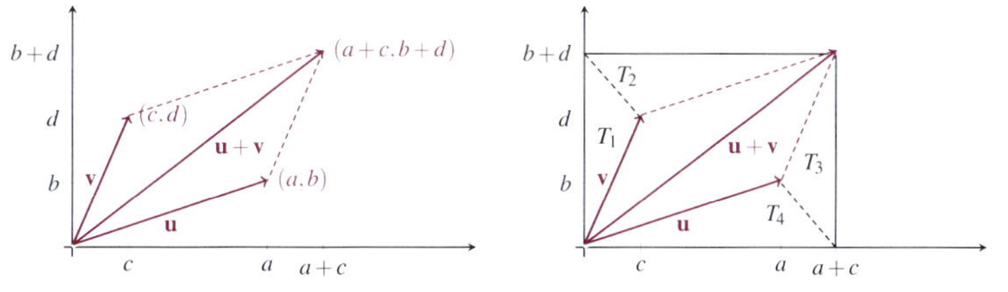

Figure 5.13: The sum of two vectors and the parallelogram spanned by them

- Let $\mathbf{A} = \begin{pmatrix} a & b \\ c & d \end{pmatrix}$. Then, $\det(\mathbf{A}) = \det\begin{pmatrix} a & b \\ c & d \end{pmatrix} = ad - bc$. Indeed, $\det(\mathbf{A})$ is the area of a parallelogram with vertices $(0,0)$, (a,b), (c,d), $(a+c,b+d)$ as shown in Figure 5.13. Specifically, to get the area of the parallelogram, which is actually the same as $\det(\mathbf{A})$, we subtract the area of four triangles T_1–T_4 from the area of the rectangle (see, the right panel of Figure 5.13.)

$$\begin{aligned} &\text{area of parallelogram} \\ &= \text{area of rectangle} - \text{area of } T_1 - \text{area of } T_2 - \text{area of } T_3 - \text{area of } T_4 \\ &= (a+c)(b+d) - c(b+d)/2 - b(a+c)/2 - a(b+d)/2 - b(a+c)/2 \\ &= ad - bc = \det(\mathbf{A}). \end{aligned}$$

■ **Example 5.29 — Convolution (revisited).** Suppose that two random variables X_1 and X_2 have a given joint p.d.f. $f_{X_1,X_2}(x_1,x_2)$ and that it is desired to find the p.d.f. of the random variable $Y = X_1 + X_2$.

- Let $Y = X_1 + X_2$ and $Z = X_2$. Then the inverse transformation is specified as

$$X_1 = Y - Z,\ X_2 = Z,\ \mathbf{A} = \begin{pmatrix} 1 & 1 \\ 0 & 1 \end{pmatrix},\ J(y_1,y_2) = \frac{1}{\det(\mathbf{A})} = 1.$$

- Then the joint p.d.f. of Y and Z at every given point (y,z) will be

$$f_{Y,Z}(y,z) = f_{X_1,X_2}(y-z,z)$$

- Therefore, the marginal p.d.f. of f_Y of Y can be obtained as

$$f_Y(y) = \int_{-\infty}^{\infty} f_{X_1,X_2}(y-z,z)dz, \quad \text{for } -\infty < y < \infty.$$

- If X_1 and X_2 are independent, then $f_Y(y) = \int_{-\infty}^{\infty} f_{X_1}(y-z) f_{X_2}(z)dz$.

■

■ **Example 5.30 — Sum of the Cauchy random variables.** Suppose X_1 and X_2 are independent and identically distributed as the Cauchy distribution, $C(0,1)$. Find the distribution of $Y = X_1 + X_2$.

- The p.d.f. of X_1 is given by $f_{X_1}(x_1) = \dfrac{1}{\pi(1+x_1^2)}$.
- Since X_1 and X_2 are independent, the p.d.f. Y can be obtained by the convolution.

$$\begin{aligned} f_Y(y) &= \int_{-\infty}^{\infty} f_{X_1}(y-z) f_{X_2}(z)\,dz = \int_{-\infty}^{\infty} \frac{1}{\pi(1+(y-z)^2)} \frac{1}{\pi(1+z^2)}\,dz \\ &= \frac{1}{\pi^2}\int_{-\infty}^{\infty} \frac{1}{1+(y-z)^2}\frac{1}{1+z^2}\,dz. \end{aligned}$$

- To integrate the above integral, we decompose the integrand using partial fraction decomposition, $\dfrac{1}{1+(y-z)^2}\dfrac{1}{1+z^2} = \dfrac{2az+b}{1+z^2} + \dfrac{2c(y-z)+d}{1+(y-z)^2}$, for which we have $a = c = \dfrac{1}{y(4+y^2)}$ and $b = d = \dfrac{1}{4+y^2}$. Thus, it follows that

$$\begin{aligned} &\frac{1}{\pi^2}\int_{-\infty}^{\infty} \frac{1}{1+(y-z)^2}\frac{1}{1+z^2}dz \\ &= \frac{1}{\pi^2 y^2(4+y^2)}\left[y\ln\left(\frac{1+z^2}{1+(y-z)^2}\right) + y^2\tan^{-1}z - y^2\tan^{-1}(y-z)\right]_{-\infty}^{\infty} \\ &= \frac{1}{\pi^2 y^2(4+y^2)}\left[0 + y^2\pi + y^2\pi\right] = \frac{2}{\pi(4+y^2)}. \end{aligned}$$

Hence the sum of two independent Cauchy random variables $Y = X_1 + X_2$ is not a Cauchy random variable.

- However, if $X_1 \sim C(0,1)$ and $X_2 \sim C(0,1)$, then it can be easily shown that the random variable $Z = \dfrac{X_1+X_2}{2}$ is again Cauchy, that is $Z \sim C(0,1)$. Hence, by induction, one can show that if $X_1,\dots,X_n$ are i.i.d. with the Cauchy distribution, then $\frac{1}{n}(X_1+\ldots+X_n)$ also has the Cauchy distribution.

∎

Example 5.31 Let X and Y be independent $N(0,1)$ random variables. Then the distribution of X/Y is the Cauchy distribution, Cauchy$(0,1)$.

- Let $U = \dfrac{X}{Y}$ and $V = Y$. First, we compute the joint p.d.f. of U and V. Since $U = \dfrac{X}{Y}$ and $V = Y$, we get by solving for X and Y, $X = UV$ and $Y = V$.

- Then, the Jacobian of the transformation is given by

$$J = \frac{\partial x}{\partial u}\frac{\partial y}{\partial v} - \frac{\partial x}{\partial v}\frac{\partial y}{\partial u} = v\cdot(1) - u\cdot(0) = v.$$

- Thus, the joint p.d.f. of U and V is obtained as

$$f_{U,V}(u,v) = |J|\, f_{X,Y}(uv,v) = |v|\,\frac{1}{2\pi}e^{-\frac{1}{2}\left[u^2v^2+v^2\right]} = |v|\,\frac{1}{2\pi}e^{-\frac{1}{2}v^2\left(u^2+1\right)}$$

for $-\infty < u < \infty$ and $-\infty < v < \infty$.

- Now, we derive the marginal p.d.f. of U by integrating the joint p.d.f. of U and V with respect to v,

$$\begin{aligned} f_U(u) &= \int_{-\infty}^{\infty} f_{U,V}(u,v)\,dv = \int_{-\infty}^{\infty} |v|\,\frac{1}{2\pi}e^{-\frac{1}{2}v^2\left(u^2+1\right)}\,dv \\ &= 2\int_0^{\infty} v\frac{1}{2\pi}e^{-\frac{1}{2}v^2\left(u^2+1\right)}\,dv = 2\frac{1}{2\pi}\left(\frac{1}{2}\right)\left[\frac{-2}{u^2+1}e^{-\frac{1}{2}v^2\left(u^2+1\right)}\right]_0^{\infty} \\ &= \frac{1}{\pi\left(u^2+1\right)}, \quad -\infty < u < \infty, \end{aligned}$$

which is a p.d.f. of Cauchy distribution. Hence, $\dfrac{X}{Y} \sim$ Cauchy$(0,1)$.

- Alternatively, we can define $W = \dfrac{X}{\sqrt{Y^2}}$, which is shown to have a Cauchy distribution, $W \sim$ Cauchy$(0,1)$. Indeed, W has a t-distribution with one degree of freedom, $W \sim t(1)$, which will be discussed in Example 5.44. Here, we compute the c.d.f. of W and obtain the p.d.f. of W by differentiation. With the polar coordinate changes with $X = R\sin\Theta$ and $Y = R\cos\Theta$, we have $W = \tan\Theta$, where $\Theta \sim$ Uniform$(-\pi/2,\pi/2)$. Thus, the c.d.f of W is given by

$$P(W \le w) = \int_0^{\infty}\int_{-\pi/2}^{\tan^{-1}(w)} \frac{r}{\pi}\exp\left(-\frac{1}{2}r^2\right)d\theta dr = \frac{1}{2} + \frac{1}{\pi}\tan^{-1}(w),$$

which is actually the c.d.f. of the Cauchy distribution with parameters 0 and 1.

Hence, $f_W(w) = \dfrac{1}{\pi(1+w^2)}$, for $-\infty < w < \infty$.

■

■ **Example 5.32** Let X and Y be independent and identically distributed as exponential distribution with parameter 1, with $f_X(x) = \exp(-x)$, $0 < x < \infty$.

- Let $U = \dfrac{X}{Y}$ and $V = Y$. First, we compute the joint p.d.f. of U and V. Since $U = \dfrac{X}{Y}$ and $V = Y$, we get by solving for X and Y, $X = UV$ and $Y = V$. Then, the Jacobian of the transformation is given by $J(u,v) = \dfrac{\partial x}{\partial u}\dfrac{\partial y}{\partial v} - \dfrac{\partial x}{\partial v}\dfrac{\partial y}{\partial u} = v$.
- Thus, the joint p.d.f. of U and V is obtained as

$$f_{U,V}(u,v) \quad = \quad f_{X,Y}(uv,v)\,|J(u,v)| = ve^{-(u+1)v}, \quad u>0,\ v>0.$$

- Now, we derive the marginal p.d.f. of U by integrating the joint p.d.f. of U and V with respect to v,

$$f_U(u) \quad = \quad \int_0^\infty f_{U,V}(u,v)\,dv = \int_0^\infty ve^{-(u+1)v}dv = \frac{1}{(u+1)^2}, \quad u>0.$$

■

■ **Example 5.33 — Difference of two exponentials.** Let X_1 and X_2 be independent and identically distributed as exponential distribution with parameter 1, with $f_{X_1}(x_1) = \exp(-x_1)$, $0 < x_1 < \infty$. Define $Y = X_1 - X_2$.

- Note that the joint p.d.f. of X_1 and X_2 is $f_{X_1,X_2}(x_1,x_2) = e^{-(x_1+x_2)}$.
- Let $Z = X_2$. Then the inverse transforms are $x_1 = y+z$ and $x_2 = z$. Thus, the Jacobian of the transformation is given by $J = \begin{vmatrix} 1 & 1 \\ 0 & 1 \end{vmatrix} = 1$. Hence, the joint p.d.f. of Y and Z is obtained as $f_{Y,Z}(y,z) = e^{-(y+2z)}$.
- Note that $-\infty < y < \infty$, $0 < z < \infty$ and $y+z>0$ since $x_1 > 0$ and $x_2 > 0$, which implies that $-y < z < \infty$ if $y < 0$, and that $0 < z < \infty$ if $y > 0$ as shown in the shaded area of Figure 5.14.

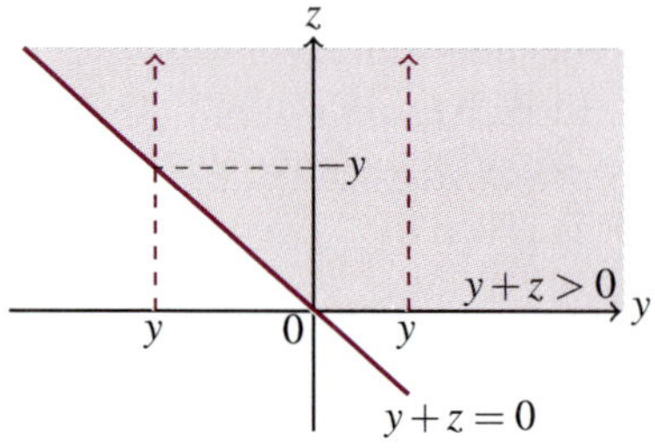

Figure 5.14: The shaded area of $\{(y,z) : 0 < z < \infty,\ -\infty < y < \infty,\ y+z>0\}$

Thus, we have

$$f_Y(y) = \int f_{Y,Z}(y,z)dz = \begin{cases} \int_{-y}^{\infty} e^{-(y+2z)}dz = \frac{1}{2}e^{y}, & \text{for } -\infty < y < 0, \\ \int_0^{\infty} e^{-(y+2z)}dz = \frac{1}{2}e^{-y}, & \text{for } 0 \le y < \infty. \end{cases}$$

Hence, $f_Y(y) = \dfrac{1}{2}e^{-|y|}$, for $-\infty < y < \infty$, which is p.d.f. of the Laplace (double exponential) distribution, DE$(0,1)$.

∎

Example 5.34 — Range of two uniforms. Let U_1 and U_2 be independent random variables, each one distributed uniformly on $[0,1]$. Let Z be the minimum, $Z = \min\{U_1, U_2\}$ and W be the maximum, $W = \max\{U_1, U_2\}$. The random variable $Y = W - Z$ is called the range of two uniforms U_1 and U_2.

- Notice that transform of (U_1, U_2) to (Z, W) is not 1-1, for which we may not use the change of variable formula. Thus, we start with the joint c.d.f. of Z and W with the support of $\{(z,w) : 0 \le z \le w \le 1\}$ and then find the joint p.d.f of Z and W by differentiation.

- Since the event $\{Z \le z\} \cap \{W \le w\}$ is the same as $\{W \le w\} \cap \{Z > z\}^c$, the joint c.d.f. of (Z,W), $F_{Z,W}(z,w) = P(Z \le z,\ W \le w)$ is computed as

$$\begin{aligned} F_{Z,W}(z,w) &= P(W \le w) - P(Z > z,\ W \le w) \\ &= P(\max(U_1,U_2) \le w) - P(\min(U_1,U_2) > z,\ \max(U_1,U_2) \le w) \\ &= P(U_1 \le w,\ U_2 \le w) - P(z < U_1 < w,\ z < U_2 < w) \\ &= P(U_1 \le w)P(U_2 \le w) - P(z < U_1 < w)P(z < U_2 < w) \\ &= w^2 - (w-z)^2, \text{ for } 0 \le z \le w \le 1. \end{aligned}$$

Therefore $f_{Z,W}(z,w) = \dfrac{\partial^2}{\partial z \partial w} F_{Z,W}(z,w) = 2$ for $0 \le z \le w \le 1$.

- Let $Y = W - Z$ and $V = Z$. Then, the inverse transform is $z = v$ and $w = y + v$ with $J = 1$ and $0 \le v \le y + v \le 1$. Hence, the marginal p.d.f. is

$$f_Y(y) = \int_0^{1-y} 2dv = 2(1-y), \quad \text{for } 0 < y < 1.$$

∎

Example 5.35 Let X_1 and X_2 be two independent random variables that have gamma distributions with parameters α, 1, β and 1, respectively. Then, $X_1 + X_2$ and $\dfrac{X_1}{X_1+X_2}$ are independent. In addition, $\dfrac{X_1}{X_1+X_2}$ has a beta distribution with parameters α and β.

- Let $Y_1 = X_1 + X_2$ and $Y_2 = X_1/(X_1+X_2)$. First, we compute the joint p.d.f. of Y_1 and Y_2. Since $Y_1 = X_1 + X_2$ and $Y_2 = X_1/(X_1+X_2)$, we get by solving for X_1 and X_2

$X_1 = Y_1Y_2$ and $X_2 = Y_1(1-Y_2)$. Thus, the Jacobian of the transformation is given by

$$J(y_1,y_2) = \det\begin{pmatrix} y_2 & y_1 \\ 1-y_2 & -y_1 \end{pmatrix} = \frac{\partial x_1}{\partial y_1}\frac{\partial x_2}{\partial y_2} - \frac{\partial x_1}{\partial y_2}\frac{\partial x_2}{\partial y_1} = -y_1.$$

- Note that $y_1 > 0$ and $0 < y_2 < 1$. Hence, the joint density of Y_1 and Y_2 is

$$\begin{aligned} f_{Y_1,Y_2}(y_1,y_2) &= f_{X_1,X_2}(y_1y_2, y_1(1-y_2))\,|J(y_1,y_2)| \\ &= \frac{\Gamma(\alpha+\beta)}{\Gamma(\alpha)\Gamma(\beta)} y_2^{\alpha-1}(1-y_2)^{\beta-1}\frac{y_1^{\alpha+\beta-1}}{\Gamma(\alpha+\beta)}e^{-y_1} = f_{Y_2}(y_2)f_{Y_1}(y_1), \end{aligned}$$

  which shows that Y_1 and Y_2 are independent. In addition, we see that Y_1 has a gamma distribution with parameters $\alpha+\beta$ and 1, $Y_1 \sim \text{Gamma}(\alpha+\beta,1)$, which was also shown in Example 5.22 using the convolution, and that Y_2 has a beta distribution with parameters α and β, $Y_2 \sim \text{Beta}(\alpha,\beta)$.

- Alternatively, we can find the c.d.f. of Y_2 and then get its p.d.f. by differentiation. For simplicity, letting $\alpha = \beta = 1$, for $0 < y_2 < 1$,

$$\begin{aligned} F_{Y_2}(y_2) &= P(Y_2 \le y_2) = P\left(X_1 \le \frac{y_2}{1-y_2}X_2\right) \\ &= \int_0^\infty \int_0^{\frac{x_2y_2}{1-y_2}} f_{X_1}(x_1)f_{X_2}(x_2)dx_1dx_2 = \int_0^\infty \int_0^{\frac{x_2y_2}{1-y_2}} e^{-x_1}e^{-x_2}dx_1dx_2 \\ &= \int_0^\infty e^{-x_2}\left(1-e^{-x_2y_2/(1-y_2)}\right)dx_2 = y_2, \\ f_{Y_2}(y_2) &= \frac{d}{dy_2}F_{Y_2}(y_2) = 1, \text{ for } 0 < y_2 < 1, \end{aligned}$$

  which implies that Y_2 has a uniform distribution between 0 and 1, $Y_2 \sim \text{Uniform}(0,1)$, in other words, a beta distribution with parameters 1 and 1, $Y_2 \sim \text{Beta}(1,1)$.

- For two independent gamma random variables, $W_1 \sim \text{Gamma}(\alpha,\lambda)$ and $W_2 \sim \text{Gamma}(\beta,\lambda)$, we consider $X_1 = \lambda W_1 \sim \text{Gamma}(\alpha,1)$ and $X_2 = \lambda W_2 \sim \text{Gamma}(\beta,1)$, which are independent. Then, $\dfrac{W_1}{W_1+W_2}$ has the same distribution with $\frac{X_1}{X_1+X_2}$ as a Beta distribution with parameters α and β.

- In addition, since $Y_1 = X_1+X_2$ and $Y_2 = \dfrac{X_1}{X_1+X_2} = \dfrac{X_1/X_2}{X_1/X_2+1}$ are independent, $\dfrac{X_1}{X_2}$ and X_1+X_2 are also independent.

■

The use of Jacobian can be understood as the change in infinitesimal volumes in moving from one coordinate system to another system to compute the integral of a two-dimensional function after a change of variables, as presented in Theorem 5.3.3.

Theorem 5.3.3 — Change of variable formula for double integrals. Let $x = s_1(u,v)$ and $y = s_2(u,v)$ define a one-to-one mapping of a region A' in the uv-plane onto a region A in the xy-plane. Assume that if $(x,y) = (s_1(u,v), s_2(u,v))$, then, the partial derivatives $\frac{\partial x}{\partial u}, \frac{\partial x}{\partial v}, \frac{\partial y}{\partial u}, \frac{\partial y}{\partial v}$ exist and are continuous. Then

$$\iint_A f_{X,Y}(x,y)\,dxdy = \iint_{A'} f_{X,Y}(s_1(u,v), s_2(u,v))\,J(u,v)\,dudv,$$

where $J(u,v)$ is the Jacobian of x and y with respect to u and v,

$$J(u,v) = \frac{\partial x \partial y}{\partial u \partial v} = \det \begin{pmatrix} \frac{\partial x}{\partial u} & \frac{\partial x}{\partial v} \\ \frac{\partial y}{\partial u} & \frac{\partial y}{\partial v} \end{pmatrix}.$$

Remark The proof of Theorem 5.3.3 is beyond the scope of the text, but we give a heuristic explanation. Consider the subrectangles that are enclosed completely within the region A. For such a subrectangle $[x_i, x_{i+1}] \times [y_j, y_{j+1}]$, we construct a cuboid with the height $f_{X,Y}(x_i, y_j)$ as shown in Figure 5.16.

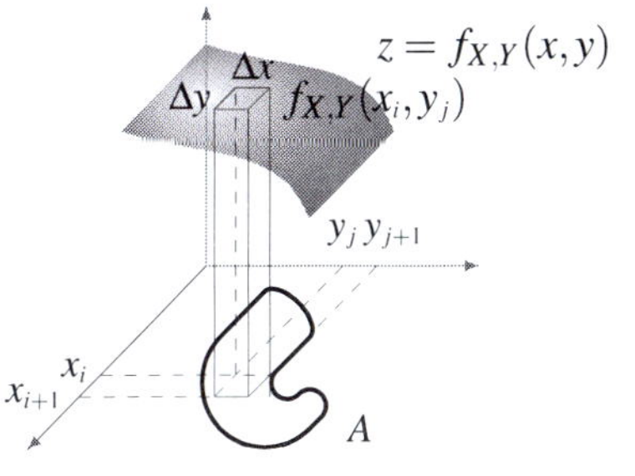

Figure 5.15: A cuboid over a subrectangle

- The volume under the surface $z = f_{X,Y}(x,y)$ over that subrectangle is approximately $f_{X,Y}(x_i, y_j)\Delta x \Delta y$, where $\Delta x = x_{i+1} - x_i$, $\Delta y = y_{j+1} - y_j$, and $f_{X,Y}(x_i, y_j)$ is the height and $\Delta x \Delta y$. Thus, the total volume under the surface is approximately the sum of the volumes of all such rectangular solids as the limit of the double summation, $\iint_A f(x,y)\,dxdy \approx \sum_j \sum_i f_{X,Y}(x_i, y_j)\Delta x \Delta y$.
- Imagine a parallelogram with four vertices (u,v), $(u+\Delta u, v)$, $(u+\Delta u, v+\Delta v)$ and $(u, v+\Delta v)$. Then, using the map $s = (s_1, s_2)$, we have a parallelogram corresponding to four vertices $(s_1(u,v), s_2(u,v))$, $(s_1(u+\Delta u, v), s_2(u+\Delta u, v))$, $(s_1(u+\Delta u, v+\Delta v), s_2(u+\Delta u, v+\Delta v))$ and $(s_1(u, v+\Delta v), s_2(u, v+\Delta v))$. Now, using the first-order Taylor series expansion in each coordinate, we have

$$\begin{aligned} s_1(u+\Delta u, v) &\approx s_1(u,v) + \frac{\partial x}{\partial u}\Delta u, \quad s_2(u+\Delta u, v) \approx s_2(u,v) + \frac{\partial y}{\partial u}\Delta u, \\ s_1(u, v+\Delta v) &\approx s_1(u,v) + \frac{\partial x}{\partial v}\Delta v, \quad s_2(u, v+\Delta v) \approx s_2(u,v) + \frac{\partial y}{\partial v}\Delta v. \end{aligned}$$

That is, two vectors **x** and **y** consisting of three vertices, one from $(s_1(u,v),s_2(u,v))$ to $(s_1(u+\Delta u,v),s_2(u+\Delta u,v))$ and the other from $(s_1(u,v),s_2(u,v))$ to $(s_1(u,v+\Delta v),s_2(u,v+\Delta v))$ are approximated as follows.

$$\begin{aligned}\mathbf{x}=(s_1(u+\Delta u,v),s_2(u+\Delta u,v))-(s_1(u,v),s_2(u,v)) &\approx \Delta u\left(\frac{\partial x}{\partial u},\frac{\partial y}{\partial u}\right),\\ \mathbf{y}=(s_1(u,v+\Delta v),s_2(u,v+\Delta v))-(s_1(u,v),s_2(u,v)) &\approx \Delta v\left(\frac{\partial x}{\partial v},\frac{\partial y}{\partial v}\right).\end{aligned}$$

- Thus, the area of the subrectangle $[x_i,x_{i+1}]\times[y_j,y_{j+1}]$ in Figure 5.15, that is, the area of parallelogram based on two vectors $\overrightarrow{x}$ and $\overrightarrow{y}$, or four vertices (x_i,y_j), $(x_i+\Delta x,y_j)$, $(x_i+\Delta x,y_j+\Delta y)$ and $(x_i,y_j+\Delta y)$ is approximated by

$$\Delta x\Delta y\approx\Delta u\Delta v\left|\det\begin{pmatrix}\frac{\partial x}{\partial u} & \frac{\partial y}{\partial u}\\ \frac{\partial x}{\partial v} & \frac{\partial y}{\partial v}\end{pmatrix}\right|=\Delta u\Delta v\left|J(u,v)\right|.$$

- Therefore, the double integral $\iint_A f(x,y)\,dxdy$ is written as

$$\begin{aligned}&\iint_A f_{X,Y}(x,y)\,dxdy\approx\sum_j\sum_i f_{X,Y}(x_i,y_j)\,\Delta x\Delta y\\ &\approx\sum_j\sum_i f_{X,Y}(s_1(u_i,v_j),s_2(u_i,v_j))\Delta u\Delta v\left|J(u,v)\right|\\ &\approx\iint_{A'} f_{X,Y}(s_1(u,v),s_2(u,v))\,J(u,v)dudv.\end{aligned}$$

Hence, the change of variables formula in Theorem 5.3.3 tells us how to use the joint p.d.f. of X and Y to get the joint p.d.f. of the transformed random variables U and V with $X=s_1(U,V)$ and $Y=s_2(U,V)$.

- Let A be a region in the xy-plane and $A'=\{(u,v):(s_1(u,v),s_2(u,v))\in A\}$ be the corresponding region in the uv-plane. Then, $(X,Y)\in A$ is the same event as $(U,V)\in A'$. Thus,

$$\begin{aligned}P((U,V)\in A')=\iint_{A'} f_{U,V}(u,v)dudv=P((X,Y)\in A)=\iint_A f_{X,Y}(x,y)dxdy\\ =\iint_{A'} f_{X,Y}(s_1(u,v),s_2(u,v))\,J(u,v)dudv,\end{aligned}$$

which implies that $f_{U,V}(u,v)=f_{X,Y}(s_1(u,v),s_2(u,v))|J(u,v)|$.

- Alternatively, the nonlinear transformation $X=s_1(U,V)$ and $Y=s_2(U,V)$ can also be understood as an approximate linear transformation $\begin{pmatrix}X\\Y\end{pmatrix}\approx\mathbf{A}\begin{pmatrix}U\\V\end{pmatrix}$, where $\mathbf{A}=\begin{pmatrix}\frac{\partial x}{\partial u} & \frac{\partial y}{\partial u}\\ \frac{\partial x}{\partial v} & \frac{\partial y}{\partial v}\end{pmatrix}$. Thus, we have

$$f_{U,V}(u,v)=f_{X,Y}(s_1(u,v),s_2(u,v))\,|\det(\mathbf{A})|=f_{X,Y}(s_1(u,v),s_2(u,v))|J(u,v)|.$$

- Note that for the one-dimensional case, we have $f_Y(y)=f_X(r^{-1}(y))\left|\frac{d}{dy}r^{-1}(y)\right|$

if $Y = r(X)$ and $r(x)$ is a one-to-one function and that $\left|\frac{d}{dy}r^{-1}(y)\right|$ becomes the absolute value of the Jacobian in the two-dimensional case.

Theorem 5.3.4 — Box-Müller transformation. The Box-Müller transformation generates two independent standard normal random variables based on two independent uniform random variables.

① Let $U \sim \text{Uniform}(0, 2\pi)$ and let $T \sim \text{Exp}(1)$ be independent of U. Define

$$Z_1 = \sqrt{2T}\cos U \quad \text{and} \quad Z_2 = \sqrt{2T}\sin U.$$

Then, Z_1 and Z_2 are independent and have the standard normal distributions.

② Let U_1 and U_2 be two independent uniform random variables on $(0,1)$, $U_1, U_2 \sim \text{Uniform}(0,1)$. Define

$$Z_1 \equiv \sqrt{-2\log U_2}\cos(2\pi U_1) \quad \text{and} \quad Z_2 = \sqrt{-2\log U_2}\sin(2\pi U_1).$$

Then, Z_1 and Z_2 are independent and have the standard normal distribution.

Proof. To prove Box-Müller transformation, we can use either ① the change of two variables or ② its construction starting with two independent standard normal random variables.

① First, note that the joint p.d.f. of U and T is $f_{U,T}(u,t) = \frac{1}{2\pi}e^{-t}$. Since $Z_1 = \sqrt{2T}\cos U$ and $Z_2 = \sqrt{2T}\sin U$, we get $T = (Z_1^2 + Z_2^2)/2$ and $U = \tan^{-1}(Z_2/Z_1)$.

$$\begin{aligned} J &= \det\begin{pmatrix} \frac{\partial t}{\partial z_1} & \frac{\partial t}{\partial z_2} \\ \frac{\partial u}{\partial z_1} & \frac{\partial u}{\partial z_2} \end{pmatrix} = \frac{\partial t}{\partial z_1}\frac{\partial u}{\partial z_2} - \frac{\partial t}{\partial z_2}\frac{\partial u}{\partial z_1} \\ &= \left[\det\begin{pmatrix} \frac{\partial z_1}{\partial t} & \frac{\partial z_1}{\partial u} \\ \frac{\partial z_2}{\partial t} & \frac{\partial z_2}{\partial t} \end{pmatrix}\right]^{-1} = \left[\det\begin{pmatrix} \frac{1}{\sqrt{2t}}\cos u & -\sqrt{2t}\sin u \\ \frac{1}{\sqrt{2t}}\sin u & \sqrt{2t}\cos u \end{pmatrix}\right]^{-1} = 1. \end{aligned}$$

Thus, for $-\infty < z_1, z_2 < \infty$, the joint p.d.f. of Z_1 and Z_2 is given by

$$f_{Z_1,Z_2}(z_1,z_2) = f_{U,T}\left(\tan^{-1}(z_2/z_1), (z_1^2+z_2^2)/2\right)|J| = \frac{1}{2\pi}e^{-\frac{1}{2}(z_1^2+z_2^2)},$$

which is the product of two standard normal densities.

② Start with two independent standard normal numbers Z_1 and Z_2 and consider (Z_1, Z_2) as a point in the plane (see, Figure 5.16).Then, in polar coordinates, $Z_1 = R\cos\Theta$ and $Z_2 = R\sin\Theta$, where $R^2 \equiv Z_1^2 + Z_2^2$, the squared distance(radius) from the origin $(0,0)$ to the point (Z_1, Z_2), has a $\chi^2(2) \stackrel{d}{=} \text{Exp}(1/2)$ distribution, as discussed in Example

5.20, and $\Theta \in [0, 2\pi)$ is the angle between Z_1 axis and the straight line that connects $(0,0)$ and (Z_1, Z_2). By symmetry, R^2 is independent of Θ, and $\Theta \sim \text{Uniform}[0, 2\pi)$. Hence, we have $\Theta = 2\pi U_1$ and $R^2 = -2\log U_2$ that follows from the probability integral transform.

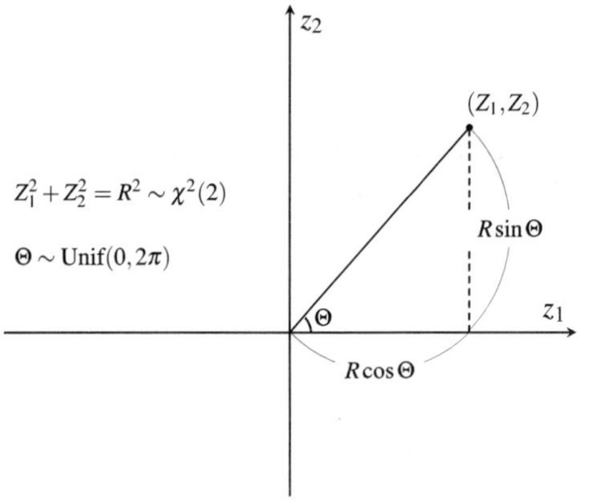

Figure 5.16: Polar coordinate change of (z_1, z_2)

Since $z_1^2 + z_2^2 = -2\log(u_1)$ and $\dfrac{z_2}{z_1} = \tan(2\pi u_2)$, the inverse transformation is

$$u_1 = e^{-(z_1^2+z_2^2)/2} \quad \text{and} \quad u_2 = \frac{1}{2\pi}\tan^{-1}\left(\frac{z_2}{z_1}\right).$$

Using the fact that $d(\tan^{-1}(x))/dx = 1/(1+x^2)$, we find the Jacobian and $f_{Z_1,Z_2}(z_1,z_2)$

$$\begin{aligned} J(z_1,z_2) &= \det\begin{pmatrix} -z_1 e^{-(z_1^2+z_2^2)/2} & -z_2 e^{-(z_1^2+z_2^2)/2} \\ -\frac{z_2}{2\pi z_1^2}\times\frac{1}{1+z_2^2/z_1^2} & \frac{1}{2\pi z_1}\times\frac{1}{1+z_2^2/z_1^2} \end{pmatrix} = -\frac{1}{2\pi}e^{-(z_1^2+z_2^2)/2}, \\ f_{Z_1,Z_2}(z_1,z_2) &= f_{U_1,U_2}(u_1,u_2)|J(z_1,z_2)| = \frac{1}{2\pi}e^{-\frac{z_1^2+z_2^2}{2}}, \end{aligned}$$

which implies that Z_1 and Z_2 are independent and identically distributed standard normals.

■

▪ **Example 5.36** If X and Y have joint density function $f_{X,Y}(x,y)$, show that the density function of $Z = XY$ is $f_Z(z) = \int_{-\infty}^{\infty} f_{X,Y}(u, z/u)|u|^{-1}du$.

- Consider the transformation $u = x$ and $z = xy$, with inverse transformation $x = u$ and $y = z/u$. Then, $J(u,z) = \det\begin{pmatrix} 1 & 0 \\ \frac{-z}{u^2} & \frac{1}{u} \end{pmatrix} = u^{-1}$.

  Thus, the joint p.d.f. of $U = X$ and $Z = XY$ is given by $f_{U,Z}(u,z) = \dfrac{1}{|u|}f_{X,Y}(u, z/u)$, and the marginal p.d.f. Z is obtained from $f_{U,Z}(u,z)$ by integration with respect to u.

- Let X and Y be independent with respective p.d.f.'s $f_X(x) = xe^{-\frac{x^2}{2}}$ for $x > 0$ and $f_Y(y) = \pi^{-1}\left(1-y^2\right)^{-\frac{1}{2}}$ for $|y| < 1$. Then, the marginal p.d.f. of $Z = XY$ is obtained as the standard normal p.d.f.,

$$f_Z(z) = \int_{-\infty}^{\infty} \frac{1}{|u|} f_X(u) f_Y\left(\frac{z}{u}\right) du = \int_{u>z} \frac{1}{|u|} ue^{-\frac{u^2}{2}} \pi^{-1}\left(1-\frac{z^2}{u^2}\right)^{-\frac{1}{2}} du$$
$$= \frac{1}{\pi}\int_z^{\infty} \frac{e^{-\frac{u^2}{2}}}{(u^2-z^2)^{\frac{1}{2}}} udu = \frac{1}{\pi} e^{-\frac{z^2}{2}} \int_0^{\infty} e^{-\frac{v^2}{2}} dv = \frac{1}{\sqrt{2\pi}} e^{-\frac{z^2}{2}},$$

where the last equality follows from the substitution $u^2 = z^2 + v^2$.

- Note that the radius R from the Box-Müller transformation in Figure 5.16 has a Rayleigh distribution as discussed in Example 5.20 with the p.d.f $re^{-r^2/2}$, which implies that X has the same distribution with R, $X \stackrel{d}{=} R$. In addition, let $W = \cos\Theta$, where Θ has a uniform distribution from $[0, 2\pi)$ from the Box-Müller transformation in Figure 5.16. Then, the p.d.f of W is

$$f_W(w) = \frac{d}{dw} P(W \le w) = \frac{d}{dw} 2\int_0^{\cos^{-1}(y)} \frac{1}{2\pi} d\theta = \frac{1}{\pi}\frac{1}{1-w^2}, \; -1 \le w \le 1,$$

which is the same as the p.d.f. of Y, $\cos\Theta \stackrel{d}{=} Y$. Hence, $Z = XY \stackrel{d}{=} R\cos\Theta$ has a standard normal distribution as shown in Theorem 5.3.4 for the Box-Müller transformation.

■

5.4 Joint moments

Suppose that the continuous random variable $W = g(X,Y)$ is a function of two continuous random variables with p.d.f. $f_W(w)$. Then, by definition, $\mathrm{E}(W) = \int_{-\infty}^{\infty} wf_W(w)dw$ provided that $\mathrm{E}(|W|) < \infty$. However, suppose we know only the joint p.d.f. $f_{X,Y}(x,y)$ of X and Y and that it is not easy to find the p.d.f. of $W = g(X,Y)$. In this case, we can still find the expectation of W, using the law of unconscious statistician for two-dimensional function. Indeed, it follows from the change of variables formula for double integrals in Theorem 5.3.3.

Theorem 5.4.1 If X and Y have joint p.d.f. $f_{X,Y}(x,y)$, then

$$\mathrm{E}(g(X,Y)) = \int_{-\infty}^{\infty}\int_{-\infty}^{\infty} g(x,y) f_{X,Y}(x,y)dxdy$$

provided that $\int_{-\infty}^{\infty}\int_{-\infty}^{\infty} |g(x,y)| f_{X,Y}(x,y)dxdy < \infty$.

Example 5.37 As a special case of Theorem 5.4.1, the linearity of expectation holds. That is, let X and Y be two continuous random variables with a joint p.d.f. $f_{X,Y}(x,y)$ and finite expectations. Then, for any constants a and b,

$$\mathrm{E}(aX+bY)=a\mathrm{E}(X)+b\mathrm{E}(Y).$$

- By Theorem 5.4.1 with $g(X,Y)=aX+bY$, we have

$$\begin{aligned}\mathrm{E}(aX+bY) &= \int_{-\infty}^{\infty}\int_{-\infty}^{\infty}(ax+by)f_{X,Y}dxdy\\ &= \int_{-\infty}^{\infty}ax\left(\int_{-\infty}^{\infty}f_{X,Y}dy\right)dx+\int_{-\infty}^{\infty}by\left(\int_{-\infty}^{\infty}f_{X,Y}dx\right)dy\\ &= a\mathrm{E}(X)+b\mathrm{E}(Y).\end{aligned}$$

■

Example 5.38 — Difference of two uniforms. Suppose that U_1 and U_2 are independent and identically distributed as Uniform$(0,1)$. Find $\mathrm{E}(|U_1-U_2|)$.

- First, we find the p.d.f. of $Y_1=U_1-U_2$ and evaluate $\mathrm{E}(|Y_1|)$. Consider change of variables by $Y_1=U_1-U_2$ and $Y_2=U_2$, then

$$\begin{cases}Y_1=U_1-U_2,\\ Y_2=U_2,\end{cases}\quad \begin{cases}U_1=Y_1+Y_2,\\ U_2=Y_2,\end{cases}\quad J(y_1,y_2)=\frac{\partial u_1\partial u_2}{\partial y_1\partial y_2}=\det\begin{pmatrix}1&1\\0&1\end{pmatrix}=1,$$

where $0<y_1+y_2<1, 0<y_2<1$. Thus, the joint p.d.f. of Y_1 and Y_2 is given by

$$f_{Y_1,Y_2}(y_1,y_2)=f_{U_1,U_2}(u_1,u_2)\cdot|J(y_1,y_2)|=1,\ \text{for } 0<y_1+y_2<1 \text{ and } 0<y_2<1.$$

To find the marginal p.d.f. of Y_1, we consider two integrals of $f_{Y_1,Y_2}(y_1,y_2)$ with respect to y_2, depending on the range of y_1,

$$f_{Y_1}(y_1)=\begin{cases}\int_{-y_1}^{1}1dy_2=1+y_1, & \text{for } -1<y_1<0,\\ \int_0^{1-y_1}1dy_2=1-y_1, & \text{for } 0<y_1<1,\end{cases}$$

which has a triangular shape, $f_{Y_1}(y_1)=1-|y_1|$, for $|y_1|<1$. Therefore,

$$\mathrm{E}(|Y_1|)=\int_{-1}^{1}|y_1|f_{Y_1}(y_1)dy=\int_{-1}^{0}-y_1(1+y_1)dy_1+\int_0^1 y_1(1-y_1)dy_1=\frac{1}{3}.$$

- On the other hand, we can directly compute $\mathrm{E}(|U_1-U_2|)$ using the law of unconscious statistician for double integrals in Theorem 5.4.1

$$\mathrm{E}(|U_1-U_2|)=\int_0^1\int_0^1|u_1-u_2|f_{U_1,U_2}(u_1,u_2)du_1du_2$$
$$=\int_0^1\int_{u_2}^1(u_1-u_2)du_1du_2+\int_0^1\int_0^{u_2}(u_2-u_1)du_1du_2=\frac{1}{3}.$$

- Since $\max\{a,b\} = \dfrac{a+b}{2} + \left|\dfrac{a-b}{2}\right|$ and $\min\{a,b\} = \frac{a+b}{2} - \left|\frac{a-b}{2}\right|$, we have $\max(U_1,U_2)+\min(U_1,U_2) = U_1+U_2$ and $\max(U_1,U_2)-\min(U_1,U_2) = |U_2-U_1|$, which is actually obvious by definition. Thus, it is also noted that $|U_2-U_1|$ has the same distribution with the range of two uniforms in Example 5.34.

■

■ **Example 5.39** Suppose that Z_1 and Z_2 are independent standard normal random variables, $Z_1, Z_2 \overset{i.i.d.}{\sim} N(0,1)$. Define $X = \max\{Z_1,Z_2\}$. To compute $\mathrm{E}(X)$,

- We can obtain the p.d.f. of X, $f_X(x) = \dfrac{d}{dx}F_X(x) = \dfrac{d}{dx}P(\max\{Z_1,Z_2\} \le x) = \dfrac{d}{dx}P(Z_1 \le x)P(Z_2 \le x) = \dfrac{d}{dx}\Phi(x)^2 = 2\Phi(x)\phi(x)$. Thus,
$$\begin{aligned}\mathrm{E}(X) &= \int_{-\infty}^{\infty} x f_X(x)dx = \int_{-\infty}^{\infty} 2x\Phi(x)\phi(x)dx \\ &= \frac{1}{\pi}\int_{-\infty}^{\infty} xe^{-\frac{x^2}{2}}\int_{-\infty}^{x} e^{-\frac{z^2}{2}}dzdx = \frac{1}{\pi}\int_{-\infty}^{\infty} e^{-\frac{z^2}{2}}\int_{z}^{\infty} xe^{-\frac{x^2}{2}}dxdz \\ &= \frac{1}{\pi}\int_{-\infty}^{\infty} e^{-\frac{z^2}{2}}e^{-\frac{z^2}{2}}dz = \frac{1}{\sqrt{\pi}}\int_{-\infty}^{\infty}\frac{1}{\sqrt{\pi}}e^{-z^2}dz = \frac{1}{\sqrt{\pi}}.\end{aligned}$$
- Alternatively, we can use the law of unconscious statistician for $\mathrm{E}(\max\{Z_1,Z_2\})$,
$$\begin{aligned}\mathrm{E}(\max\{Z_1,Z_2\}) &= \int_{-\infty}^{\infty}\int_{-\infty}^{\infty}\max\{z_1,z_2\}\phi(z_1)\phi(z_2)dz_1dz_2 \\ &= 2\int_{-\infty}^{\infty}\int_{z_2}^{\infty} z_1\phi(z_1)dz_1\phi(z_2)dz_2 = 2\int_{-\infty}^{\infty}\phi(z_2)^2dz_2 = \frac{1}{\sqrt{\pi}}.\end{aligned}$$
- On the other hand, note that $\max\{Z_1,Z_2\} = \frac{Z_1+Z_2}{2} + \frac{|Z_1-Z_2|}{2}$ as given in Example 5.38, where $Z_1+Z_2 \sim N(0,2)$ and $Z_1-Z_2 \sim N(0,2)$. Thus, $\mathrm{E}(\max\{Z_1,Z_2\}) = \frac{\mathrm{E}(|Y|)}{\sqrt{2}} = \frac{1}{\sqrt{\pi}}$ as shown in Example 4.23, where $Y \sim N(0,1)$.

■

When we have more than one random variable, we would like to consider a summary of the joint distribution and a measure of dependence between the random variables. It would be nice if it were 0 when the random variables were independent. The *covariance* tries to measure how two random variables vary together.

Definition 5.4.1 — Covariance. If two random variables X and Y have finite variances, their *covariance* $\mathrm{Cov}(X,Y)$ is $\mathrm{E}[(X-\mu_X)(Y-\mu_Y)]$, where μ_X and μ_Y are the means.

Remark If $\mathrm{Cov}(X,Y) > 0$, it means that $X-\mu_X$ and $Y-\mu_Y$ tend to have the same sign. This means that X and Y are both above their means together and they are both below their means together to a greater extent than they are on opposite sides of their means.

■ **Example 5.40** Suppose that X and Y have the following joint p.d.f.

$$f_{X,Y}(x,y) = \begin{cases} x+y, & \text{if } 0 < x,y < 1, \\ 0, & \text{otherwise.} \end{cases}$$

To find $\text{Cov}(X,Y)$, we need the means, variances, and mean of the product. The marginal p.d.f.'s are $f_1(x) = x + 1/2$ for $0 < x < 1$ and $f_2(y) = y + 1/2$ for $0 < y < 1$. So,

$$\text{E}(X) = \text{E}(Y) = \int_0^1 x(x+1/2)dx = \frac{7}{12}, \ \text{E}(XY) = \int_0^1 \int_0^1 xy(x+y)dxdy = \frac{1}{3},$$
$$\text{Cov}(X,Y) = \frac{1}{3} - \frac{7}{12} \cdot \frac{7}{12} = -\frac{1}{144}.$$

■

■ **Example 5.41** Assume that (X,Y) is a random point in the right triangle $T = \{(x,y) : x,y \geq 0,\ x+y \leq 1\}$. Compute $\text{Cov}(X,Y)$.

- Note that the area of the right triangle T is $1/2$, so the joint p.d.f. is 2 on the triangle.
- By symmetry and the law of unconscious statistician, we have

$$\text{E}(X) = \text{E}(Y) = \int_0^1 dx \int_0^{1-x} 2x dy = \int_0^1 2x(1-x)dx = \frac{1}{3},$$
$$\text{E}(XY) = \int_0^1 dx \int_0^{1-x} 2xy dy = \int_0^1 2x \frac{(1-x)^2}{2} dx = \frac{1}{12},$$
$$\text{Cov}(X,Y) = \frac{1}{12} - \left(\frac{1}{3}\right)^2 = -\frac{1}{36}.$$

■

Theorem 5.4.2 Let X and Y be two random variables. Then, it is easy to see that $\text{Cov}(X,Y) = \text{E}(XY) - \text{E}(X)\text{E}(Y)$, $\text{Cov}(X,Y) = \text{Cov}(Y,X)$, and that $\text{Var}(X) = \text{Cov}(X,X)$. Also, covariance is *bilinear* in that $\text{Cov}(aX,bY) = ab\text{Cov}(X,Y)$. Further, we have the following properties of covariance :

- $\text{Cov}(X,c) = 0$, for any constant c.
- $\text{Cov}(X+c,Y) = \text{Cov}(X,Y)$.
- $\text{Cov}(X+Z,Y) = \text{Cov}(X,Y) + \text{Cov}(Z,Y)$.
- $\text{Cov}(aX+b,cY+d) = ac\text{Cov}(X,Y)$ for any constants a, b, c, and d.
- $\text{Var}(X \pm Y) = \text{Var}(X) + \text{Var}(Y) \pm 2\text{Cov}(X,Y)$.
- $\text{Var}\left(\sum_{i=1}^n a_i X_i\right) = \sum_{i=1}^n a_i^2 \text{Var}(X_i) + 2\sum\sum_{i<j} a_i a_j \text{Cov}(X_i,X_j)$.

Proof. Let $\mathrm{E}(X) = \mu_X$ and $\mathrm{E}(Y) = \mu_Y$. Since the expectation is linear, it follows that

$$\begin{aligned}
\mathrm{Cov}(X,Y) &= \mathrm{E}((X-\mu_X)(Y-\mu_Y)) = \mathrm{E}(XY - \mu_X Y - \mu_Y X + \mu_X \mu_Y) \\
&= \mathrm{E}(XY) - \mathrm{E}(X)\mathrm{E}(Y) = \mathrm{Cov}(Y,X), \\
\mathrm{Cov}(X,X) &= \mathrm{E}(XX) - \mathrm{E}(X)\mathrm{E}(X) = \mathrm{Var}(X), \\
\mathrm{Cov}(aX,bY) &= \mathrm{E}(aXbY) - \mathrm{E}(aX)\mathrm{E}(bY) = ab\mathrm{Cov}(X,Y).
\end{aligned}$$

Also, the properties of covariance are easily verified :

- $\mathrm{Cov}(X,c) = \mathrm{E}(Xc) - \mathrm{E}(X)\mathrm{E}(c) = 0.$
- $\mathrm{Cov}(X+c,Y) = \mathrm{E}(XY+cY) - (\mathrm{E}(X)+c)\mathrm{E}(Y) = \mathrm{Cov}(X,Y).$
- If X, Y, and Z are three random variables, then

$$\begin{aligned}
&\mathrm{Cov}(X+Y,Z) = \mathrm{E}((X+Y)Z) - \mathrm{E}(X+Y)\mathrm{E}(Z) \\
= \ &\mathrm{E}(XZ+YZ) - \mathrm{E}(X)\mathrm{E}(Z) - \mathrm{E}(Y)\mathrm{E}(Z) \\
= \ &\mathrm{E}(XZ) - \mathrm{E}(X)\mathrm{E}(Z) + \mathrm{E}(YZ) - \mathrm{E}(Y)\mathrm{E}(Z) = \mathrm{Cov}(X,Z) + \mathrm{Cov}(Y,Z).
\end{aligned}$$

- To see covariance is bilinear, consider $\mathrm{Cov}(aX+b, cY+d)$, where a, b, c, and d are constants. Then,

$$\begin{aligned}
\mathrm{Cov}(aX+b,\, cY+d) &= \mathrm{E}((aX+b)(cY+d)) - \mathrm{E}(aX+b)\mathrm{E}(cY+d) \\
&= \mathrm{E}(acXY + adX + bcY + bd) - (a\mathrm{E}(X)+b)\,(c\mathrm{E}(Y)+d) \\
&= ac\,[\mathrm{E}(XY) - \mathrm{E}(X)\mathrm{E}(Y)] = ac\mathrm{Cov}(X,\,Y).
\end{aligned}$$

- Since $\mathrm{Var}(X) = \mathrm{Cov}(X,X)$ and covariance is bilinear, we have

$$\begin{aligned}
\mathrm{Var}(X \pm Y) &= \mathrm{Cov}(X \pm Y, X \pm Y) = \mathrm{Cov}(X \pm Y, X) \pm \mathrm{Cov}(X \pm Y, Y) \\
&= \mathrm{Cov}(X,X) \pm \mathrm{Cov}(Y,X) \pm \mathrm{Cov}(X,Y) \pm \mathrm{Cov}(Y,Y) \\
&= \mathrm{Var}(X) + \mathrm{Var}(Y) \pm 2\mathrm{Cov}(X,Y).
\end{aligned}$$

  Furthermore, it follows from bilinearity of covariance that

$$\begin{aligned}
\mathrm{Var}\left(\sum_{i=1}^{n} a_i X_i\right) &= \mathrm{Cov}\left(\sum_{i=1}^{n} a_i X_i, \sum_{i=1}^{n} a_i X_i\right) = \sum_{i=1}^{n}\sum_{j=1}^{n} \mathrm{Cov}\left(a_i X_i, a_j X_j\right) \\
&= \sum_{i=1}^{n} a_i^2 \mathrm{Var}(X_i) + 2\sum\sum_{i<j} a_i a_j \mathrm{Cov}(X_i, X_j).
\end{aligned}$$

■

■ **Example 5.42** Let $X = I_A$ be the indicator of event A, and $Y = I_B$ the indicator of event B. Then, $XY = I_A I_B = I_{A\cap B}$ is the indicator of the intersection of two events A and B. Thus,

$$\mathrm{E}(I_A) = P(A),\ \mathrm{E}(I_B) = P(B),\ \mathrm{E}(I_A I_B) = P(A\cap B),\ \mathrm{Cov}(X,Y) = P(A\cap B) - P(A)P(B).$$

- $\text{Cov}(X,Y) > 0$ if and only if $P(A \cap B) > P(A)P(B)$, when A and B are called *positively dependent*. That is, additional information that one event has occurred would increase the chance of the other event, which implies that $P(A|B) > P(A)$ and $P(B|A) > P(B)$, where $P(A) > 0$ and $P(B) > 0$.

- $\text{Cov}(X,Y) = 0$ if and only if $P(A \cap B) = P(A)P(B)$, when A and B are *independent*.

- $\text{Cov}(X,Y) < 0$ if and only if $P(A \cap B) < P(A)P(B)$, when A and B are called *negatively dependent*. In other words, $P(A|B) < P(A)$ and $P(B|A) < P(B)$.

■

The mean of the product $\text{E}(XY)$ is a special case of $\text{E}(X^kY^m)$, the moment of the pair of two random variables X and Y, which is called the *joint moment* or *product moment of order* (k,m) with $g(x,y) = x^ky^m$ in Theorem 5.4.1 if $\text{E}(|X|^k|Y|^m) < \infty$. Indeed, we have the joint moment generating function in Definition 5.4.2, which can also be used to compute the joint moments.

Definition 5.4.2 — Joint moment generating function. The joint moment generating function (m.g.f.) $M_{X,Y}(t_1,t_2)$ is defined by

$$M_{X,Y}(t_1,t_2) \equiv \text{E}\left[e^{t_1X+t_2Y}\right], \text{ for } |t_1| < h_1 \text{ and } |t_2| < h_2.$$

Then, the joint m.g.f. provides the joint moment, $\text{E}(X^kY^m) = \left.\dfrac{\partial^{k+m}}{\partial t_1^k \partial t_2^m} M_{X,Y}(t_1,t_2)\right|_{t_1=t_2=0}$.

In general, let $\mathbf{t}$ denote an element of R^k. The m.g.f. of k-dimensional random vector $\mathbf{X}$ is $M_{\mathbf{X}}(\mathbf{t}) = \text{E}\left(\exp\left\{\mathbf{t}^\top \mathbf{X}\right\}\right)$ if it exists for $\|\mathbf{t}\| = \sqrt{\sum_{i=1}^k t_i^2} < h$, and $h > 0$.

■ **Example 5.43** Suppose that X and Y have a joint p.d.f. $f_{X,Y}(x,y) = e^{-y}I_{(0<x<y<\infty)}$. We would like to compute $\text{Var}(X-Y)$.

- First, we can compute the joint m.g.f.

$$M_{X,Y}(t_1,t_2) = \int_{-\infty}^{\infty}\int_{-\infty}^{\infty} e^{t_1x+t_2y}e^{-y}I_{(0<x<y<\infty)}dydx = \int_0^{\infty}\int_x^{\infty} e^{-(1-t_2)y}dye^{t_1x}dx$$
$$= (1-t_2)^{-1}(1-t_1-t_2)^{-1}, \ t_2 < 1, \ t_1+t_2 < 1.$$

Then, we evaluate its joint moments

$$\text{E}(X) = \left.\frac{\partial}{\partial t_1}M_{X_1,X_2}(t_1,t_2)\right|_{t_1=t_2=0} = 1, \ \text{E}(Y) = \left.\frac{\partial}{\partial t_2}M_{X_1,X_2}(t_1,t_2)\right|_{t_1=t_2=0} = 2,$$
$$\text{E}(XY) = \left.\frac{\partial^2}{\partial t_1 \partial t_2}M_{X_1,X_2}(t_1,t_2)\right|_{t_1=t_2=0} = 3.$$

Thus, $\text{Cov}(X,Y) = \text{E}(XY) - \text{E}(X)\text{E}(Y) = 3-2 = 1$.

- Alternatively, we can get marginal distributions of X and Y and $\mathrm{E}(XY)$.

$$\begin{aligned}
f_Y(y) &= \int_0^y e^{-y}dx = ye^{-y}, \ Y \sim \text{Gamma}(2,1), \ \mathrm{E}(Y) = 2, \\
f_X(x) &= \int_x^\infty e^{-y}dy = e^{-x}, \ X \sim \text{Exp}(1), \ \mathrm{E}(X) = 1, \\
\mathrm{E}(XY) &= \int_0^\infty \int_0^y xye^{-y}dxdy = \int_0^\infty \frac{y^3}{2}e^{-y}dy = \frac{\Gamma(4)}{2} = 3.
\end{aligned}$$

Hence, $\mathrm{Var}(X-Y) = \mathrm{Var}(X) + \mathrm{Var}(Y) - 2\mathrm{Cov}(X,Y) = 1+2-2 = 1$.

■

Theorem 5.4.3 Suppose that two continuous random variables X and Y are independent and $\mathrm{E}[g(X)]$ and $\mathrm{E}[h(Y)]$ are finite. Then,

- $\mathrm{E}[g(X)h(Y)] = \mathrm{E}[g(X)]\mathrm{E}[h(Y)]$.
- $\mathrm{E}(XY) = \mathrm{E}(X)\mathrm{E}(Y)$ and hence $\mathrm{Cov}(X,Y) = 0$.
- If X and Y are independent, then the joint m.g.f. of X and Y is the product of two marginal m.g.f, $M_{X,Y}(t_1,t_2) = M_X(t_1)M_Y(t_2)$. In addition, the m.g.f. of $X+Y$ exists and is the product of two m.g.f.'s of X and Y, $M_{X+Y}(t) = M_X(t)M_Y(t)$.

Proof. Recall that X and Y are independent if and only if $f_{X,Y}(x,y) = f_X(x)f_Y(y)$. Then,

$$\mathrm{E}[g(X)h(Y)] = \int_{-\infty}^\infty \int_{-\infty}^\infty g(x)h(y)f_{X,Y}(x,y)dxdy = \int_{-\infty}^\infty \int_{-\infty}^\infty g(x)h(y)f_X(x)f_Y(y)dxdy$$
$$= \left(\int_{-\infty}^\infty g(x)f_X(x)dx\right)\left(\int_{-\infty}^\infty h(y)f_Y(y)dy\right) = \mathrm{E}(g(X))\mathrm{E}(h(Y)).$$

- Letting $g(X) = X$ and $h(Y) = Y$, we have $\mathrm{E}(XY) = \mathrm{E}(X)\mathrm{E}(Y)$ and $\mathrm{Cov}(X,Y) = 0$.
- Let $g(X) = e^{t_1X}$ and $h(Y) = e^{t_2Y}$. Then, we have

$$M_{X,Y}(t_1,t_2) \equiv \mathrm{E}\left[e^{t_1X+t_2Y}\right] = \mathrm{E}\left[e^{t_1X}\right]\mathrm{E}\left[e^{t_2Y}\right] = M_X(t_1)M_Y(t_2).$$

In fact, if the joint m.g.f. is the product of two marginal m.g.f.'s, then X and Y are independent. In addition, since X and Y are independent, it is obvious that

$$M_{X+Y}(t) = \mathrm{E}\left(e^{t(X+Y)}\right) = \mathrm{E}\left(e^{tX}\right)\mathrm{E}\left(e^{tY}\right) = M_X(t)M_Y(t).$$

■

■ **Example 5.44 — t-distribution.** Let X and Y be two independent random variables distributed as $X \sim N(0,1)$ and $Y \sim \chi^2(r)$. Define a random variable T by $T = \dfrac{X}{\sqrt{Y/r}}$.

- The joint p.d.f. of X and Y is the product of two marginal p.d.f.'s of X and Y

$$f_{X,Y}(x,y)=\begin{cases}\frac{1}{\sqrt{2\pi}}e^{-x^2/2}\frac{1}{\Gamma(r/2)2^{r/2}}y^{r/2-1}e^{-y/2}, & -\infty<x<\infty, 0<y<\infty\\ 0, & \text{otherwise.}\end{cases}$$

- Define two new variables $t=\dfrac{x}{\sqrt{y/r}}$ and $u=y$ and their inverse transformations $x=t\sqrt{u}/\sqrt{r}$ and $y=u$. Then, $J(t,u)=\det\begin{pmatrix}\sqrt{\frac{u}{r}} & \frac{t}{2\sqrt{ru}}\\ 0 & 1\end{pmatrix}=\sqrt{\dfrac{u}{r}}.$

- Thus, the joint p.d.f. of T and U is given by

$$f_{T,U}(t,u)=f_{X,Y}\left(\frac{t\sqrt{u}}{\sqrt{r}},u\right)|J(t,u)|$$

$$=\begin{cases}\frac{u^{r/2-1}}{\sqrt{2\pi}\Gamma(r/2)2^{r/2}}\exp\left[-\frac{u}{2}\left(1+\frac{t^2}{r}\right)\right]\frac{\sqrt{u}}{\sqrt{r}}, & -\infty<t<\infty, 0<u<\infty,\\ 0, & \text{otherwise.}\end{cases}$$

- Hence, the marginal p.d.f. of T is obtained by

$$f_T(t)=\int_{-\infty}^{\infty}f_{T,U}(t,u)du=\int_0^{\infty}\frac{1}{\sqrt{2\pi r}\Gamma(r/2)2^{r/2}}u^{(r+1)/2-1}\exp\left[-\frac{u}{2}\left(1+\frac{t^2}{r}\right)\right]du$$

$$=\int_0^{\infty}\frac{1}{\sqrt{2\pi r}\Gamma(r/2)2^{r/2}}\left(\frac{2z}{1+t^2/r}\right)^{(r+1)/2-1}e^{-z}\left(\frac{2}{1+t^2/r}\right)dz$$

$$=\frac{\Gamma\left(\frac{r+1}{2}\right)}{(r\pi)^{1/2}\Gamma\left(\frac{r}{2}\right)}\left(1+\frac{t^2}{r}\right)^{-(r+1)/2}, \quad \text{for } -\infty<t<\infty,$$

where we use a change of variable formula with $z=u\left[1+(t^2/r)\right]/2$. The distribution of T is called a *t-distribution with r degrees of freedom, denoted by $T\sim t(r)$.*

- To compute the kth moment of T, we use the representation of $T^k=g(X)h(Y)=X^k\left(\frac{Y}{r}\right)^{-k/2}$. Then, from the independence of X and Y and the general result of the moment of a gamma random variable, it follows that

$$\mathrm{E}\left(T^k\right)=\mathrm{E}\left(X^k\right)\mathrm{E}\left[\left(\frac{Y}{r}\right)^{-k/2}\right]=\mathrm{E}\left(X^k\right)\frac{2^{-k/2}\Gamma\left(\frac{r}{2}-\frac{k}{2}\right)}{\Gamma\left(\frac{r}{2}\right)r^{-k/2}}, \quad \text{if } k<r.$$

Hence, for the mean of T with $k=1$, we have $\mathrm{E}(T)=0$ since $\mathrm{E}(X)=0$, as long as $r>1$. For the variance of T, we use $k=2$ with the condition $r>2$. Since $\mathrm{E}(X^2)=1$, the variance of T is given by $\mathrm{Var}(T)=\mathrm{E}\left(T^2\right)=\dfrac{2^{-1}\Gamma\left(\frac{r-2}{2}\right)}{\Gamma\left(\frac{r}{2}\right)r^{-2/2}}=\dfrac{r}{r-2}$, for $r>2$.

■

■ **Example 5.45 — F-distribution.** Let X and Y be two independent random variables distributed as $X\sim\chi^2(r_1)$ and $Y\sim\chi^2(r_2)$ such that r_1 and r_2 are given positive integers.

Define a random variable W as $W = \dfrac{X/r_1}{Y/r_2} = \dfrac{r_2 X}{r_1 Y}$. Then, the distribution of W is called the *F-distribution with r_1 and r_2 degrees of freedom, denoted by $W \sim F(r_1, r_2)$.*

- The joint p.d.f. of X and Y is the product of two marginal p.d.f.'s of X and Y

$$f_{X,Y}(x,y) = \begin{cases} \frac{x^{r_1/2-1} y^{r_2/2-1}}{\Gamma(r_1/2)\Gamma(r_2/2)2^{(r_1+r_2)/2}} e^{-(x+y)/2}, & \text{for } 0 < x, y < \infty, \\ 0, & \text{elsewhere .} \end{cases}$$

- We consider two new variables $w = \dfrac{x/r_1}{y/r_2}$ and $z = y$ and their inverse transformations

$$x = (r_1/r_2)wz, \; y = z, \; J = \det \begin{pmatrix} \frac{r_1}{r_2} z & \frac{r_1}{r_2} w \\ 0 & 1 \end{pmatrix} = \frac{r_1}{r_2} z.$$

Thus, the joint p.d.f. of W and Z is given by

$$f_{W,Z}(w,z) = \frac{1}{\Gamma\left(\frac{r_1}{2}\right)\Gamma\left(\frac{r_2}{2}\right)2^{(r_1+r_2)/2}} \left(\frac{r_1 z w}{r_2}\right)^{\frac{r_1-2}{2}} z^{\frac{r_2-2}{2}} \exp\left[\ \frac{z}{2}\left(\frac{r_1 w}{r_2}+1\right)\right] \frac{r_1 z}{r_2},$$

where $0 < w, z < \infty$, and zero otherwise.

- Hence, the marginal p.d.f. $f_W(w)$ of W is then

$$\begin{aligned} f_W(w) &= \int_{-\infty}^{\infty} f_{W,Z}(w,z)dz \\ &= \int_0^{\infty} \frac{(r_1/r_2)^{r_1/2}(w)^{r_1/2-1}}{\Gamma(r_1/2)\Gamma(r_2/2)2^{(r_1+r_2)/2}} z^{(r_1+r_2)/2-1} \exp\left[-\frac{z}{2}\left(\frac{r_1 w}{r_2}+1\right)\right] dz \\ &= \int_0^{\infty} \frac{(r_1/r_2)^{r_1/2}(w)^{r_1/2-1}}{\Gamma(r_1/2)\Gamma(r_2/2)2^{(r_1+r_2)/2}} \left(\frac{2y}{r_1 w/r_2+1}\right)^{(r_1+r_2)/2-1} e^{-y} \times \left(\frac{2}{r_1 w/r_2+1}\right) dy \\ &= \begin{cases} \frac{\Gamma[(r_1+r_2)/2](r_1/r_2)^{r_1/2}}{\Gamma(r_1/2)\Gamma(r_2/2)} \frac{(w)^{r_1/2-1}}{(1+r_1 w/r_2)^{(r_1+r_2)/2}}, & \text{for } 0 < w < \infty, \\ 0, & \text{elsewhere,} \end{cases} \end{aligned}$$

where we use the change of variable formula with $y = \dfrac{z}{2}\left(\dfrac{r_1 w}{r_2}+1\right)$.

- The kth moment of F can be obtained using the representation of $W^k = g(X)h(Y) = \left(\dfrac{r_2}{r_1}\right)^k X^k Y^{-k}$. Thus, from the independence of X and Y, it follows that

$$\mathrm{E}(W) = \frac{r_2}{r_1} r_1 \mathrm{E}(X)\mathrm{E}(Y^{-1}) = \frac{r_2}{r_1} r_1 \frac{2^{-1}\Gamma\left(\frac{r_2}{2}-1\right)}{\Gamma\left(\frac{r_2}{2}\right)} = \frac{r_2}{r_2-2}, \text{ if } r_2 \geq 3.$$

- By definition, it is obvious that if $W \sim F_{r_1,r_2}$, then $1/W \sim F_{r_2,r_1}$. In addition, if $T \sim t(r)$, then $T^2 \sim F(1,r)$.

- Note that $X \sim \text{Gamma}\left(\frac{r_1}{2}, \frac{1}{2}\right)$, $Y \sim \text{Gamma}\left(\frac{r_2}{2}, \frac{1}{2}\right)$ and that $W = \frac{r_2}{r_1}\frac{X}{Y}$. Thus, it follows that $\dfrac{\frac{r_1}{r_2}W}{1+\frac{r_1}{r_2}W} = \dfrac{X/Y}{1+X/Y} = \dfrac{X}{X+Y}$, which implies that $\dfrac{\frac{r_1}{r_2}W}{1+\frac{r_1}{r_2}W}$ has the beta distribution with parameters $\frac{r_1}{2}$ and $\frac{r_2}{2}$.

■

Remark A moment-generating function uniquely determines the distribution, which is known as the *uniqueness of m.g.f.* That is, let X and Y have the same m.g.f. in some neighborhood of 0, i.e., $M_X(t) = \mathrm{E}(e^{tX}) = M_Y(t) = E(e^{tY})$ for $|t| < h$, where $h > 0$. Then, X and Y have the same distribution, $X \stackrel{d}{=} Y$.

■ **Example 5.46 — Sum of two independent normals.** Let X_1 and X_2 be independent normal random variables $X_1 \sim N(\mu_1, \sigma_1^2)$ and $X_2 \sim N(\mu_2, \sigma_2^2)$. Let $Y = aX_1 + bX_2$. Find the distribution of Y.

- Because X_1 and X_2 are independent and normally distributed, the m.g.f. of $Y = aX_1 + bX_2$ is given by

$$\begin{aligned}M_Y(t) &= \mathrm{E}\left(e^{t(aX_1+bX_2)}\right) = M_{X_1}(at)M_{X_2}(bt)\\ &= \exp\left((a\mu_1 + b\mu_2)t + \frac{1}{2}(a^2\sigma_1^2 + b^2\sigma_2^2)t^2\right),\end{aligned}$$

which implies that Y has a normal distribution, $N(a\mu_1 + b\mu_2, a^2\sigma_1^2 + b^2\sigma_2^2)$.

- Suppose that $\mu_1 = \mu_2 = \mu$ and $\sigma_1^2 = \sigma_2^2 = \sigma^2$. Let $Y = X_1 + X_2$ and $W = X_1 - X_2$. Then, $Y \sim N(2\mu, 2\sigma^2)$ and $W \sim N(0, 2\sigma^2)$. Also, the joint m.g.f. of Y and W is

$$\begin{aligned}M_{Y,W}(t_1,t_2) &= \mathrm{E}\left(e^{t_1Y+t_2W}\right) = \mathrm{E}\left(e^{(t_1+t_2)X_1}\right)\mathrm{E}\left(e^{(t_1-t_2)X_2}\right)\\ &= \exp\left(\mu(t_1+t_2) + \frac{1}{2}\sigma^2(t_1+t_2)^2\right)\exp\left(\mu(t_1-t_2) + \frac{1}{2}\sigma^2(t_1-t_2)^2\right)\\ &= \exp\left(2\mu t_1 + \sigma^2 t_1^2\right)\exp\left(\sigma^2 t_2^2\right) = M_Y(t_1)M_W(t_2),\end{aligned}$$

which implies that $X_1 + X_2$ and $X_1 - X_2$ are independent and that $\mathrm{Cov}(X_1 + X_2, X_1 - X_2) = 0$. Thus, $\overline{X}_2 = \frac{X_1+X_2}{2}$ is independent of $X_1 - \overline{X}_2 = \frac{X_1-X_2}{2} = \overline{X}_2 - X_2$. Hence, $\overline{X}_2$ and $S_2^2 = (X_1 - \overline{X}_2)^2 + (X_2 - \overline{X}_2)^2$ are independent, which is the version of independence of sample mean and sample variance with the sample size $n = 2$. Further discussion with the general $n \geq 2$ will be given in Chapter 6.

■

■ **Example 5.47 — Sum of two independent gammas.** Let X_1 and X_2 be independent gamma random variables $X_1 \sim \text{Gamma}(\alpha_1, \lambda)$ and $X_2 \sim \text{Gamma}(\alpha_2, \lambda)$.

- The m.g.f. of $Y = X_1 + X_2$ is

$$M_Y(t) = \mathrm{E}\left(e^{t(X_1+X_2)}\right) = M_{X_1}(t)M_{X_2}(t) = \left(\frac{\lambda}{\lambda-t}\right)^{\alpha_1}\left(\frac{\lambda}{\lambda-t}\right)^{\alpha_2} = \left(\frac{\lambda}{\lambda-t}\right)^{\alpha_1+\alpha_2},$$

which implies that Y has a gamma distribution, Gamma$(\alpha_1+\alpha_2,\lambda)$, which we obtained using the convolution in Example 5.22.

- Suppose that $\lambda = 1/2$. Then, $X_1 \sim \chi^2(2\alpha_1)$ and $X_2 \sim \chi^2(2\alpha_2)$. Then, Y has a chi-squared distribution with $2(\alpha_1+\alpha_2)$ degrees of freedom.

- Suppose that $\alpha_1 = \alpha_2 = \lambda = 1$. Let $Y = X_1 + X_2$ and $W = X_1 - X_2$. Then, $Y \sim$ Gamma$(2,1)$, and the m.g.f. of W is

$$M_W(t) = \mathrm{E}\left(e^{tW}\right) = M_{X_1}(t)M_{X_2}(-t) = \frac{1}{1-t}\cdot\frac{1}{1+t} = \frac{1}{1-t^2},$$

which is the m.g.f. of a Laplace distribution in Example 4.32. Thus, $W \sim \mathrm{DE}(0,1)$.

- Further, the joint m.g.f. of Y and W is given by

$$\begin{aligned} M_{Y,W}(t_1,t_2) &= \mathrm{E}\left(e^{t_1Y+t_2W}\right) = \mathrm{E}\left(e^{(t_1+t_2)X_1}\right)\mathrm{E}\left(e^{(t_1-t_2)X_2}\right) \\ &= \left(\frac{1}{1-t_1-t_2}\right)\left(\frac{1}{1-t_1+t_2}\right) \neq \frac{1}{1-t_1}\cdot\frac{1}{1-t_2^2} = M_Y(t_1)M_W(t_2), \end{aligned}$$

which implies that X_1+X_2 and X_1-X_2 are not independent.

∎

∎ **Example 5.48** Suppose that X_1 and X_2 are independent and identically distributed with $M_{X_1}(t) = M_{X_2}(t)$.

- If X_1+X_2(or X_1-X_2) is normally distributed $N(0,\sigma^2)$, then its m.g.f. is

$$\begin{aligned} M_{X_1+X_2}(t) &= e^{\frac{1}{2}\sigma^2t^2} = M_{X_1}(t)M_{X_2}(t) = M_{X_1}(t)^2, \\ M_{X_1-X_2}(t) &= e^{\frac{1}{2}\sigma^2t^2} = M_{X_1}(t)M_{X_2}(-t) = M_{X_1}(t)^2, \end{aligned}$$

which implies that $M_{X_1}(t) = e^{\frac{1}{2}\frac{\sigma^2}{2}t^2}$ and that $X_1, X_2 \overset{i.i.d.}{\sim} N\left(0,\frac{1}{2}\sigma^2\right)$.

- If X_1+X_2 has a chi-squared distribution with 2α degrees of freedom, $X_1+X_2 \sim \chi^2(2\alpha)$, then its m.g.f. is

$$M_{X_1+X_2}(t) = \left(\frac{1}{1-2t}\right)^{\alpha} = M_{X_1}(t)M_{X_2}(t) = M_{X_1}(t)^2,$$

which implies that $M_{X_1}(t) = \left(\frac{1}{1-2t}\right)^{\alpha/2}$ and that $X_1, X_2 \overset{i.i.d.}{\sim} \chi^2(\alpha)$.

∎

■ **Example 5.49** Note that if X and Y are independent, then $\mathrm{Cov}(X,Y)=0$. However, the converse is not true. In other words, if the covariance of X and Y is zero, that does not imply they are independent. Such counter examples can often be found in bivariate random variables symmetric around zero :

- Let X have a distribution that is symmetric around 0 and let $Y=|X|$. Then $\mathrm{E}(X)=0$ and $\mathrm{E}(XY)=0$, so $\mathrm{Cov}(X,Y)=0$. As a specific example, let $f_X(x)=1/3$ for $x=-1,0,1$. Then the joint p.f. of X and Y is

$$f_{X,Y}(x,y)=\begin{cases}\frac{1}{3} & \text{if } x=-1, y=1,\\ \frac{1}{3} & \text{if } x=0, y=0,\\ \frac{1}{3} & \text{if } x=1, y=1,\\ 0 & \text{otherwise.}\end{cases}$$

- Let X have a standard normal distribution, $X\sim N(0,1)$ and let $Y=X^2$. Then, $\mathrm{E}(X)=0$ and $\mathrm{E}(XY)=0$, so $\mathrm{Cov}(X,Y)=0$. But X and Y are not independent.

- Let X have a Laplace distribution, $X\sim \mathrm{DE}(0,1)$ and let $Y=|X|$. Then, $\mathrm{E}(X)=0$ and $\mathrm{E}(XY)=\mathrm{E}(X|X|)=0$, so $\mathrm{Cov}(X,Y)=0$. But X and Y are not independent.

- Let X and Y have a joint p.d.f.

$$f_{X,Y}(x,y)=\begin{cases}1/\pi, & x^2+y^2\le 1\\ 0, & \text{otherwise.}\end{cases}$$

The marginal density of X is given by $f_X(x)=(2/\pi)(1-x^2)^{1/2}$ for $-1<x<1$. By symmetry, the marginal density f_Y has the same form. Thus,

$$\mathrm{E}(X)=\mathrm{E}(Y)=\int_{-1}^{1}x\frac{2}{\pi}(1-x^2)^{1/2}dx=0$$

and, it follows by skew symmetry with no integrations required that

$$\mathrm{E}(XY)=\frac{1}{\pi}\int\int_{x^2+y^2<1}xydxdy=0.$$

Alternatively, we can evaluate the integral

$$\begin{aligned}\mathrm{E}(XY) &= \frac{1}{\pi}\int\int_{x^2+y^2<1}xydxdy=\frac{1}{\pi}\int_{-1}^{1}\int_{-\sqrt{1-y^2}}^{\sqrt{1-y^2}}xydxdy\\ &= \frac{1}{\pi}\int_{-1}^{0}y\int_{-\sqrt{1-y^2}}^{\sqrt{1-y^2}}xdxdy+\frac{1}{\pi}\int_{0}^{1}y\int_{-\sqrt{1-y^2}}^{\sqrt{1-y^2}}xdxdy\\ &= -\frac{1}{\pi}\int_{0}^{1}u\int_{-\sqrt{1-u^2}}^{\sqrt{1-u^2}}xdxdu+\frac{1}{\pi}\int_{0}^{1}y\int_{-\sqrt{1-y^2}}^{\sqrt{1-y^2}}xdxdy=0,\end{aligned}$$

where the last equality follows from the transformation $u=-y$ and the fact $g(y)=\int_{-\sqrt{1-y^2}}^{\sqrt{1-y^2}}xdx=g(-y)$. Hence $\mathrm{Cov}(X,Y)=0$. However, X and Y are not independent. On the other hand, let R and Θ denote the polar coordinates of the point (X,Y),

$X = R\cos\Theta$ and $Y = R\sin\Theta$. Then, the joint p.d.f. is

$$f_{R,\Theta}(r,\theta) = f_{X,Y}(r\cos\theta, r\sin\theta)|J(r,\theta)| = \frac{r}{\pi}, \quad 0 \le r \le 1,\ 0 \le \theta < 2\pi,$$

which implies that R and Θ are independent and that $\text{Cov}(R,\Theta) = 0$.

- More generally, suppose that X and Y are i.i.d. with finite second moments. Then,

$$\text{Cov}(X+Y, X-Y) = \text{Var}(X) + \text{Cov}(X,Y) - \text{Cov}(X,Y) - \text{Var}(Y) = 0.$$

However, $X+Y$ need not be independent of $X-Y$. For example, when X and Y are independent standard exponential random variables, $X+Y$ and $X-Y$ are not independent as given in Example 5.47.

- Hence, $\text{Cov}(X,Y) = 0$ does not imply that X and Y are independent. However, $\text{Cov}(X,Y) = 0$ indicates that two random variables X and Y, more correctly, two centered random variables $X - \text{E}(X)$ and $Y - \text{E}(Y)$, are *orthogonal or perpendicular* to each other, by viewing $\text{Cov}(X,Y)$ as an inner product of $X - \text{E}(X)$ and $Y - \text{E}(Y)$ defined on the set of centered random variables.

■

Remark Let $\mathbf{u} = (a,b)$ and $\mathbf{v} = (c,d)$ be two-dimensional vectors. The *inner product* or the dot product of $\mathbf{u}$ and $\mathbf{v}$ is defined $\langle \mathbf{u},\ \mathbf{v} \rangle = \mathbf{u} \cdot \mathbf{v} = ac + bd$. An inner product $\langle \mathbf{u},\ \mathbf{v} \rangle$ maps two vectors to a real number that satisfies four properties.

① $\langle \mathbf{u}+\mathbf{v},\ \mathbf{w} \rangle = \langle \mathbf{u},\ \mathbf{w} \rangle + \langle \mathbf{v},\ \mathbf{w} \rangle$.

② $\langle \alpha\mathbf{v},\ \mathbf{w} \rangle = \alpha \langle \mathbf{v},\ \mathbf{w} \rangle$.

③ $\langle \mathbf{v},\ \mathbf{w} \rangle = \langle \mathbf{w},\ \mathbf{v} \rangle$.

④ $\langle \mathbf{v},\ \mathbf{v} \rangle \ge 0$ where equality holds if and only if $\mathbf{v} = \mathbf{0}$.

Remark A *norm* of a vector measures the size (length) of the vector. We denote the norm of $\mathbf{v}$ as $||\mathbf{v}||$, which is defined to be $||\mathbf{v}|| = \langle \mathbf{v},\ \mathbf{v} \rangle^{1/2}$. Then a norm satisfies the following rules.

① For any $c \in \mathbb{R}$ and vector $\mathbf{v}$, $||c\mathbf{v}|| - |c|\,||\mathbf{v}||$.

② For any vectors $\mathbf{v}$ and $\mathbf{w}$, $||\mathbf{v}+\mathbf{w}|| \le ||\mathbf{v}|| + ||\mathbf{w}||$, known as the *triangle* inequality.

③ For any vector $\mathbf{v}$, $||\mathbf{v}|| \ge 0$, and only equals 0 if $\mathbf{v}$ is the zero vector.

The distance between two vectors $\mathbf{u}$ and $\mathbf{v}$ is $||\mathbf{u} - \mathbf{v}|| = \langle \mathbf{u} - \mathbf{v},\ \mathbf{u} - \mathbf{v} \rangle^{1/2}$.

Theorem 5.4.4 — Covariance as an inner product. For a random variable X whose mean is finite, then a centered random variable $\widetilde{X} = X - \mathrm{E}(X)$ has mean 0 and centered random variables form a vector space. Then, $\mathrm{Cov}(\widetilde{X}, \widetilde{Y}) = \mathrm{E}[\widetilde{X}\widetilde{Y}]$, and covariance is an inner product. The norm is the standard deviation.

Remark For any inner product and vectors $\mathbf{x}$ and $\mathbf{y}$, $|\mathbf{x} \cdot \mathbf{y}| \leq ||\mathbf{x}|| \cdot ||\mathbf{y}||$, which is known as the *Cauchy-Schwarz inequality*.

- For any real numbers $a_1, a_2, \ldots, a_n$ and $b_1, b_2, \ldots, b_n$,
$$(a_1b_1 + a_2b_2 + \cdots + a_nb_n)^2 \leq (a_1^2 + a_2^2 + \cdots + a_n^2)(b_1^2 + b_2^2 + \cdots + b_n^2).$$
- If θ is the angle between two vectors $\mathbf{x}$ and $\mathbf{y}$, by the Pythagorean theorem, it follows that $||\mathbf{x}+\mathbf{y}||^2 = ||\mathbf{x}\sin\theta||^2 + (||\mathbf{y}|| + ||\mathbf{x}\cos\theta||^2)$ as shown in Figure 5.17. Thus, we have $\cos\theta = \dfrac{\mathbf{x} \cdot \mathbf{y}}{||\mathbf{x}|| \cdot ||\mathbf{y}||}$.
- Two nonzero vectors are orthogonal (perpendicular) if the angle between them is 90°. In other words, two nonzero vectors $\mathbf{x}$ and $\mathbf{y}$ are perpendicular if and only if $\mathbf{x} \cdot \mathbf{y} = 0$, which is denoted as $\mathbf{x} \perp \mathbf{y}$. Thus, $\mathbf{x}$ and $\mathbf{y}$ form a right triangle if and only if $\mathbf{x} \cdot \mathbf{y} = 0$. In other words, $\cos\theta = 0$ and thus $\theta = \pi/2$.

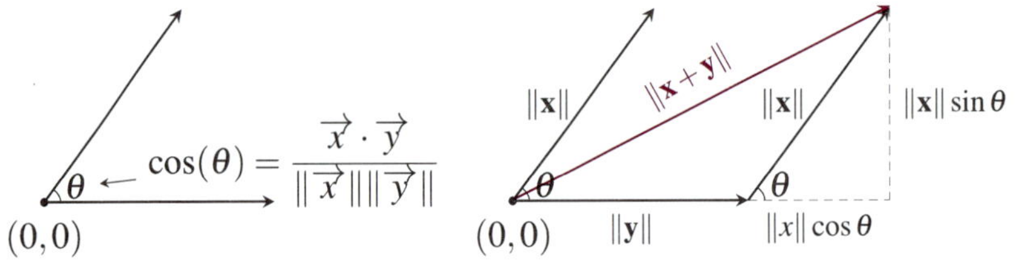

Figure 5.17: Two nonzero vectors and the angle between them

■ **Example 5.50** Let X and Y be uniformly distributed over the ellipse as follows.

$$f_{X,Y}(x,y) = \begin{cases} c, & (x,y) \in E = \left\{(x,y) : \frac{x^2}{a^2} + \frac{y^2}{b^2} = 1\right\}, \\ 0, & \text{otherwise}. \end{cases}$$

Find the covariance between X and Y.

- For $f_{X,Y}(x,y)$ to be the joint p.d.f., $\int\int_E f_{X,Y}(x,y)dxdy = 1$. Thus, c should be the reciprocal of $|E|$, where $|E|$ denotes the size of the ellipse E. Using the change of variables with $\frac{x}{a} = r\cos\theta$, $\frac{y}{b} = r\sin\theta$ and the Jacobian $J(r,\theta) = abr$, where $0 \leq r \leq 1$ and $0 < \theta < 2\pi$, we evaluate the integral,
$$1 = \int\int_E f_{X,Y}(x,y)dxdy = \int_0^{2\pi}\int_0^1 cJ(r,\theta)drd\theta = \int_0^{2\pi}\int_0^1 cabrdrd\theta = cab\pi,$$
which implies that $|E| = \pi ab$ and $f_{X,Y} = \dfrac{1}{\pi ab}$ for $(x,y) \in E$.

- From the joint p.d.f. of X and Y, we obtain the marginal p.d.f.'s of X and Y,

$$
\begin{aligned}
f_X(x) &= \int_{-b\sqrt{1-(x^2/a^2)}}^{b\sqrt{1-(x^2/a^2)}} \frac{1}{\pi ab} dy = 2\frac{\sqrt{1-\frac{x^2}{a^2}}}{\pi a}, \text{ for } -a \le x \le a, \\
f_Y(y) &= \int_{-a\sqrt{1-(y^2/b^2)}}^{a\sqrt{1-(y^2/b^2)}} \frac{1}{\pi ab} dx = 2\frac{\sqrt{1-\frac{y^2}{b^2}}}{\pi b}, \text{ for } -b \le y \le b,
\end{aligned}
$$

which implies that X and Y are symmetric about 0, respectively and that $\mathrm{E}(X) = \mathrm{E}(Y) = 0$.

- The covariance between X and Y is $\mathrm{Cov}(X,Y) = \mathrm{E}(XY)$ since $\mathrm{E}(X) = \mathrm{E}(Y) = 0$. Thus, using the change of variables with $x = ar\cos\theta$ and $y = br\sin\theta$

$$
\begin{aligned}
\mathrm{Cov}(X,Y) = \mathrm{E}(XY) &= \int\int_E xy f_{X,Y}(x,y) dx dy \\
&= \int_0^{2\pi}\int_0^1 a^2b^2r^3\cos\theta\sin\theta \frac{1}{\pi ab} dr d\theta = \int_0^{2\pi} \frac{ab}{4\pi}\cos\theta\sin\theta d\theta = 0,
\end{aligned}
$$

since $\sin\theta\cos\theta$ are symmetric about the origin. Alternatively, define $U = X/a$ and $V = Y/b$. Then, U and V are random points on the unit circle with $\mathrm{Cov}(U,V) = 0$ as given in Example 5.49. Thus $\mathrm{Cov}(X,Y) = \mathrm{Cov}(aU,bV) = ab\mathrm{Cov}(U,V) = 0$. However, note that $f_{X,Y}(x,y) \neq f_X(x)f_Y(y)$. Hence, X and Y are not independent but uncorrelated.

- On the other hand, using the change of variables $X = aR\cos\Theta$ and $Y = bR\sin\Theta$, we can get the joint p.d.f. of R and Θ,

$$
f_{R,\Theta}(r,\theta) = \frac{1}{\pi ab} J(r,\theta) = \frac{r}{\pi}, \text{ for } 0 \le r \le 1, 0 < \theta < 2\pi,
$$

which implies that R and Θ are independent as also discussed in Example 5.49.

■

The problem with interpreting the size of covariance as a measure of how much X and Y vary together is the fact that $\mathrm{Cov}(X,Y)$ is also affected by the variances. Thus, the magnitude (not the sign) of the covariance is a function of the standard deviations of the random variables, and we remove the effect of the standard deviations by defining the correlation.

We define two random variables $X' = (X - \mathrm{E}(X))/\sqrt{\mathrm{Var}(X)}$ and $Y' = (Y - \mathrm{E}(Y))/\sqrt{\mathrm{Var}(Y)}$, centered and rescaled to standard units so that $\mathrm{E}(X') = \mathrm{E}(Y') = 0$ and $\mathrm{Var}(X') = \mathrm{Var}(Y') = 1$. Then, we consider covariance between X' and Y', $\mathrm{Cov}(X',Y') = \mathrm{E}(X'Y')$, which measures the linear dependence between X and Y, not depending on their units.

Definition 5.4.3 The *correlation* between X and Y is

$$\rho(X,Y) = \frac{\mathrm{Cov}(X,Y)}{\sqrt{\mathrm{Var}(X)\mathrm{Var}(Y)}} = \mathrm{Cov}\left(\frac{X-\mathrm{E}(X)}{\sqrt{\mathrm{Var}(X)}}, \frac{Y-\mathrm{E}(Y)}{\sqrt{\mathrm{Var}(Y)}}\right).$$

- If $\rho(X,Y) = 0$, we say that X and Y are *uncorrelated.*
- If $\rho(X,Y) > 0$, we say that X and Y are *positively correlated.*
- If $\rho(X,Y) < 0$, we say that X and Y are *negatively correlated.*

Example 5.51 Suppose that the joint p.d.f. of X and Y is

$$f_{X,Y}(x,y) = \begin{cases} e^{-y}, & (0 < x < y < \infty), \\ 0, & \text{otherwise.} \end{cases}$$

- In Example 6.27, we obtained $X \sim \mathrm{Exp}(1)$, $Y \sim \mathrm{Gamma}(2,1)$, and $\mathrm{Cov}(X,Y) = 1$. Thus, $\rho(X,Y) = \frac{\mathrm{Cov}(X,Y)}{\sqrt{\mathrm{Var}(X)\mathrm{Var}(Y)}} = \frac{1}{\sqrt{2}}$.
- Let $U = Y - X$ and $V = X$. Then, the joint p.d.f. of U and V is

  $$f_{U,V}(u,v) = f_{X,Y}(v,u+v)|J(u,v)| = e^{-(u+v)} I_{(0<v<u+v<\infty)}.$$

  and the marginal p.d.f.'s of U and V are $f_U(u) = \int_0^\infty e^{-(u+v)}dv = e^{-u}$, $f_V(v) = \int_0^\infty e^{-(u+v)}du = e^{-v}$, which implies that U and V are independent and $\mathrm{Cov}(U,V) = 0$. Indeed, we have $\mathrm{Cov}(Y-X,X) = \mathrm{Cov}(Y,X) - \mathrm{Cov}(X,X) = 0$.

■

Example 5.52 Suppose that the joint p.d.f. of X and Y is given by

$$f_{X,Y}(x,y) = \frac{1}{2\pi} e^{-(x^2+(y-x)^2)/2}, \quad -\infty < x,y < \infty.$$

- Let $W = X - Y$ and $V = X$. Then, the joint p.d.f. of V and W is given by

  $$f_{V,W}(v,w) = \frac{1}{2\pi} e^{-(v^2+w^2)/2},$$

  which implies that V and W are independent and identically distributed as a standard normal distribution.

- Note that $0 = \mathrm{Cov}(X, X-Y) = \mathrm{Var}(X) - \mathrm{Cov}(X,Y)$ and $1 = \mathrm{Var}(W) = \mathrm{Var}(X) - 2\mathrm{Cov}(X,Y) + \mathrm{Var}(Y)$. Thus, $\mathrm{Cov}(X,Y) = 1$ and $\mathrm{Var}(Y) = 2$. Hence, $\rho(X,Y) = \frac{\mathrm{Cov}(X,Y)}{\sqrt{\mathrm{Var}(X)\mathrm{Var}(Y)}} = \frac{1}{\sqrt{2}}$. Indeed, this is an example of a bivariate normal distribution, which will be discussed in Chapter 6.

■

■ **Example 5.53** Let X and Y have a joint p.d.f.

$$f_{X,Y}(x,y) = \begin{cases} 2, & \text{for } 1-y < x < 2-2y,\ 0 < y < 1, \\ 0, & \text{otherwise.} \end{cases}$$

- Note that (X,Y) is a random point in a triangle as in Examples 5.4 and 5.7. As given in Example 5.7, the marginal p.d.f. of X is given by $f_X(x) = \begin{cases} x, & \text{for } 0 < x < 1, \\ 2-x, & \text{for } 1 \leq x < 2, \end{cases}$ and the marginal p.d.f. of Y is $f_Y(y) = 2-2y$, for $0 < y < 1$.

- Thus, the joint moments are computed

$$\begin{aligned}
\mathrm{E}(X) &= \int_0^1 \int_{1-y}^{2-y} 2x dx dy = \int_0^1 (2-y)^2 - (1-y)^2 dy = \int_0^1 3 - 2y dy = 2, \\
\mathrm{E}(Y) &= \int_0^1 \int_{1-y}^{2-y} 2y dx dy = \int_0^1 2y dy = 1, \\
\mathrm{E}(XY) &= \int_0^1 \int_{1-y}^{2-y} 2xy dx dy - \int_0^1 y(3-2y) dy - \frac{3}{2} - \frac{2}{3} = \frac{5}{6}, \\
\mathrm{Cov}(X,Y) &= \mathrm{E}(XY) - \mathrm{E}(X)\mathrm{E}(Y) = -\frac{7}{6},
\end{aligned}$$

which implies that $\rho(X,Y) < 0$ and that X and Y are negatively correlated. Indeed, Figures 5.5 and 5.7 indicate that as X increases, Y decreases, which we also expect that they would be negatively correlated.

■

Theorem 5.4.5 Let X and Y be two random variables with positive variances. Then,

- $-1 \leq \rho(X,Y) \leq 1$.
- Let a and b be real constants with $a \neq 0$. Then, $\rho(aX+b,Y) = \frac{a}{|a|}\rho(X,Y)$.
- $\rho(aX+b,cY+d) = \rho(X,Y)$ for any constants a,b,c, and d with $ac > 0$.

Proof. From the Cauchy-Schwarz inequality, we have $\langle \widetilde{X}, \widetilde{Y} \rangle \leq \|\widetilde{X}\|\|\widetilde{Y}\|$, which implies that $\mathrm{Cov}(X,Y) \leq \sqrt{\mathrm{Var}(X)\mathrm{Var}(Y)}$. In other words, since $\cos\theta = \frac{\mathrm{Cov}(X,Y)}{\sqrt{\mathrm{Var}(X)\mathrm{Var}(Y)}} = \rho(X,Y)$, $\rho(X,Y)$ is in between -1 and 1.

① Alternatively, note that $\mathrm{Var}(aX+Y) = a^2\mathrm{Var}(X) + 2a\mathrm{Cov}(X,Y) + \mathrm{Var}(Y) \geq 0$ for any real number a, which is a quadratic function of a. Thus, it is nonnegative everywhere if and only if the coefficient of a^2 is positive and the discriminant D is nonpositive, that is,

$$D = [\mathrm{Cov}(X,Y)]^2 - \mathrm{Var}(X)\mathrm{Var}(Y) \leq 0.$$

Therefore, we have $\mathrm{Cov}(X,Y)^2 \leq \mathrm{Var}(X)\mathrm{Var}(Y)$. Or, assuming that X and Y have variance 1, we have $\mathrm{Var}(X \pm Y) = \mathrm{Var}(X) + \mathrm{Var}(Y) \pm 2\mathrm{Cov}(X,Y) = 2 \pm 2\rho \geq 0$, which implies that $-1 \leq \rho \leq 1$.

- On the other hand, let $X' = (X - \mathrm{E}(X))/\sqrt{\mathrm{Var}(X)}$ and let $Y' = (Y - \mathrm{E}(Y))/\sqrt{\mathrm{Var}(Y)}$. Then, we see that

$$0 \leq \mathrm{Var}(X' - Y') = 1 + 1 - 2\mathrm{Cov}(X', Y'),\ 0 \leq \mathrm{Var}(X' + Y') = 1 + 1 + 2\mathrm{Cov}(X', Y'),$$

which implies that $-1 \leq \mathrm{Cov}(X', Y') = \rho(X,Y) \leq 1$.

- From the properties of the variance and covariance, we have

$$\rho(aX + b, Y) = \frac{\mathrm{Cov}(aX + b, Y)}{\sqrt{\mathrm{Var}(aX + b)\mathrm{Var}(Y)}} = \frac{a\mathrm{Cov}(X,Y)}{\sqrt{a^2\mathrm{Var}(X)\mathrm{Var}(Y)}} = \frac{a}{|a|}\rho(X,Y).$$

That is, for $a \neq 0$, $\rho(aX + b, Y)$ is $\rho(X,Y)$ or $-\rho(X,Y)$, depending on the sign of a.

- Using the bilinearity of covariance, we have

$$\rho(aX + b, cY + d) = \frac{\mathrm{Cov}(aX + b, cY + d)}{\sqrt{\mathrm{Var}(aX + b)\mathrm{Var}(cY + d)}} = \frac{ac\mathrm{Cov}(X,Y)}{\sqrt{a^2\mathrm{Var}(X)c^2\mathrm{Var}(Y)}} = \rho(X,Y)$$

if a and c have the same sign, i.e. $ac > 0$.

■

■ **Example 5.54** Let $X = I_A$ be the indicator of event A, and $Y = I_B$ the indicator of event B. Then, note that $X \sim \mathrm{Bernoulli}(P(A))$ and $Y \sim \mathrm{Bernoulli}(P(B))$.

- Thus, we have $\mathrm{E}(I_A) = P(A)$, $\mathrm{E}(I_B) = P(B)$, and $\mathrm{Cov}(X,Y) = P(A \cap B) - P(A)P(B)$ as given in Example 5.42. In addition, we have $\mathrm{Var}(I_A) = P(A) - P(A)^2$, and $\mathrm{Var}(I_B) = P(B) - P(B)^2$. Hence, we have

$$\rho(X,Y) = \frac{\mathrm{Cov}(X,Y)}{\sqrt{\mathrm{Var}(X)}\sqrt{\mathrm{Var}(Y)}} = \frac{P(A \cap B) - P(A)P(B)}{\sqrt{(P(A)(1 - P(A))P(B)(1 - P(B))}}.$$

- Note that $I_{A^c} = 1 - I_A = 1 - X$ and $I_{B^c} = 1 - I_B = 1 - Y$. Thus, by Theorem 5.4.5, it follows that $\rho(I_A, I_B) = \rho(I_{A^c}, I_{B^c}) = -\rho(I_A, I_{B^c}) = -\rho(I_{A^c}, I_B)$, which implies that if A and B are positively correlated, then A^c and B^c are positively correlated, but A and B^c are negatively dependent, as are A^c and B.

■

■ **Example 5.55** Suppose that there are r red balls, b blue balls and g green balls in a box, where $r + b + g = n$. Consider drawing a ball n times with replacement. Let X be the number of red balls out of n trials, and let Y be the number of blue balls out of n trials. Find $\rho(X,Y)$.

- When there is no green ball ($g = 0$), it is obvious that X and $Y = n - X$ are negatively correlated, and actually $\rho(X,Y) = -1$. Even if $g > 0$, it is intuitively true that $\rho(X,Y) < 0$ since $X + Y + Z = n$, where Z is the number of green balls out of n trials.

- Note that $X \sim \text{Binomial}(n, p_r)$, where $p_r = r/(r+b+g)$, and $Y \sim \text{Binomial}(n, p_b)$, where $p_b = b/(r+b+g)$. Thus,

$$\text{E}(X) = np_r,\ \text{Var}(X) = np_r(1-p_r),\ \text{E}(Y) = np_b,\ \text{Var}(Y) = np_b(1-p_b).$$

- In addition, since $X + Y \sim \text{Binomial}(n, p_r + p_b)$, we have

$$\text{Var}(X+Y) = \text{Var}(X) + \text{Var}(Y) + 2\text{Cov}(X,Y) = n(p_r + p_b)(1 - p_r - p_b),$$

which implies that $\text{Cov}(X,Y) = -np_rp_b$. Hence,

$$\rho(X,Y) = \frac{\text{Cov}(X,Y)}{\sqrt{\text{Var}(X)\text{Var}(Y)}} = \frac{-np_rp_b}{n\sqrt{p_r(1-p_r)p_b(1-p_b)}}$$
$$= -\sqrt{\frac{p_rp_b}{(1-p_r)(1-p_b)}} < 0.$$

■

■ **Example 5.56** Suppose that three random variables X, Y, and Z have finite variances and covariances such that $\text{Var}(X) = \text{Var}(Y) = \text{Var}(Z) = 1$, and $\rho(X,Y) = \rho(X,Z) = \rho$.

- Consider $\text{Var}(\alpha X + Y + Z)$. Then, by Theorem 5.4.2, we have

$$\begin{aligned}\text{Var}(\alpha X + Y + Z) &= \alpha^2\text{Var}(X) + \text{Var}(Z) + \text{Var}(Y)\\ &+2\alpha\left[\text{Cov}(X,Y) + \text{Cov}(X,Z)\right] + 2\text{Cov}(Y,Z) \geq 0.\end{aligned}$$

Since $\text{Var}(X) = \text{Var}(Y) = \text{Var}(Z) = 1$ and $\rho(X,Y) = \rho(X,Z) = \rho$, we have $\text{Cov}(X,Y) = \text{Cov}(X,Z) = \rho$ and $\text{Cov}(Y,Z) = \rho(Y,Z)$. Thus, we have the quadratic equation of α, $\alpha^2 + 4\rho\alpha + 2 + 2\rho(Y,Z) \geq 0$ for all α. Hence, the discriminant D of the equation must be nonpositive, $D = 4\rho^2 - 2 - 2\rho(Y,Z) \leq 0$, which implies that $\rho(Y,Z) \geq 2\rho^2 - 1$.

- More generally, suppose that $X_1, X_2, \ldots, X_n$ have the same variance $\text{Var}(X_i) = 1$, $i = 1, \ldots, n$, and the same correlation $\rho(X_i, X_j) = \rho$, $i \neq j$. Then, by Theorem 5.4.2,

$$\text{Var}\left(\sum_{i=1}^{n} X_i\right) = \sum_{i=1}^{n} \text{Var}(X_i) + 2\sum\sum_{i<j} \text{Cov}(X_i, X_j) \geq 0.$$

Since $\text{Var}(X_i) = 1$ for all i, $\text{Cov}(X_i, X_j) = \rho(X_i, X_j) = \rho$, $i \neq j$, and there are $\binom{n}{2}$ ways of choosing X_i and X_j, $i \neq j$, we have $n + n(n-1)\rho \geq 0$. Hence, $\rho \geq -1/(n-1)$.

■

■ **Example 5.57** Suppose that Z_1 and Z_2 are independent and identically distributed standard normal random variables, $N(0,1)$. Define $U = \max\{Z_1, Z_2\}$ and $V = \min\{Z_1, Z_2\}$. Find the correlation between U and V.

- Note that $\max\{Z_1, Z_2\} = -\min\{-Z_1, -Z_2\}$, $U = \frac{Z_1+Z_2}{2} + \frac{|Z_1-Z_2|}{2}$, and $V = \frac{Z_1+Z_2}{2} - \frac{|Z_1-Z_2|}{2}$. Thus, we have $U+V = Z_1+Z_2$, $U-V = |Z_1-Z_2|$, and $UV = Z_1Z_2$.

- As discussed in Example 5.28, Z_1+Z_2 and Z_1-Z_2 are independent and identically distributed as $N(0,2)$. Then,

$$\begin{aligned}
&\mathrm{E}(|Z_1-Z_2|) = 2\int_0^\infty t\frac{1}{\sqrt{4\pi}}\exp\left(-\frac{1}{4}t^2\right)dt = \frac{2}{\sqrt{\pi}},\\
&\mathrm{Var}(|Z_1-Z_2|) = \mathrm{E}\left(|Z_1-Z_2|^2\right) - \left(\mathrm{E}\left(|Z_1-Z_2|\right)\right)^2 = 2-\frac{4}{\pi},\\
&\mathrm{E}(U)+\mathrm{E}(V) = 0,\ \mathrm{E}(U)-\mathrm{E}(V) = \frac{2}{\sqrt{\pi}},\ \mathrm{E}(UV) = \mathrm{E}(Z_1Z_2) = 0,\\
&\mathrm{Cov}(U,V) = \mathrm{E}(UV)-\mathrm{E}(U)\mathrm{E}(V) = \frac{1}{\pi},\\
&\mathrm{Var}(U) = \mathrm{Var}(V) = \mathrm{Var}\left(\frac{Z_1+Z_2}{2}\right)+\mathrm{Var}\left(\frac{|Z_1-Z_2|}{2}\right) = \frac{1}{4}\left(4-\frac{4}{\pi}\right) = 1-\frac{1}{\pi}.
\end{aligned}$$

Thus, $\rho(U,V) = \dfrac{1/\pi}{1-1/\pi} = \dfrac{1}{\pi-1}$. If X and Y are independent normal random variables with mean μ and variance σ^2, then $X = \sigma Z_1+\mu$ and $Y = \sigma Z_2+\mu$. Hence, $\rho(\max\{X,Y\}, \min\{X,Y\}) = \rho(U,V)$.

■

■ **Example 5.58** Suppose that Z_1 and Z_2 are independent and identically distributed standard exponential random variables, Exp(1). Define $U = \max\{Z_1, Z_2\}$ and $V = \min\{Z_1, Z_2\}$. Find the correlation between U and V.

- As discussed in Example 5.34, the joint c.d.f. of U and V is given by

$$F_{U,V}(u,v) = \{F_{Z_1}(u)\}^2 - \{F_{Z_1}(u)-F_{Z_1}(v)\}^2.$$

Thus, the joint p.d.f. of U and V is obtained

$$f_{U,V}(u,v) = \frac{\partial^2}{\partial u\partial v}F_{U,V}(u,v) = 2f_{Z_1}(u)f_{Z_1}(v) = 2e^{-(u+v)},\ 0<v<u<\infty.$$

- Define $W_1 = U-V$ and $W_2 = V$. Then, the joint p.d.f. of W_1 and W_2 is

$$f_{W_1,W_2}(w_1,w_2) = f_{U,V}(w_1+w_2,w_2)|J(w_1,w_2)| = 2e^{-w_1-2w_2},\ \ 0<w_1,w_2<\infty,$$

which implies that $U-V$ and V are independent and that $U-V$ and $2V$ are identically distributed as an exponential distribution with parameter 1, Exp(1).

- Note that $\mathrm{Var}(V) = 1/4$, $\mathrm{Var}(U-V) = 1$, $0 = \mathrm{Cov}(U-V,V) = \mathrm{Cov}(U,V)-\mathrm{Var}(V)$, and $\mathrm{Var}(U) = \mathrm{Var}(U-V)+\mathrm{Var}(V)$. Thus, $\mathrm{Cov}(U,V) = 1/4$, $\mathrm{Var}(V) = 1/4$, and $\mathrm{Var}(U) = 5/4$. Hence, $\rho(U,V) = \dfrac{\mathrm{Cov}(U,V)}{\sqrt{\mathrm{Var}(U)\mathrm{Var}(V)}} = \dfrac{1/4}{\sqrt{1/4\cdot 5/4}} = \dfrac{1}{\sqrt{5}}$.

■

Note that in Examples 5.57 and 5.58, the correlation between $\max\{Z_1, Z_2\}$ and $\min\{Z_1, Z_2\}$ is positive. We expect that $\max\{Z_1, Z_2\}$ and $\min\{Z_1, Z_2\}$ are positively correlated since the maximum of two numbers would increase as the minimum of two numbers increases and that the maximum of two numbers is at least as large as the minimum of two numbers. We can check it using Monte Carlo experiments. That is, we first generate sequences of two i.i.d. random numbers $\{(z_{i1}, z_{i2})\}_{i=1}^n$ and evaluate $u_{i1} = \max\{z_{i1}, z_{i2}\}$ and $v_{i1} = \min\{z_{i1}, z_{i2}\}$ for $i = 1, \ldots, n$. Then, we can have a Monte Carlo approximation of $\rho(U, V)$ by a sample correlation coefficient based on $\{(u_{i1}, u_{i2})\}_{i=1}^n$.

- (Normal max and min) We generate n pairs of two independent standard normal random variables $\{Z_{i1}, Z_{i2}\}_{i=1}^n$ using a Box-Müller transformation in Theorem 5.3.4

$$Z_{i1} = \sqrt{-2\log(U_{i1})}\cos(2\pi U_{i2}),\ Z_2 = \sqrt{-2\log(U_{i1})}\sin(2\pi U_{i2}),$$

where $\{U_{i1}, U_{i2}\} \overset{i.i.d.}{\sim} \text{Uniform}(0,1)$.

```
nsim<-10^5; z<-matrix(0,nsim/2,2)
for (i in 1:(nsim/2)){
 u<-runif(2)
 z[i,1]<-sqrt(-2*log(u[1]))*cos(2*pi*u[2])
 z[i,2]<-sqrt(-2*log(u[1]))*sin(2*pi*u[2])
}
u<-pmax(z[,1],z[,2]); v<-pmin(z[,1],z[,2])
cor(u,v)
> cor(u,v)
[1] 0.4706403
> 1/(pi-1)
[1] 0.4669422
```

- (Exponential max and min) We generate n pairs of two independent exponential random variables $\{Z_{i1}, Z_{i2}\}_{i=1}^n$ using probability integral transform in Theorem 4.1.3

$$Z_{i1} = -\log(1-U_{i1}),\ Z_2 = -\log(1-U_{i2}),\ \{U_{i1}, U_{i2}\} \overset{i.i.d.}{\sim} \text{Uniform}(0,1).$$

```
nsim<-10^5; z<-matrix(0,nsim/2,2)
for (i in 1:(nsim/2)){
 u<-runif(2)
 z[i,1]<- -log(1-u[1]); z[i,2]<- -log(1-u[2])
}
u<-pmax(z[,1],z[,2]); v<-pmin(z[,1],z[,2])
cor(u,v)
> cor(u,v)
[1] 0.4439405
> 1/sqrt(5)
[1] 0.4472136
```

5.5 Exercises

Problem 5.1 Let X have p.d.f. $f_1(x) = 2x$ for $0 < x < 1$ and let Y have p.d.f. $f_2(y) = y/2$ for $0 < y < 2$. Find the p.d.f of $Z = X + Y$.

Problem 5.2 Suppose that $f_{X_1,X_2}(x_1,x_2) = \begin{cases} 2, & \text{for } 0 < x_2 < x_1 < 1 \\ 0, & \text{otherwise.} \end{cases}$ Find the p.d.f. $Y = X_1 + X_2$.

Problem 5.3 A point is chosen at random uniformly in the triangular area defined by the points $(1,0), (2,0)$ and $(0,1)$. Let X be the x-coordinate of the chosen point and Y be the y-coordinate of the chosen point. Let W be the area of the rectangle with vertices $(0,0)$, $(X,0)$, $(0,Y)$ and (X,Y). Find $\mathrm{E}(W)$.

Problem 5.4 Let the joint p.d.f of X and Y be

$$f_{X,Y}(x,y) = \begin{cases} 2xe^{x^2-y}, & \text{if } 0 < x < 1 \text{ and } y > x^2 \\ 0, & \text{otherwise.} \end{cases}$$

① Find the marginal p.d.f.'s of X and Y, respectively.

② Compute $P(Y < 2X^2)$.

Problem 5.5 Suppose that the joint p.d.f. of X and Y is

$$f_{X,Y}(x,y) = \begin{cases} 3x^2 & \text{for } 0 < x < 1, 0 < y < 1 \\ 0 & \text{otherwise} \end{cases}$$

① Calculate $P(X/Y \le w)$ for $w \in R$.

② Calculate $P(X + Y \le w)$ for $w \in R$.

Problem 5.6 Suppose that the joint p.d.f. of X and Y is given by

$$f_{X,Y}(x,y) = \begin{cases} 3x, & \text{for } 0 \le y \le x \le 1, \\ 0, & \text{otherwise.} \end{cases}$$

Compute $\mathrm{Cov}(X,Y)$.

Problem 5.7 Jaden and Joanne agree to meet at KU cafe at noon. Assume that the times that they actually arrive are independent random variables, identically distributed with a triangular distribution on the interval $(-5,5)$, where time is measured in minutes relative to noon. That is, we denote X and Y to be the times that the Jaden and Joanne arrive respectively with p.d.f. $f_X(x) = f_Y(x) = \begin{cases} \frac{1}{5}\left(1 - \frac{|x|}{5}\right), & -5 < x < 5, \\ 0, & \text{otherwise.} \end{cases}$ Determine the probability that the first person to arrive will wait more than 5 minutes before the other person arrives.

Problem 5.8 Suppose that a joint p.d.f. of X and Y is $f(x,y) = \lambda_1\lambda_2 e^{-(\lambda_1 x+\lambda_2 y)}$, $x,y > 0$, and $f(x,y) = 0$ otherwise, where λ_1 and λ_2 are positive constants.

① Find the distribution of $Z = X/Y$. Compute the covariance of Z and Y, i.e. $\text{Cov}(Z,Y)$.

② Let $W = \min(X,Y)$. Determine the distribution of W. Compute the variance of W.

Problem 5.9 Let X_1 and X_2 denote i.i.d. standard exponential random variables, Exp(1). Define $Y_1 = \sqrt{X_1X_2}$ and $Y_2 = \sqrt{X_1/X_2}$. Find the joint p.d.f. of Y_1 and Y_2.

Problem 5.10 A point (X,Y) inside the unit circle can be constructed as follows. A number R is randomly chosen between 0 and 1 and an angle Θ is chosen at random between 0 and 2π. Then, a point (X,Y) is constructed as $X = R\cos\Theta$ and $Y = R\sin\Theta$. Find the joint density of X and Y. Compute $\text{Cov}(X,Y)$ if they are not independent.

Problem 5.11 Let U and V be independent, $U \sim \text{Beta}(\alpha,\beta)$ and $V \sim \text{Gamma}(\alpha+\beta,\lambda)$. Define $X = UV$ and $Y = (1-U)V$. Find the joint p.d.f. of X and Y. In addition, compute the covariance between X and Y, $\text{Cov}(X,Y)$.

Problem 5.12 Let X have p.d.f. $f_X(x) = 2x$ for $0 < x < 1$ and let Y have p.d.f. $f_Y(y) = y/2$ for $0 < y < 2$. Let $Z = X - Y$. Find the p.d.f. of Z.

Problem 5.13 Suppose that the joint p.d.f. of X and Y is

$$f_{X,Y}(x,y) = \begin{cases} \frac{4}{\sqrt{\pi}} y e^{-y^2}, & \text{for } 0 < x < y < \infty, \\ 0, & \text{otherwise.} \end{cases}$$

Are X and Y independent? Compute $\text{Cov}(X,Y)$ if they are not independent.

Problem 5.14 Let $X \sim \text{Gamma}(\alpha,\lambda)$, $(\alpha \ge 3)$ and let $Y \sim \text{Gamma}(\beta,\lambda)$, $(\beta \ge 3)$ with X and Y independent.

① Is the ratio X/Y independent of $X+Y$?

② Compute the expected value of X/Y.

Problem 5.15 Suppose X and Y are independent uniform variables on $[0,1]$. Let $W = \max\{X,Y\}$. Show that W has the same distribution as $Z_1/(Z_1+Z_2)$ where $Z_1 \sim \text{Gamma}(2,1)$, $Z_2 \sim \text{Exp}(1)$, and two random variables Z_1 and Z_2 are independent.

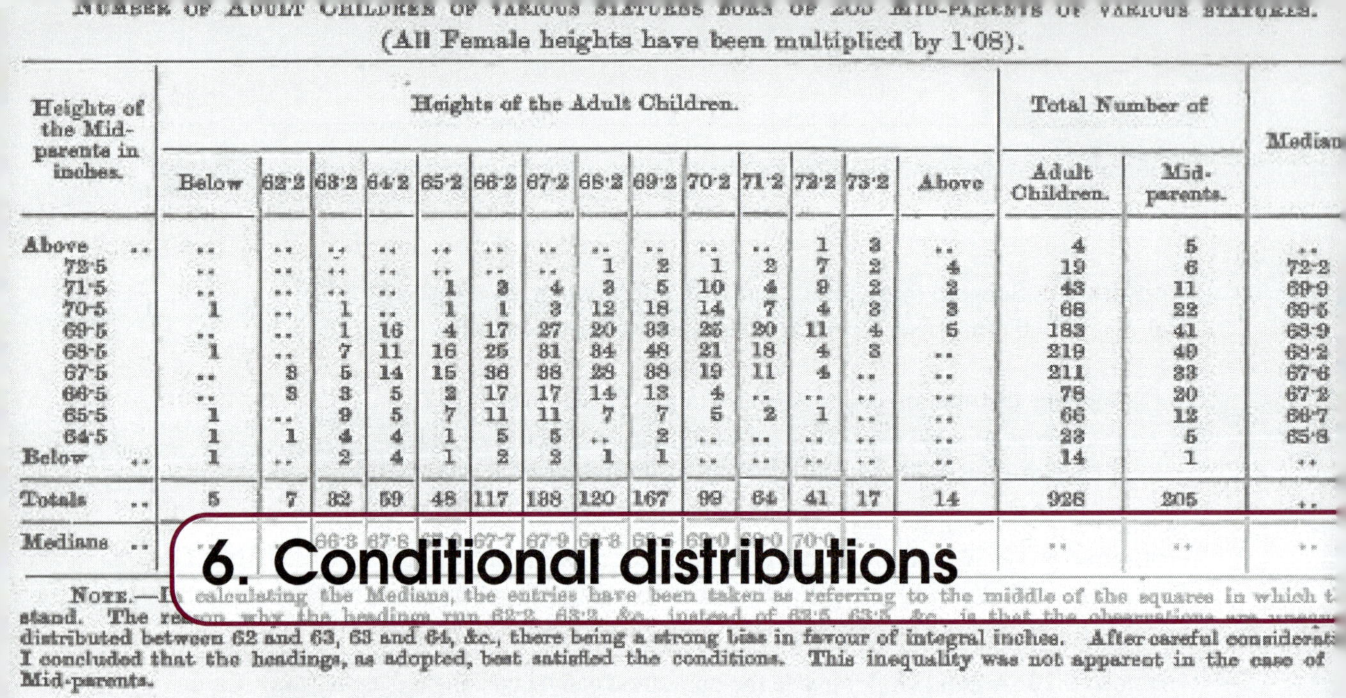

NUMBER OF ADULT CHILDREN OF VARIOUS STATURES BORN OF 205 MID-PARENTS OF VARIOUS STATURES.
(All Female heights have been multiplied by 1·08).

Heights of the Mid-parents in inches.	Heights of the Adult Children.														Total Number of		Medians.
	Below	62·2	63·2	64·2	65·2	66·2	67·2	68·2	69·2	70·2	71·2	72·2	73·2	Above	Adult Children.	Mid-parents.	
Above ..	..	..	..	..	..	..	..	..	..	..	..	1	3	..	4	5	..
72·5	..	..	..	..	..	..	..	1	2	1	2	7	2	4	19	6	72·2
71·5	..	..	..	..	1	3	4	3	5	10	4	9	2	2	43	11	69·9
70·5	1	..	1	..	1	1	3	12	18	14	7	4	3	3	68	22	69·5
69·5	..	..	1	16	4	17	27	20	33	25	20	11	4	5	183	41	68·9
68·5	1	..	7	11	16	25	31	34	48	21	18	4	3	..	219	49	68·2
67·5	..	3	5	14	15	36	38	28	38	19	11	4	..	..	211	33	67·6
66·5	..	3	3	5	2	17	17	14	13	4	..	..	..	..	78	20	67·2
65·5	1	..	9	5	7	11	11	7	7	5	2	1	..	..	66	12	66·7
64·5	1	1	4	4	1	5	5	..	2	..	..	..	..	..	23	5	65·8
Below ..	1	..	2	4	1	2	2	1	1	..	..	..	..	..	14	1	..
Totals ..	5	7	32	59	48	117	138	120	167	99	64	41	17	14	928	205	..
Medians ..	[illegible]	[illegible]	66·3	67·8	[illegible]	67·7	67·9	[illegible]	[illegible]	[illegible]	[illegible]	70·0	..	..	..	..	..

NOTE.—In calculating the Medians, the entries have been taken as referring to the middle of the squares in which they stand. The reason why the headings run 62·2, 63·2, &c., instead of 62·5, 63·5, &c., is that the observations are unequally distributed between 62 and 63, 63 and 64, &c., there being a strong bias in favour of integral inches. After careful consideration I concluded that the headings, as adopted, best satisfied the conditions. This inequality was not apparent in the case of Mid-parents.

6. Conditional distributions

6.1 Conditional density

The conditional distribution or conditioning is a very important tool in probability and statistics. Roughly speaking, conditioning is the procedure by which we learn from one random variable Y, incorporating new information of the other random variable X into the distribution of Y. The conditional distribution of Y given X improves our predictions about one measurement with Y with the help of our knowledge of another measurement with X, and it is generalized from the concept of conditional probability of one event given the other event. Thus, the conditional distribution of Y given X is understood as the distribution that we would use for Y after we learn the value of X.

Definition 6.1.1 — Conditioning a discrete random variable. The conditional distribution of Y given X is easiest to understand when X has a discrete distribution. In this case, for each x such that $P(X=x)>0$,

$$P(Y \in A|X=x) = \frac{P(Y \in A, X=x)}{P(X=x)}.$$

- If Y also has a discrete distribution, then $P(Y \in A, X=x) = \sum_{y \in A} f(x,y)$ and

$$P(Y \in A|X=x) = \frac{\sum_{y \in A} f_{X,Y}(x,y)}{f_X(x)} = \sum_{y \in A} \frac{f_{X,Y}(x,y)}{f_X(x)},$$

where $f_{Y|X}(y|x) = \dfrac{f_{X,Y}(x,y)}{f_X(x)}$ is called the *conditional p.m.f. of Y given $X=x$*.

- If Y has a continuous distribution, and X has a discrete distribution then it is still true that

$$P(Y \in A|X = x) = \frac{P(Y \in A, X = x)}{P(X = x)} = \frac{\int_A f_{X,Y}(x,y)dy}{f_X(x)} = \int_A \frac{f_{X,Y}(x,y)}{f_X(x)}dy$$

where $f_{Y|X}(y|x) = \dfrac{f_{X,Y}(x,y)}{f_X(x)}$ is called the *conditional p.d.f. of Y given $X = x$*.

Example 6.1 — Customer arrivals. Let X be the number of customers who arrive at a store during a specific time period, and let Y be the rate at which customers arrive in general. Suppose that our joint p.f./p.d.f. of X and Y is

$$f_{X,Y}(x,y) = \begin{cases} \frac{y^{x+1}}{x!}\exp(-2y) & \text{for } y > 0 \text{ and } x = 0, 1, \ldots, \\ 0 & \text{otherwise.} \end{cases}$$

Find the conditional p.d.f. of Y given $X = 1$.

- We are asked to find $f_{Y|X}(y|1)$. We need $f_X(1)$ for this.

$$\begin{aligned} f_X(1) &= \int_0^\infty y^2\exp(-2y)dy = -\frac{1}{2}y^2\exp(-2y)\Big|_{y=0}^\infty + \int_0^\infty y\exp(-2y)dy \\ &= -\frac{1}{2}y\exp(-2y)\Big|_{y=0}^\infty + \frac{1}{2}\int_0^\infty \exp(-2y)dy = \frac{1}{4}. \end{aligned}$$

Then $f_{Y|X}(y|1) = \dfrac{f_{X,Y}(1,y)}{f_X(1)} = 4y^2\exp(-2y)$, for $y > 0$.

■

The difficult cases arise when X has a continuous distribution. In these cases, we continue to define $f_{Y|X}(y|x) = f_{X,Y}(x,y)/f_X(x)$. This $f_{Y|X}$ will either be a conditional p.d.f. of Y given X or a conditional p.m.f. of Y given X depending on whether Y has a continuous or discrete distribution. The problem is that these conditional p.f.'s and p.d.f.'s cannot be interpreted strictly as being conditioned on events of the form $\{X = x\}$. Instead, think of this conditioning as being on events of the form $\{X \text{ is very close to } x\}$. This will be fine so long as $f_X(\cdot) > 0$ in an interval that contains x.

Definition 6.1.2 — Conditioning a continuous random variable. Suppose that X and Y have joint density $f_{X,Y}(x,y)$, and we are given the value of Y. By analogy with the conditional p.m.f., we make the following definition.

- If X and Y have a joint p.d.f. $f_{X,Y}(x,y)$, then the conditional p.d.f. of Y given $X = x$ is given by $f_{Y|X}(y|x) = \begin{cases} \frac{f_{X,Y}(x,y)}{f_X(x)}, & \text{if } 0 < f_X(x) < \infty \\ 0, & \text{otherwise.} \end{cases}$
- $f_{Y|X}(y|x) \geq 0$, $\forall y$, and $\int_{-\infty}^{\infty} f_{Y|X}(y|x)dy = 1$.

Remark Conditional p.d.f $f_{Y|X}(y|x)$ is a function of y (for a fixed value of $x : f_X(x) > 0$).

- Consider the conditional probability $P(Y \in A|x \le X \le x+\Delta x)$

$$
\begin{aligned}
P(Y \in A|x \le X \le x+\Delta x) &= \frac{P(Y \in A, x \le X \le x+\Delta x)}{P(x \le X \le x+\Delta x)} \\
&= \int_A \frac{1}{\Delta x}\left(\int_x^{x+\Delta x} f_{X,Y}(t,y)dt \Big/ \frac{1}{\Delta x}\int_x^{x+\Delta x} f_X(z)dz\right)dy \\
&\approx \int_A \frac{f_{X,Y}(x,y)}{f_X(x)}dy = \int_A f_{Y|X}(y|x)dy.
\end{aligned}
$$

- If X and Y are continuous with joint p.d.f. $f_{X,Y}(x,y)$, then

$$
\begin{aligned}
F_{Y|X}(y|x) &= \int_{-\infty}^{y} f_{Y|X}(t|x)dt = \lim_{h\to 0^+} P(Y \le y|x \le X \le x+h) \\
&= \lim_{h\to 0^+} \frac{\frac{1}{h}\int_{-\infty}^{y}\int_x^{x+h} f_{X,Y}(s,t)dsdt}{\frac{1}{h}\int_x^{x+h} f_X(s)ds} = \frac{\int_{-\infty}^{y} f(x,t)dt}{f_X(x)}, \\
f_{Y|X}(y|x) &= \frac{d}{dy}F_{Y|X}(y|x) = \frac{f_{X,Y}(x,y)}{f_X(x)}.
\end{aligned}
$$

- For each x, $f_{Y|X}(y|x)$ is a function of y, and it is obvious that

$$
\int_{-\infty}^{\infty} f_{Y|X}(y|x)dy = \int_{-\infty}^{\infty} \frac{f_{X,Y}(x,y)}{f_X(x)}dy = \frac{\int_{-\infty}^{\infty} f_{X,Y}(x,y)dy}{f_X(x)} = 1.
$$

- If it is known that $X = x_0$, then the point (x,y) must be on the line $X = x_0$ in the xy-plane with the height $f_{X,Y}(x_0,y)$ as shown in Figure 6.1,and $f_{Y|X}(y|x_0)$ is proportional to $f_{X,Y}(x_0,y)$.

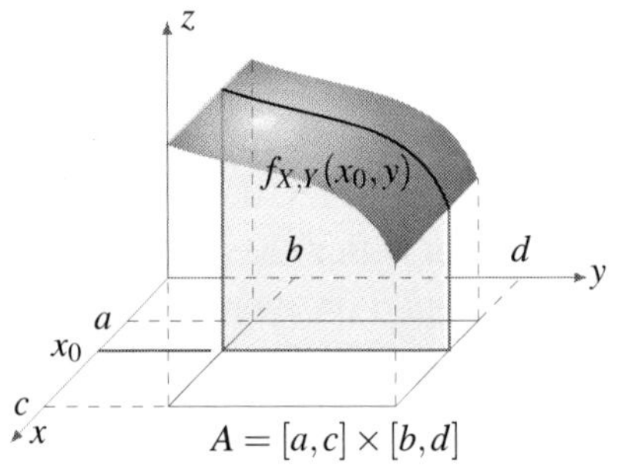

Figure 6.1: $f_{Y|X}(y|x_0)$ proportional to $f_{X,Y}(x_0,y)$ over a line $X = x_0$

- Similarly, for each value of y such that $f_Y(y) > 0$, the conditional p.d.f. of X given $Y = y$ is defined as $f_{X|Y}(x|y) = \dfrac{f_{X,Y}(x,y)}{f_Y(y)}$.

Example 6.2 Consider the uniform distribution on the circle of radius 1. The joint p.d.f. of X and Y is

$$
f_{X,Y}(x,y) = \begin{cases} \dfrac{1}{\pi}, & \text{for } x^2+y^2 \le 1, \\ 0, & \text{otherwise.} \end{cases}
$$

- The marginal p.d.f. of X is computed as $f_X(x) = \int_{-\sqrt{1-x^2}}^{\sqrt{1-x^2}} \frac{1}{\pi} dy = \frac{2\sqrt{1-x^2}}{\pi}$, for $-1 \le x \le 1$ as in Example 5.8.

- The conditional p.d.f. of Y given $X = x$ is then

$$f_{Y|X}(y|x) = \frac{f_{X,Y}(x,y)}{f_X(x)} = \begin{cases} \frac{1}{2\sqrt{1-x^2}}, & \text{for } -\sqrt{1-x^2} < y < \sqrt{1-x^2}, \\ 0, & \text{otherwise.} \end{cases}$$

- Hence, the conditional distribution of Y given $X = x$ is the uniform distribution on the interval $[-\sqrt{1-x^2}, \sqrt{1-x^2}]$, $Y|X = x \sim \text{Uniform}[-\sqrt{1-x^2}, \sqrt{1-x^2}]$ as displayed in Figure 6.2.

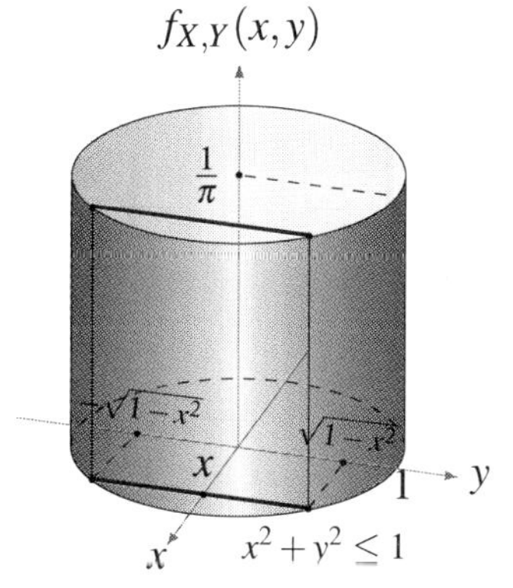

Figure 6.2: Uniformly distributed over a line $X = x$ in a cylinder of radius 1 and height $\frac{1}{\pi}$.

■

■ **Example 6.3** Let (X,Y) be a random point in the triangle $\{(x,y) : x,y \ge 0, x+y \le 1\}$ with a joint p.d.f. $f_{X,Y}(x,y) = \begin{cases} 2, & \text{for } x,y \ge 0 \text{ and } x+y \le 1, \\ 0, & \text{otherwise.} \end{cases}$ Find $f_{X|Y}(x|y)$.

- For a given $y \in [0,1]$, we know that, if $Y = y$, X is between 0 and $1-y$. Thus, the marginal p.d.f. of Y is $f_Y(y) = \int_0^{1-y} 2dx = 2(1-y)$.

- Therefore,

$$f_{X|Y}(x|y) = \begin{cases} \frac{1}{1-y}, & 0 \le x \le 1-y, \\ 0 & \text{otherwise} \end{cases}$$

In other words, given $Y = y$, X is distributed uniformly on $[0, 1-y]$.

- For example, we can compute the probability that $P(X > 0.5|Y = y)$

$$P(X > 0.5|Y = y) = \begin{cases} \int_{0.5}^{1-y} \frac{1}{1-y} dx = \frac{0.5-y}{1-y}, & \text{for } 0.5 \le y \le 1, \\ 0, & \text{otherwise.} \end{cases}$$

■

Theorem 6.1.1 The multiplication theorem for conditional probabilities says that $P(A \cap B) = P(A|B)P(B)$. For random variables, we have the analogous :

- $f_{X,Y}(x,y) = f_{X|Y}(x|y)f_Y(y) = f_{Y|X}(y|x)f_X(x)$.
- $f_X(x) = \int_{-\infty}^{\infty} f_{X|Y}(x|y)f_Y(y)dy$, $f_Y(y) = \int_{-\infty}^{\infty} f_{Y|X}(y|x)f_X(x)dx$.
- X and Y are independent if and only if
$$f_{Y|X}(y|x) = f_Y(y), \text{ for all } y \text{ and all } x \text{ such that } f_X(x) > 0.$$
- Bayes' Theorem : $f_{Y|X}(y|x) = \dfrac{f_{X|Y}(x|y)f_Y(y)}{f_X(x)}$.

Example 6.4 Suppose that a point X is chosen from the uniform distribution on the interval $(0,1)$, $X \sim \text{Uniform}(0,1)$. After obtaining a value $X = x$, a point Y is then chosen from the uniform distribution on the interval $(x,1)$, $Y|X = x \sim \text{Uniform}(x,1)$. What is the marginal distribution of Y?

- The joint p.d.f. of X and Y is given by
$$f_{X,Y}(x,y) = f_{Y|X}(y|x)f_X(x) = \begin{cases} \frac{1}{1-x}, & \text{for } 0 < x < y < 1, \\ 0, & \text{otherwise.} \end{cases}$$
- Thus, for $0 < y < 1$, the marginal p.d.f of Y is given by
$$f_Y(y) = \int_0^y \frac{1}{1-x}dx = -\log(1-y).$$
- Furthermore, the conditional p.d.f of X given $Y = y$ is given by
$$f_{X|Y}(y) = \frac{f_{X,Y}(x,y)}{f_Y(y)} = \begin{cases} \frac{-1}{(1-x)\log(1-y)}, & \text{for } 0 < x < y, \\ 0, & \text{otherwise.} \end{cases}$$

■

Example 6.5 Suppose that X has a gamma distribution with parameters 2 and λ, $X \sim \text{Gamma}(2,\lambda)$, and given $X = x$, Y has a uniform distribution on $(0,x)$, $Y|X = x \sim \text{Uniform}(0,x)$,

$$f_X(x) = \begin{cases} \lambda^2 xe^{-\lambda x}, & \text{for } x > 0, \\ 0, & \text{otherwise,} \end{cases} \quad f_{Y|X}(y|x) = \begin{cases} \frac{1}{x}, & \text{for } 0 < y < x, \\ 0, & \text{otherwise.} \end{cases}$$

- The joint p.d.f. of X and Y is obtained as
$$f_{X,Y}(x,y) = f_X(x)f_{Y|X}(y|x) = \begin{cases} \lambda^2 e^{-\lambda x}, & \text{for } 0 < y < x, \\ 0, & \text{otherwise.} \end{cases}$$

- The marginal p.d.f of Y is obtained by integrating out x in the joint p.d.f. of X and Y

$$f_Y(y) = \int_{-\infty}^{\infty} f_{X,Y}(x,y)dx = \int_y^{\infty} \lambda^2 e^{-\lambda x}dx = \lambda e^{-\lambda y},$$

which implies that Y has an exponential distribution with parameter λ, $Y \sim \text{Exp}(\lambda)$.

- Note that in the Poisson process with rate λ, the waiting time until the first arrival, T_1, has exponential distribution with parameter λ as discussed in Chapter 4, which implies that $Y \stackrel{d}{=} T_1$, and the waiting time until the second arrival (or two arrivals), T_2, has gamma distribution with parameters 2 and λ, that is, $X \stackrel{d}{=} T_2$. Thus, we can interpret that T_1 given $T_2 = t_2$ is uniformly distributed on $(0, t_2)$.

■

■ **Example 6.6** Suppose that the joint p.d.f. of X and Y is given by

$$f_{X,Y}(x,y) = \begin{cases} 3x, & \text{if } 0 \le y \le x \le 1 \\ 0, & \text{otherwise.} \end{cases}$$

- Since the joint p.d.f. is a constant with respect to y, the conditional distribution of Y given $X = x$ is expected to be uniformly distributed. Indeed, we compute the marginal p.d.f. of X and the conditional p.d.f. of Y given $X = x$

$$f_X(x) = \int_0^x 3x dy - 3x^2, \ f_{Y|X}(y|x) = \frac{f_{X,Y}(x,y)}{f_X(x)} = \frac{3x}{3x^2} = \frac{1}{x}, \text{ for } 0 \le y \le x,$$

which implies that $Y|X = x \sim \text{Uniform}[0,x]$ for $0 \le x \le 1$.

- Note that the marginal p.d.f. of Y is obtained as

$$f_Y(y) = \int_y^1 3x dx = \frac{3}{2}(1 - y^2), \text{ for } 0 \le y \le 1.$$

Hence $f_Y(y) \neq f_{Y|X}(y|x)$, which implies that X and Y are not independent.

■

■ **Example 6.7** Suppose that the joint density of X and Y is given by

$$f_{X,Y}(x,y) = \begin{cases} e^{-y}, & 0 < x < y < \infty \\ 0, & \text{otherwise.} \end{cases}$$

- $f_X(x) = \int_x^{\infty} e^{-y}dy = e^{-x}, x > 0$. Thus, $X \sim \text{Exp}(1)$.

- $f_{Y|X}(y|x) = \frac{f_{X,Y}(x,y)}{f_X(x)} = \begin{cases} e^{-(y-x)}, & \text{for } y \ge x, \\ 0, & \text{otherwise.} \end{cases}$ That is, the conditional distribution of Y given $X = x$ is a shifted exponential distribution, $Y - x|X = x \sim \text{Exp}(1)$.

■

Example 6.8 — Customer arrivals (revisited). Let X be the number of customers who arrive at a store during a specific time period, and let Y be the rate at which customers arrive in general. Suppose that the conditional p.f. of X given $Y = y$ is

$$f_X(x|y) = \frac{y^x}{x!}\exp(-y), \text{ for } x = 0, 1, \ldots.$$

Suppose that the marginal p.d.f. of Y is $f_Y(y) = y\exp(-y)$, for $y > 0$. Find the conditional p.d.f. of Y given $X = x$ for general x.

- $X|Y = y \sim \text{Poisson}(y)$ and $Y \sim \text{Gamma}(2, 1)$.
- The joint p.f./p.d.f. is obtained by the multiplication theorem that $f_{X,Y}(x,y) = f_{X|Y}(x|y)f_Y(y)$, namely $y^{x+1}\exp(-2y)/x!$, for $y > 0$ and $x = 0, 1, \ldots$.
- We can find $f_X(x)$ by integrating the joint p.f./p.d.f. with respect to y,

$$f_X(x) = \int_0^\infty \frac{y^{x+1}}{x!}\exp(-2y)dy = \frac{(x+1)!}{x!2^{x+2}} = \frac{x+1}{2^{x+2}}, \; x = 0, 1, \ldots.$$

- Hence $f_Y(y|x) = \dfrac{y^{x+1}\exp(-2y)/x!}{(x+1)/2^{x+2}} = \dfrac{2^{x+2}y^{x+1}}{(x+1)!}\exp(-2y)$, for $y > 0$.

■

Example 6.9 Suppose that the conditional distribution of Y given $X = x$ is an exponential with parameter x and that the marginal distribution of X is an exponential distribution with parameter 1. Find the conditional distribution of X given $Y = y$.

- We can construct the joint p.d.f. of X and Y

$$f_{X,Y}(x,y) = f_{Y|X}(y|x)f_X(x) = \begin{cases} xe^{-x(y+1)}, & \text{for } x > 0,\ y > 0, \\ 0, & \text{otherwise.} \end{cases}$$

- Since $f_{X|Y}(x|y) \propto f_{X,Y}(x,y)$, the conditional distribution of X given $Y = y$ is a gamma distribution with parameters 2 and $y+1$. Indeed, we can obtain $f_{X|Y}(x|y)$ using Bayes' theorem for random variables given in Theorem 6.1.1

$$f_{X|Y}(x|y) = \frac{f_{X,Y}(x,y)}{f_Y(x)} = \frac{xe^{-x(y+1)}}{\int_0^\infty xe^{-x(y+1)}dx} = (y+1)^2xe^{-yx}, \; X|Y = y \sim \text{Gamma}(2, y+1).$$

■

Example 6.10 — Scale mixture of normal : *t*-distribution. Let X and Y be two independent random variables distributed as $X \sim N(0,1)$ and $Y \sim \chi^2(r)$. Consider the random variable $T = \dfrac{X}{\sqrt{Y/r}}$ as discussed in Example 5.44. Here, we revisit t-distribution as a *scale mixture of normal distributions*.

- Suppose that $T|Y=y \sim N(0, r/y)$ and $Y \sim \text{Gamma}(\alpha, \lambda)$. Then, we have

$$f_{T|Y}(w|y) = \frac{\sqrt{y}}{\sqrt{2\pi r}} \exp\left(-\frac{y}{2r}t^2\right), \quad f_Y(y) = \frac{\lambda^\alpha}{\Gamma(\alpha)} y^{\alpha-1} \exp(-\lambda y),$$

$$f_{T|Y}(t|y) f_Y(y) = \left[\frac{\sqrt{y}}{\sqrt{2\pi r}} \exp\left(-\frac{y}{2r}t^2\right)\right] \left[\frac{\lambda^\alpha}{\Gamma(\alpha)} y^{\alpha-1} \exp(-\lambda y)\right],$$

$$f_T(t) = \int_0^\infty f_{T|Y}(t|y) f_Y(y) dy = \int_0^\infty \frac{y^{\alpha+1/2-1}\lambda^\alpha}{\sqrt{2\pi r}\Gamma(\alpha)} \exp\left[-y\left(\frac{t^2}{2r}+\lambda\right)\right] dy$$

$$= \frac{\Gamma\left(\alpha+\frac{1}{2}\right)\lambda^\alpha}{\sqrt{2\pi r}\Gamma(\alpha)} \left(\frac{2}{2\lambda+(t/r)^2}\right)^{\alpha+1/2}.$$

- Thus, if $\alpha = r/2$ and $\lambda = 1/2$, then $f_T(t)$ is simplified as

$$f_T(t) = \frac{\Gamma\left(\frac{r}{2}+\frac{1}{2}\right)(1/2)^{r/2}}{\sqrt{2\pi r}\Gamma(r/2)} \left(\frac{2}{1+(t/r)^2}\right)^{r/2+1/2}, \quad \text{for } -\infty < t < \infty,$$

which implies that T has a t-distribution with r degrees of freedom.

- In addition, since Z and Y are independent,

$$F_T(t) = P(T \le t) = P\left(\frac{Z}{\sqrt{Y/r}} \le t\right) = \int_0^\infty \int_{-\infty}^{t\sqrt{y/r}} f_Z(z)dz f_Y(y)dy,$$

$$f_T(t) = \frac{d}{dt}F_T(t) = \int_0^\infty \left\{\frac{1}{\sqrt{2\pi}} e^{-\frac{y}{2r}t^2} \frac{\sqrt{y}}{\sqrt{r}}\right\} \frac{1}{\Gamma(r/2)2^{r/2}} y^{r/2-1} e^{-x/2} dy,$$

which also shows the scale mixture of normal distributions.

- If Y has a gamma distribution with parameters α and β, $Y \sim \text{Gamma}(\alpha, \beta)$, then, the distribution of $W = 1/Y$ is called the *inverse gamma distribution with parameters α and β*, $W \sim \text{IG}(\alpha, \beta)$ with p.d.f.

$$f_W(w) = f_Y(1/w)\left|\frac{d}{dw}\frac{1}{w}\right| = \frac{\lambda^\alpha}{\Gamma(\alpha)} w^{-(\alpha+1)} \exp(-\lambda/w).$$

Hence, t-distribution is also understood as a marginal distribution with a normal-inverse gamma mixture, $X|w \sim N(\mu, rw)$ and $W \sim \text{IG}(r/2, 1/2)$.

■

Remark Many distributions can be represented as mixtures of other distributions.

$$f_X(x) = \int f_{X|Y}(x|y) f_Y(y) dy \quad \text{or} \quad f_X(x) = \sum_y f_{X|Y}(x|y) P(Y=y)$$

where the mixing distribution of a random variable Y can be continuous with a p.d.f. $f_Y(y)$ or discrete with a p.m.f. $P(Y=y)$. To generate a random variable X, we can first generate a mixing variable Y from the mixing distribution and then generate X from the selected conditional distribution $f_{X|Y}(x|y)$.

- If $Y \sim f_Y(y)$ (if continuous) and $X \sim f_{X|Y}(x|y)$, then $X \sim f_X(x)$. For example, as in Example 6.10, $X \sim t(r)$ is a continuous mixture of $X|y \sim N(0, r/y)$ and $Y \sim \chi^2(r)$.

- If $Y \sim P(Y=y)$ (if discrete) and $X \sim f_{X|Y}(x|y)$, then $X \sim f(x)$. For example, if X is a negative binomial random variable, $X \sim \text{Negbin}(r,p)$, then X has the mixture representation

$$X|y \sim \text{Poisson}(y) \ \ \textit{and} \ \ Y \sim \text{Gamma}(r,\lambda),\ \lambda = \frac{p}{1-p}.$$

That is, we have $f_X(x) = \displaystyle\int_0^\infty \left[\frac{y^x e^{-y}}{x!}\right]\left[\frac{\lambda^r}{\Gamma(r)} y^{r-1} e^{-\lambda y}\right] dy = \frac{\Gamma(r+x)\lambda^r}{\Gamma(r)x!(1+\lambda)^{r+x}}$.
When $p = \lambda/(1+\lambda)$, the marginal p.d.f. of X becomes

$$f_X(x) = \frac{(r+x-1)!}{(r-1)!x!} p^r (1-p)^x,\ x = 0, 1, 2, \ldots,$$

which is a p.m.f. of the negative binomial distribution, $X \sim \text{Negbin}(r,p)$, where X is the number of failures until the rth success is obtained.

Example 6.11 — Accept-Reject algorithm (1). Let X and U be independent random variables such that X has p.d.f $f_X(x)$ and U is uniformly distributed on $(0,1)$. Suppose that there exists a constant M such that for all x a p.d.f $f_Z(x)$ satisfies $0 \le f_Z(x) \le M f_X(x)$. Then, $P(X \le x | MUf_X(X) \le f_Z(X)) = \displaystyle\int_{-\infty}^x f_Z(z)dz$.

- We compute the conditional probability of $\{X \le x\}$ given that $MUf_X(X) \le f_Z(X)$. Since $f_{X,U}(x,u) = f_X(x) \cdot 1$,

$$\begin{aligned} &F_{X|A}(x) = P(X \le x | MUf_X(X) \le f_Z(X)) \\ = \ & \frac{P(X \le x,\ MUf_X(X) \le f_Z(X))}{P(MUf_X(X) \le f_Z(X))} = \frac{\int_{-\infty}^x \int_0^{f_Z(t)/(Mf_X(t))} f_X(t)dudt}{\int_{-\infty}^\infty \int_0^{f_Z(t)/(Mf_X(t))} f_X(t)dudt} \\ = \ & \int_{-\infty}^x \frac{f_Z(t)}{Mf_X(t)} f_X(t)dt \Big/ \int_{-\infty}^\infty \frac{f_Z(t)}{Mf_X(t)} f_X(t)dt = \int_{-\infty}^x f_Z(t)dt. \end{aligned}$$

- In other words, suppose we have independent random variables X and U. Then, the conditional density of X given the event $A = \{MUf_X(X) \le f_Z(X)\}$, is $f_Z(x)$, in that

$$f_{X|A}(x) = \frac{d}{dx} F_{X|A}(x) = f_Z(x).$$

- Now, suppose we have a sequence $\{(X_n, U_n)\}_{n=1}^N$, $(N \ge 1)$, of random variables that have the same distributions as (X, U). For every pair (X_n, U_n) for which A^c occurs, we reject and only accept the pair $(X_n, U_n) \in A$. Then, we obtain $Z = X_n$, which has a p.d.f. $f_Z(z)$. This is known as the *Accept-Reject* algorithm or the *Rejection* method.

■

■ **Example 6.12 — Accept-Reject algorithm (2).** Suppose that $f_Z(z)$ is a pdf that is non-zero only on an interval $[a,b]$ for some numbers $a < b$. Suppose also that the maximum value of $f_Z(z)$ is less than $M > 0$.

To generate a random variable Z with p.d.f. $f_Z(z)$, consider the following procedure:

⓪ Initialize an index of iteration n, setting $n = 1$.
① Generate (X_n, U_n) where X_n and U_n are independent,

$$X_n \sim \text{Uniform}[a,b], \text{ and } U_n \sim \text{Uniform}[0,1].$$

② If $MU_n \leq f_Z(X_n)$, we "accept" the pair, and go to step ④.
③ Otherwise, (X_n, U_n) is "rejected". Increment i by one and go to step ①.
④ Set $X = X_n$ and return the random value X.

Then, X, the value returned by the above procedure has p.d.f. $f_Z(z)$.

- In other words, we generate an independent sequence of random points (X_n, Y_n) uniformly in the rectangle $(a,b) \times (0,M)$, where $Y_n = MU_n$, $U_n \sim \text{Uniform}(0,1)$. Return the first component of the first "accepted" pair, as illustrated in Figure 6.3.

- Let $A = \{(X,Y) \ : \ Y \leq f_Z(X)\}$. Let X be the returned value or the horizontal coordinate of the accepted pair $(X_n, Y_n) \in A$. Then,

$$P(\{X \leq z\}|A) = P(\{X_n \leq z | Y_n \leq f_Z(X_n)\}) = \frac{P(\{X_n \leq z,\ Y_n \leq f_Z(X_n)\})}{P(\{Y_n \leq f_Z(X_n)\})}$$

$$= \frac{\int_a^z \int_0^{f_Z(x)} \frac{1}{M(b-a)} dxdy}{\int_a^b \int_0^{f_Z(x)} \frac{1}{M(b-a)} dxdy} = \frac{\int_a^z \frac{f_Z(x)}{M(b-a)} dx}{\int_a^b \frac{f_Z(x)}{M(b-a)} dx} = F_Z(z) - F_Z(a) = F_Z(z).$$

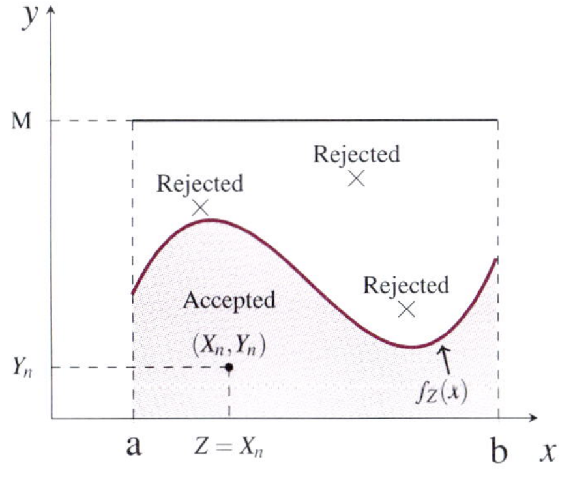

Figure 6.3: An illustration of Accept-Reject algorithm

■

Remark Let $f(x)$ be a p.d.f of a continuous random variable with a support S, $S \subset R$ and let $A = \{(x,y) \in S \times [0,\infty) \ : \ 0 \le y \le f(x)\}$. Note that the area of A, $|A|$ is obtained by integrating the height $f(x)$ over S. Thus, $|A| = \int_S f_X(x)dx = 1$.

① If X and Y are uniformly distributed on A, then X is distributed with p.d.f. f on S and $U = Y/f_X(X)$ has a uniform distribution, where $U \sim$ Uniform$[0,1]$. In addition, X and U are independent.

- Define $U = Y/f(X)$. $V = X$. Then, the inverse transforms are $X = V$ and $Y = f(V)U$ with $J(u,v) = f(v)$. Thus, the joint p.d.f. of U and V is given by $f_{U,V}(u,v) = 1 \cdot f(v)$, for $0 \le u \le 1$ and $V \in S$. Hence, the marginal p.d.f. of V is given by $f_V(v) = \int_0^1 f_{U,V}(u,v)du = f(v)$ and the marginal p.d.f. of U is $f_U(u) = \int_S f_{U,V}(u,v)dv = 1$, a uniform distribution, $U \sim$ Unif$(0,1)$. In addition, X and U are independent since $f_{U,V}(u,v) = f_U(u)f_V(v)$.

② Suppose that X is distributed with p.d.f. f on S and $U \sim$ Uniform$[0,1]$ and that X and U are independent. Then, X and $Y = f_X(X)U$ are uniformly distributed on A.

- Let $Y = f_X(X)U$ and $V = X$. Then, the inverse transformations are $X = V$ and $U = Y/f_X(V)$ with $J(y,v) = 1/f_X(v)$. Thus, the joint p.d.f. of Y and V is $f_{Y,V}(y,v) = f_{X,U}(v, y/f_X(v))|J(y,v)| = f_X(v)1/f_X(v) = 1$, for $0 \le y \le f_X(v)$ and $v \in S$. Hence, the joint p.d.f. of X and Y is a uniform distribution on $A = \{(x,y) \ : \ x \in S, \ 0 \le y \le f_X(x)\}$,

$$f_{X,Y} = \begin{cases} 1, & \text{for } x \in S \text{ and } 0 \le y \le f_X(x), \\ 0, & \text{otherwise.} \end{cases}$$

R Consider generating standard normal random numbers using accept/reject algorithm (rejection method), $Z \sim N(0,1)$ with p.d.f $f_Z(z) = \frac{1}{\sqrt{2\pi}} e^{-\frac{1}{2}z^2}$, $-\infty < z < \infty$.

- We start by generating the absolute value of Z, that is $X = |Z|$, with p.d.f. $f_X(x) = \frac{2}{\sqrt{2\pi}} e^{-x^2/2}$, $0 < x < \infty$.
- We use an exponential random variables with p.d.f. $f_Y(y) = e^{-y}$, $y > 0$ as a candidate random variable Y. To apply the rejection method, we need to bound the ratio of $f_X(x)$ and $f_Y(y)$ with a constant M,

$$\frac{f_X(x)}{f_Y(x)} = \sqrt{\frac{2}{\pi}} e^{-\frac{x^2-2x}{2}} = \sqrt{\frac{2e}{\pi}} e^{-\frac{(x-1)^2}{2}} \le \sqrt{\frac{2e}{\pi}} = M$$

and $\frac{f_X(x)}{Mf_Y(x)} e^{-\frac{(x-1)^2}{2}}$. Then, we proceed the rejection method as follows:

① Generate Y, an exponential random variable with rate 1, $Y \sim$ Exp(1).
② Generate a uniform random number U, $U \sim$ Uniform$(0,1)$.

③ If $U \le \exp\{-(Y-1)^2/2\}$, set $X = Y$. Otherwise, return to ①.

- Once we have simulated a random variable X having density function of $|Z|$, we can then obtain a standard normal Z by letting Z be equally likely to be either X or $-X$. That is, we define $W = (2S-1)X$, where $S \sim$ Bernoulli$(1/2)$ assuming that S and X are independent. Then, the marginal p.d.f. of W is

$$f_W(w) = \sum_{i=0}^{1} f_{W|S}(w|S=i)P(S=i)$$
$$= \begin{cases} \frac{1}{2}f_X(-w) = \frac{1}{\sqrt{2\pi}}e^{-w^2/2}, & \text{for } -\infty < w \le 0, \\ \frac{1}{2}f_X(w) = \frac{1}{\sqrt{2\pi}}e^{-w^2/2}, & \text{for } 0 < w < \infty, \end{cases}$$

which implies that W has the standard normal distribution. Alternatively, we can compute the m.g.f. of W using the law of total probability for expectations

$$M_W(t) = \mathrm{E}\left(e^{tW}\right) = \mathrm{E}\left[\mathrm{E}\left(e^{t(2S-1)|X|}|S\right)\right] = \frac{1}{2}\left[\mathrm{E}\left(e^{t|X|}\right) + \mathrm{E}\left(e^{-t|X|}\right)\right]$$
$$= \frac{1}{2}\int_{-\infty}^{\infty}(e^{t|x|}+e^{-t|x|})\frac{1}{2\pi}e^{-\frac{1}{2}x^2}dx = \frac{1}{2}\int_{-\infty}^{\infty}(e^{tx}+e^{-tx})\frac{1}{2\pi}e^{-\frac{1}{2}x^2}dx = e^{\frac{1}{2}t^2},$$

Thus, we conclude that W has the standard normal distribution by the uniqueness of the m.g.f.

- Hence, we generate V a uniform random variable between $(0,1)$ and generate X according to the rejection algorithm described above. If $V \in (0,1/2)$ set $Z = X$, otherwise set $Z = -X$. More generally, we can generate $X \sim N(\mu, \sigma^2)$ by the linear transformation $X = \sigma Z + \mu$.

```
fx <- function(x) {sqrt(2/pi)*exp(-x^2/2)}
norm.sim <- function(mu,sigma) {
  fx <- function(x)  sqrt(2/pi)*exp(-x^2/2)
  fy <- function(y) exp(-y)
  M <- sqrt(2*exp(1)/pi)
  while (TRUE) {
    X <- -log(runif(1));  U <- runif(1); V <- runif(1)
    if (U < fx(X)/(M*fy(X))) return(sigma*(2*(V < 0.5)-1)*X+mu)
  }
}
norm.sim(0,1)
z <- replicate(1000,norm.sim(0,1))
#generating 1000 standard normal numbers
```

6.2 Conditional expectation

Because conditional distributions behave just like distributions and $f_{X|Y}(x|y)$ is a p.d.f. of X given $Y = y$, we can define the mean, median, variance, and moments of conditional distributions. *Conditional expectations* are just expectations computed from conditional distributions. Similarly, conditional variances, conditional medians, and conditional covariances are computed from conditional distributions and conditional joint distributions.

Definition 6.2.1 — Conditional Expectation. For a random variable Y with conditional p.d.f. $f_{Y|X}(y|x)$ given $X=x$, the *conditional mean* of Y given $X=x$ is

$$\mathrm{E}(Y|x)=\mathrm{E}(Y|X=x)=\int yf_{Y|X}(y|x)dy \text{ if } \int |y|f_{Y|X}(y|x)dy<\infty.$$

If Y is discrete with conditional p.m.f. $f_{Y|X}(y|x)$, then $\mathrm{E}(Y|x)=\mathrm{E}(Y|X=x)=\sum_{\text{all } y} yf_{Y|X}(y|x)$ if $\sum_{y}|y|f_{Y|X}(y|x)<\infty$.

Example 6.13 Suppose that the pair (X,Y) has a uniform distribution on the triangle $\{(x,y):0<y<x<1\}$. The area of this triangle is 1/2, so the joint p.d.f. is

$$f_{X,Y}(x,y)=\begin{cases} 2, & \text{if } 0<y<x<1, \\ 0, & \text{otherwise.} \end{cases}$$

The marginal p.d.f. of X is $f_X(x)=\int_0^x 2dy=2x$, for $0<x<1$. The conditional p.d.f. of Y given $X=x$ is $f_{Y|X}(y|x)=\frac{2}{2x}=\frac{1}{x}$, for $0<y<x$.

- The conditional mean of Y given $X=x$ is $\mathrm{E}(Y|x)=\int_0^x y\frac{1}{x}dy=\left.\frac{y^2}{2x}\right|_{y=0}^{x}=\frac{x}{2}$.
- Notice that the conditional distribution of Y given $X=x$ is the uniform distribution on the interval $[0,x]$. The mean of this distribution is $x/2$, hence the conditional mean of Y given $X=x$ is $x/2$. We could reason this way without doing the integral.

■

Example 6.14 Let X and Y have the joint p.d.f.

$$f_{X,Y}(x,y)=\begin{cases} Cx(y-x)e^{-y}, & 0\le x\le y<\infty, \\ 0, & \text{otherwise.} \end{cases}$$

- We first find the marginal p.d.f of X and Y respectively. Note that their marginal supports are both $x>0$ and $y>0$. Thus,

$$\begin{aligned} f_Y(y) &= \int_0^y Cx(y-x)e^{-y}dx=\frac{C}{6}y^3e^{-y}, \text{ for } y>0, \\ f_X(x) &= \int_x^\infty Cx(y-x)e^{-y}dy=Cx\left[\int_x^\infty ye^{-y}dy-x\int_x^\infty e^{-y}dy\right] \\ &= Cx\left[xe^{-x}+e^{-x}-xe^{-x}\right]=Cxe^{-x}, \text{ for } x>0. \end{aligned}$$

We can find the normalizing constant C but may not need to since it is canceled. Indeed, the distributions of X and Y are easily identified as gamma distributions, $X\sim\text{Gamma}(2,1)$, and $Y\sim\text{Gamma}(4,1)$ with $C=1$.

- Given $Y = y$, the support of X is $0 < X < y$, and the conditional p.d.f. of X given $Y = y$ is obtained as $f_{X|Y}(x|y) = \dfrac{f_{X,Y}(x,y)}{f_Y(y)} = \dfrac{6x(y-x)}{y^3}$ for $0 < x < y$. Hence, we have $\mathrm{E}(X|Y = y) = \int_0^y x f_{X|Y}(x|y)dx = \int_0^y \dfrac{6x^2(y-x)}{y^3}dx = \dfrac{1}{2}y$.

- To find $\mathrm{E}(Y|X = x)$, we first evaluate the conditional p.d.f. of Y given $X = x$ $f_{Y|X}(y|x) = \dfrac{f_{X,Y}(x,y)}{f_X(x)} = (y-x)e^{x-y}$ for $x \le y < \infty$. Then, we have $\mathrm{E}(Y|X = x) = \int_x^\infty y f_{Y|X}(y|x)dy = \int_x^\infty y(y-x)e^{x-y}dy = x+2$.

■

Remark Notice that $\mathrm{E}(Y|x)$ is a function of x. If we call it $h(x)$, then we can plug the random variable X into this function. The result, $h(X)$ is called $\mathrm{E}(Y|X)$. In Example 6.12, $\mathrm{E}(Y|X) = X/2$, and $\mathrm{E}(Y|X) = X+2$ in Example 6.13, which are random variables as some functions of the random variable X.

■ **Example 6.15 — Clinical trials (revisited).** Consider a random variable Y that tells us how likely patients are to respond positively. Let X be the actual number of patients who respond positively in a sample of size n. The following function is a joint p.f./p.d.f.:

$$f_{X,Y}(x,y) = \begin{cases} \binom{n}{x} y^x(1-y)^{n-x}, & \text{if } x \in \{0, \ldots, n\} \text{ and } 0 < y < 1, \\ 0, & \text{otherwise.} \end{cases}$$

- Note that the marginal distribution of Y is a uniform on the interval $[0,1]$, $Y \sim \mathrm{Uniform}[0,1]$, and that the marginal distribution of X is a discrete uniform on $\{0,1,\ldots,n\}$ as shown in Example 5.9.

- Then, the conditional distribution of X given $Y = y$ is the binomial distribution with parameters n and y. That is $f_{X|Y}(x|y) = \binom{n}{x}y^x(1-y)^{n-x}$ for $x = 0,1,\ldots,n$. Also, we can compute the conditional distribution of Y given $X = x$

$$f_{Y|X}(y|x) = \frac{\binom{n}{x}y^x(1-y)^{n-x}}{1/(n+1)} = \frac{(n+1)!}{x!(n-x)!}y^x(1-y)^{n-x}, \text{ for } 0 < y < 1,$$

which is the beta distribution with parameters $x+1$ and $n-x+1$.

- Hence, the conditional mean of X, the number of positive responses, given $Y = y$ is $\mathrm{E}(X|y) = ny$, and the conditional mean of Y given $X = x$ is $\mathrm{E}(Y|x) = \dfrac{x+1}{n+2}$.

■

Remark Let A be an event with positive probability. The conditional expectation $\mathrm{E}(Y|A)$ given an event is defined by $\mathrm{E}(Y|A) = \mathrm{E}(Y|I_A = 1) = \dfrac{\mathrm{E}(YI_A)}{P(A)}$.

- Because the expected value of an indicator of an event is the probability of that event, $P(B|A) = \mathrm{E}(I_B|A) = \dfrac{\mathrm{E}(I_B \cdot I_A)}{P(A)} = \dfrac{\mathrm{E}(I_{B\cap A})}{P(A)} = \dfrac{P(B\cap A)}{P(A)}$.

- If Y is a discrete random variable, then the conditional expectation of Y given A is $\mathrm{E}(Y|A) = \sum_y yP(Y = y|A)$, where the sum is over the support of Y.

- If Y is a continuous random variable with p.d.f. $f_Y(y)$, then $\mathrm{E}(Y|A) = \int_{-\infty}^{\infty} yf_Y(y|A)dy$, where the conditional p.d.f $f_Y(y|A)$ is defined as the derivative of the conditional c.d.f. $F_Y(y|A) = P(Y \le y|A)$.

- When X is a discrete random variable, $\mathrm{E}(Y|X = x)$ is equivalent to $\mathrm{E}(Y|A)$, where A is the event of $\{X = x\} = \{\omega : X(\omega) = x\}$.

- When X is a continuous random variable, we take $A = \{x \le X \le x+\Delta_x)$. Then,

$$\begin{aligned} \mathrm{E}(YI_A) &= \int_A \int_{-\infty}^{\infty} yf_{X,Y}(x,y)dydx \approx \Delta_x \int_{-\infty}^{\infty} yf_{X,Y}(x,y)dy, \\ P(A) &\approx \Delta_x f_X(x) = \Delta_x \int_{-\infty}^{\infty} f_{X,Y}(x,y)dy. \end{aligned}$$

Hence, it follows that

$$\mathrm{E}(Y|A) = \frac{\mathrm{E}(YI_A)}{P(A)} = \frac{\int_{-\infty}^{\infty} yf_{X,Y}(x,y)dy}{f_X(x)} = \int_{-\infty}^{\infty} yf_{Y|X}(y|x)dy.$$

- $\mathrm{E}(X)$ can be regarded as a conditional expectation in that $\mathrm{E}(X) = \mathrm{E}(X|\Omega)$.

Example 6.16 Suppose that $X \sim \mathrm{Exp}(\lambda)$. Find $\mathrm{E}(X|A)$, where $A = \{X \ge t\}$.

- Since $F_X(x) = 1 - e^{-\lambda x}$, we have

$$\begin{aligned} \mathrm{E}(X|A) &= \frac{\mathrm{E}(XI_A)}{P(A)} = \frac{\int_t^{\infty} x\lambda e^{-\lambda x}dx}{e^{-\lambda t}} = \frac{\int_0^{\infty}(y+t)e^{-\lambda(y+t)}dy}{e^{-\lambda t}} \\ &= 1/\lambda + t = \mathrm{E}(X) + t, \end{aligned}$$

which also follows from the memoryless property of exponential distribution that $X - t|X \ge t \overset{d}{=} X \sim \mathrm{Exp}(\lambda)$ as discussed in Section 4.4.

■

Example 6.17 Suppose that a random point (X,Y) is chosen uniformly from the unit disk $\{(x,y) \mid x^2+y^2 \le 1\}$. Find $\mathrm{E}(X|\{Y > 0\})$ and $\mathrm{E}(Y|\{Y > 0\})$.

- Intuitively, we expect that $\mathrm{E}(X|\{Y > 0\}) = 0$, since the upper half-disk is symmetric about 0 in the horizontal coordinate, as discussed in Example 6.2,

$$\mathrm{E}(X|\{Y>0\}) = \frac{\mathrm{E}\left(XI_{\{Y>0\}}\right)}{P\{Y>0\}} = 2\mathrm{E}\left(XI_{\{Y>0\}}\right) = 2\int_0^1 \int_{-\sqrt{1-y^2}}^{\sqrt{1-y^2}} \frac{x}{\pi}dxdy = 0,$$

where the second equality occurs because $P\{Y>0\}=1/2$.

- Also, $\mathrm{E}(Y|\{Y>0\})$ is computed as

$$\mathrm{E}(Y|\{Y>0\})=\frac{\mathrm{E}\left(YI_{\{Y>0\}}\right)}{P\{Y>0\}}=2\int_0^1\int_{-\sqrt{1-y^2}}^{\sqrt{1-y^2}}y\frac{1}{\pi}dxdy=\frac{4}{\pi}\int_0^1 y\sqrt{1-y^2}dy=\frac{4}{3\pi}.$$

■

$\mathrm{E}(\cdot|X=x)$ can also be understood as an operator taking a single input as a random variable and returning a real number, or a real-valued function of x, which has the same properties with the unconditional expectation (operator) $\mathrm{E}(\cdot)$. In other words, since conditional expectations are expectations, they have the same properties as unconditional expectations, such as linearity and the law of the unconscious statistician. In addition, there are additional properties specific to conditional expectation as presented in Theorem 6.2.1.

Theorem 6.2.1 Let X, Y, Y_1 and Y_2 be random variables, let g be a function, and c a constant. Then,

① $\mathrm{E}(c|X=x)=c$.

② $\mathrm{E}(cY_1+dY_2|X=x)=c\mathrm{E}(Y_1|X=x)+d\mathrm{E}(Y_2|X=x)$.

③ For any function $g(x)$, $\mathrm{E}(g(X)Y|X=x)=g(x)\mathrm{E}(Y|X=x)$.

④ $\mathrm{E}(Y|X=x)=\mathrm{E}(Y)$ if X and Y are independent.

Proof. Note that ① and ② are obvious by constancy and linearity property of the unconditional expectation. For ③, it follows from the law of unconscious statistician for the unconditional expectation

$$\mathrm{E}(g(X)Y|X=x)=\int g(x)yf_{Y|X}(y|x)dy=g(x)\int yf_{Y|X}(y|x)dy=g(x)\mathrm{E}(Y|X=x),$$

which can also be understood as linearity of the conditional expectation since $g(X)$ is regarded as a constant $g(x)$ given $X=x$. If X and Y are independent, then $f_{Y|X}(y|x)=f_Y(y)$. Thus, $\mathrm{E}(Y|X=x)=\int yf_{Y|X}(y|x)dy=\int yf_Y(y)dy=\mathrm{E}(Y)$, which proves ④. ■

■ **Example 6.18** Suppose that two continuous random variables X and Y have a joint p.d.f given by $f_{X,Y}(x,y)=\begin{cases}\frac{e^{-x/y}e^{-y}}{y}, & \text{if x > 0 and y > 0,}\\ 0, & \text{otherwise.}\end{cases}$. To find $\mathrm{E}\left(\frac{X}{Y}\middle|Y=y\right)$,

- We start by finding $f_{X|Y}(x|y)$. For $x,y>0$, we get

$$f_{X|Y}(x|y)\quad=\quad\frac{f_{X,Y}(x,y)}{f_Y(y)}=\frac{\frac{e^{-x/y}e^{-y}}{y}}{\int_0^\infty\frac{e^{-x/y}e^{-y}}{y}dx}=\frac{1}{y}e^{-x/y}.$$

- Thus, $\mathrm{E}(X|Y=y)=\int_0^\infty \frac{x}{y}e^{-x/y}dx=y$. Indeed, $X|y \sim \mathrm{Exp}(1/y)$. Hence, it follows from Theorem 6.2.1 that $\mathrm{E}\left(\frac{X}{Y}\middle|Y=y\right)=\frac{1}{y}\cdot \mathrm{E}(X|Y=y)=1$. Notice that the joint p.d.f. of $U=X/Y$ and $V=Y$ is $f_{U,V}(u,v)=f_{X,Y}(uv,v)|J(u,v)|=e^{-u}e^{-v}$, which implies that U and V are independent and identically distributed as $\mathrm{Exp}(1)$. Then, we can also get the same result $\mathrm{E}(U|V=v)=\mathrm{E}(U)=1$ by Theorem 6.2.1.

■

■ **Example 6.19** Let X and Y be independent and exponential with parameter λ. Then, the density of X conditional on $X+Y=v$ is uniform on $(0,v)$.

$$\mathrm{E}(X|X+Y=v)=\mathrm{E}(Y|X+Y=v)=\frac{1}{2}v.$$

- Note that the joint p.d.f. of X and Y is $f_{X,Y}(x,y)=\lambda^2e^{-\lambda(x+y)}$, $x,y>0$ and that $X+Y\sim \mathrm{Gamma}(2,\lambda)$. To compute the joint p.d.f. of $U=X$ and $V=X+Y$, we consider the transformation $u=x$ and $v=x+y$, with inverse transform $x=u$ and $y=v-u$, so that the Jacobian $J=1$. Hence, $f_{U,V}(u,v)=\lambda^2e^{-\lambda v}$ for $0<u<v<\infty$.

- Then, the marginal p.d.f. of V is given by $f_V(v)=\int_0^v \lambda^2e^{-\lambda v}du=\lambda^2ve^{-\lambda v}$, and by definition of conditional density, we have $f_{U|V}(u|v)=\frac{f(u,v)}{f_V(v)}=\frac{1}{v}$, for $0<u<v$. This is the required uniform density, $U|V=v\sim \mathrm{Uniform}(0,v)$, and the conditional expectation is $v/2$.

- Notice that $X/(X+Y)$ and $X+Y$ are independent and that $X/(X+Y)$ has a beta distribution with parameters 1 and 1, which implies that $X/(X+Y)|X+Y=v\sim \mathrm{Uniform}(0,1)$ and that $X|X+Y=v\sim \mathrm{Uniform}(0,v)$. Hence $\mathrm{E}(X|X+Y=v)=\frac{1}{2}v$. Alternatively, by Theorem 6.2.1, we have

$$\mathrm{E}(X|X+Y=v)=\mathrm{E}\left(\frac{X}{X+Y}(X+Y)\middle|X+Y=v\right)=v\mathrm{E}\left(\frac{X}{X+Y}\right)=\frac{v}{2}.$$

■

■ **Example 6.20** Suppose that X and Y have a joint p.d.f.

$$f_{X,Y}(x,y)=\begin{cases}\frac{1}{2\pi\sqrt{x(y-x)}}e^{-y/2}, & \text{for } 0<x<y,\\ 0, & \text{otherwise.}\end{cases}\qquad \text{Find } \mathrm{E}(X|Y=1).$$

- To find the marginal p.d.f. of Y, we integrate $f_{X,Y}(x,y)$ with respect to x using the substitution $x=yt$ and $dx=ydt$,

$$f_Y(y)=\int_0^y \frac{1}{2\pi\sqrt{x(y-x)}}e^{-y/2}dx=\frac{1}{2}e^{-y/2}\int_0^1\frac{1}{\pi}\frac{1}{\sqrt{t(1-t)}}dt=\frac{1}{2}e^{-y/2},$$

since the integrand is the p.d.f. of arcsine distribution, that is, beta distribution with parameters $1/2$ and $1/2$, whose integral equals 1. Thus, the conditional p.d.f. of $X|Y=y$ is given by $f_{X|Y}(x|y)=\dfrac{f_{X,Y}(x,y)}{f_Y(y)}=\dfrac{1}{\pi\sqrt{x(y-x)}}$, for $0<x<y$. Hence, when $y=1$, it follows that $X|Y=y\sim\text{Beta}(1/2,1/2)$, which implies that $\text{E}(X|Y=1)=\frac{1/2}{1/2+1/2}=\frac{1}{2}$.

- More generally, we compute the conditional expectation $\text{E}(X|Y=y)$

$$\text{E}(X|Y=y)=\int_0^y x\frac{1}{\pi\sqrt{x(y-x)}}dx=y\int_0^1 t\frac{\Gamma(\frac{1}{2}+\frac{1}{2})}{\Gamma(\frac{1}{2})\Gamma(\frac{1}{2})}t^{1/2-1}(1-t)^{1/2-1}dt=\frac{y}{2}.$$

Indeed, note that $X|Y=y$ has the same distribution with yW, where W has the arcsine distribution, $W\sim\text{Beta}(1/2,1/2)$ and that $\text{E}(X|Y=y)=\text{E}(yW|Y=y)=\frac{y}{2}$.

- Note that X and Y are not independent, easily seen from the support $\{(x,y):0<x<y\}$, which is not rectangular. On the other hand, let $Z=Y-X$ and $W=X$. Then, the joint p.d.f. of Z and W is

$$f_{Z,W}(z,w)=f_{X,Y}(w,z+w)|J(z,w)|=\frac{1}{2\pi\sqrt{wz}}e^{-(z+w)/2}$$

$$=\frac{1}{\Gamma(\frac{1}{2})2^{\frac{1}{2}}}z^{\frac{1}{2}-1}e^{-\frac{z}{2}}\frac{1}{\Gamma(\frac{1}{2})2^{\frac{1}{2}}}w^{\frac{1}{2}-1}e^{-\frac{w}{2}}=f_Z(z)f_W(w),\ z>0,\ w>0,$$

which implies that $Y-X$ and X are independent and identically distributed as a gamma distribution with parameters 1/2 and 1/2. Thus, the conditional p.d.f of $Y|X=x$ is given by $f_{Y|X}(y|x)=\dfrac{f_{X,Y}(x,y)}{f_X(x)}=\dfrac{1}{\sqrt{2\pi(y-x)}}e^{-(y-x)/2}$, for $0<x<y$, and the conditional expectation is computed

$$\text{E}(Y|X=x)=\int_x^\infty y\frac{1}{\sqrt{2\pi(y-x)}}e^{-(y-x)/2}=\int_0^\infty(t+x)\frac{1}{\sqrt{2\pi t}}e^{-t/2}dt=1+x.$$

Note that $Y-x|X=x\sim\text{Gamma}(1/2,1/2)$. Thus, we have $\text{E}(Y|X=x)=\text{E}(Y-X+X|X=x)=\text{E}(Y-x|X=x)+x=1+x$. Alternatively, since X and $Z=Y-X$ are independent, we also have $\text{E}(Y|X=x)=\text{E}(Z+X|X=x)=\text{E}(Z)+x=1+x$. ■

Notice that $\text{E}(Y|X=x)$ is a function of x, $\text{E}(Y|X=x)=h(x)$ for some function h. Thus, we can plug the random variable X into $\text{E}(Y|X=x)$. The result, $h(X)$ is called $\text{E}(Y|X)$.

Example 6.21 Let X and Y have a joint p.d.f. $f_{X,Y}(x,y)=\begin{cases}8xy, & \text{for } 0<y<x<1,\\ 0, & \text{otherwise.}\end{cases}$

- The marginal p.d.f.s of X and Y are obtained by
$f_X(x)=\int_0^x 8xydy=4x^3$ and $f_Y(y)=\int_y^1 8xydx=4y\left(1-y^2\right)$.

- Thus, the conditional p.d.f of X given $Y=y$ is obtained as $f_{X|Y}(x|y)=2x/\left(1-y^2\right)$ for $y<x<1$. Hence $\text{E}(X|y)=\displaystyle\int_y^1\frac{2x^2}{(1-y)^2}dx=\frac{2}{3}\left(\frac{1-y^3}{1-y^2}\right)=h(y)$, which implies that $\mathrm{E}(X|Y) = h(Y) = \frac{2}{3}\left(\frac{1-Y^3}{1-Y^2}\right)$.

- Similarly, we have $f_{Y|X}(y|x) = 2y/x^2$ for $0 < y < x$. Thus, $\mathrm{E}(Y|x) = \int_0^x \frac{2y^2}{x^2}dy = \frac{2}{3}x$. Hence, $\mathrm{E}(Y|X) = 2X/3$ is a random variable as a function of X.

■

In general $\mathrm{E}(Y|X)$ is a random variable, since it is a function of X. It has a distribution which may or may not be easy to determine. But its mean (expectation), $\mathrm{E}(\mathrm{E}(Y|X))$ is always as easy to determine as the mean of Y, $\mathrm{E}(Y)$, which is known as the *law of total probability for expectations.*

Theorem 6.2.2 — Law of total probability for expectations : Discrete cases. The law of total probability for expectations for discrete random variables can be regarded as averaging the conditional averages :

- Let $A_1,\dots,A_n$ be a partition of a sample space, with $P(A_i) > 0$ for all i, and let Y be a random variable on this sample space. Then

$$\mathrm{E}(Y) = \sum_{i=1}^{n} \mathrm{E}(Y|A_i)P(A_i).$$

- Let $Y = I_B$ for an event B; then we have

$$P(B) = \mathrm{E}(I_B) = \sum_{i=1}^{n} \mathrm{E}(I_B|A_i)P(A_i) = \sum_{i=1}^{n} P(B|A_i)P(A_i).$$

- Let X and Y be discrete random variables; then we have

$$\mathrm{E}(Y) = \sum_{x} \mathrm{E}(Y|X=x)P(X=x) = \mathrm{E}(\mathrm{E}(Y|X)).$$

Proof. By the law of total probability, we have $P(Y=y) = \sum_{i=1}^{n} P(Y=y|A_i)P(A_i)$. Thus,

$$\begin{aligned}\mathrm{E}(Y) &= \sum_y yP(Y=y) = \sum_y y\sum_{i=1}^{n} P(Y=y|A_i)P(A_i) \\ &= \sum_{i=1}^{n}\left[\sum_y yP(Y=y|A_i)\right]P(A_i) = \sum_{i=1}^{n}\mathrm{E}(Y|A_i)P(A_i).\end{aligned}$$

■

■ **Example 6.22** Let $X \sim \mathrm{Geo}(p)$. Interpret X as the number of tails(failures) before the first head(success) in a sequence of coin flips with a success probability p. To get $\mathrm{E}(X)$, we condition on the event (H or T) containing the outcome (head or tail) of the first toss:

$$\mathrm{E}(X)=\mathrm{E}(X|H)\cdot p+\mathrm{E}(X|T)\cdot q=0\cdot p+(1+\mathrm{E}(X))\cdot q,$$

which gives $\mathrm{E}(X)=q/p$. ∎

∎ **Example 6.23** Messages transmitted by a computer in KUstat 433 lab through a data network are destined for KUstat 207 lab with probability 0.5, for KUstat 508 lab with probability 0.3, and for KUstat 509 lab with probability 0.2. The transit time Y of a message is random. Its mean is 0.05 secs if it is destined for KUstat 207 lab, 0.1 secs if it is destined for KUstat 508 lab, and 0.3 secs if it is destined for KUstat 509 lab. Then, compute the expected transit time $\mathrm{E}[Y]$.

- It is easily calculated using the law of total probability for expectations as
$$\mathrm{E}(Y)=\mathrm{E}(\mathrm{E}(Y|X))=\sum_{i=1}^{3}\mathrm{E}(Y|X=i)P(X=i)=0.5\cdot 0.05+0.3\cdot 0.1+0.2\cdot 0.3,$$
where X denotes three labs for which a data network is destined.

∎

∎ **Example 6.24** Suppose that the number N of customers entering a shopping mall on a given day has a Poisson distribution with parameter λ. Suppose further that each of customers buys something independently with probability p.

- We can compute the probability that exactly k customers buy something. Let X be the number of customers who buy something. Then, note that $N\sim\text{Poisson}(\lambda)$ and $X|N=n\sim\text{Binomial}(n,p)$. Thus,
$$P(X=k)=\sum_{n=k}^{\infty}P(X=k|N=n)P(N=n)=\sum_{n=k}^{\infty}\binom{n}{k}p^k(1-p)^{n-k}\frac{\lambda^n e^{-\lambda}}{n!}$$
$$=e^{-\lambda}\sum_{n=k}^{\infty}\frac{n!}{k!(n-k)!}p^k(1-p)^{n-k}\frac{\lambda^n}{n!}=\frac{e^{-\lambda}p^k\lambda^k}{k!}\sum_{n=k}^{\infty}\frac{(1-p)^{n-k}\lambda^{n-k}}{(n-k)!}$$
$$=\frac{e^{-\lambda}(p\lambda)^k}{k!}\sum_{\ell=0}^{\infty}\frac{((1-p)\lambda)^{\ell}}{\ell!}=\frac{e^{-\lambda}(p\lambda)^k}{k!}e^{(1-p)\lambda}=\frac{e^{-p\lambda}(p\lambda)^k}{k!},$$
which is the p.m.f. of a Poisson distribution with parameter λp, $X\sim\text{Poisson}(\lambda p)$.

- We can compute the expected number of customers who buy something by using either its marginal distribution, $X\sim\text{Poisson}(\lambda p)$ or the law of total probability for expectations, $\mathrm{E}(X)=\lambda p=\mathrm{E}[\mathrm{E}(X|N)]=\mathrm{E}(Np)=\lambda p$.

∎

∎ **Example 6.25** When Jaden and Joanne go to the gym, they will either run on treadmills, take bicycles, or play basketball together. They will choose one of the physical activities equally likely. And having chosen an activity, the amount of time (in minutes) they spend exercising is exponentially distributed with respective parameters λ_1, λ_2, and λ_3. Jaden excises the activity he chooses without a break, whereas when Joanne chooses taking a bicycle, she only does for 10 minutes, stops to take a drink of water, and starts all over, choosing one of the three activities at random as if she had just walked in the gym. Find the expectation of Jaden and Joanne's exercise time, respectively.

- Let Y_1 be Jaden's exercise time and Y_2 be Joanne's exercise time. Let X be a random variable that takes values 1, 2, and 3 corresponding to their choices of running, bicycling, and basketball. The conditional distribution of exercise time given the choice is exponentially distributed with $\text{Exp}(\lambda_i)$, $i = 1,2,3$. Thus, the conditional expected time of exercise given the choice is the $1/\lambda_i$, $i = 1,2,3$.

- By the law of total probability for expectations, we have

$$\mathrm{E}(Y_1) = \mathrm{E}[\mathrm{E}(Y_1|X)] = \sum_{i=1}^{3} \mathrm{E}[Y_1|X = i]P(X = i).$$

Thus, Jaden's exercise is $\mathrm{E}(Y_1) = \sum_{i=1}^{3} \frac{1}{3\lambda_i}$.

- The conditional expectation given that Joanne picks playing basketball is $\mathrm{E}[Y_2|X = 3] = 10 + \mathrm{E}(Y_2)$ since after 10 minutes of bicycling, Joanne's subsequent exercise time has the same distribution as if she had just walked into the gym and started anew. Hence,

$$\mathrm{E}(Y_2) = \mathrm{E}[\mathrm{E}(Y_2|X)] = \frac{1}{3\lambda_1} + \frac{1}{3\lambda_2} + \frac{1}{3}(\mathrm{E}(Y_2) + 10).$$

Solving for $\mathrm{E}(Y_2)$ gives the expected amount of time for her exercise.

■

Theorem 6.2.3 — Law of total probability for expectations : Continuous case. The expected value of the random variable $\mathrm{E}(Y|X)$ is equal to the expected value of Y.

$$\mathrm{E}[\mathrm{E}(Y|X)] = \mathrm{E}(Y)$$

Proof. For the continuous random variables X and Y with the joint p.d.f $f_{X,Y}(x,y)$,

$$\begin{aligned}
\mathrm{E}[\mathrm{E}(Y|X)] &= \int_{-\infty}^{\infty} \mathrm{E}(Y|x) f_X(x)dx = \int_{-\infty}^{\infty} \left(\int_{-\infty}^{\infty} y f_{Y|X}(y|x)dy \right) f_X(x)dx \\
&= \int_{-\infty}^{\infty}\int_{-\infty}^{\infty} y \frac{f_{X,Y}(x,y)}{\cancel{f_X(x)}} \cancel{f_X(x)} dydx = \int_{-\infty}^{\infty}\int_{-\infty}^{\infty} y f_{X,Y}(x,y)dydx \\
&= \int_{-\infty}^{\infty} y \left(\int_{-\infty}^{\infty} f_{X,Y}(x,y)dx \right) dy = \int_{-\infty}^{\infty} y f_Y(y)dy = \mathrm{E}(Y).
\end{aligned}$$

■

Remark The conditional expectation $E(Y|X)$ can be written as $\mathrm{E}_X\left[\mathrm{E}_{Y|X}(Y|X)\right] = \mathrm{E}_Y(Y)$, where $\mathrm{E}_X(X)$ stands for the expectation of X with respect to the distribution of X and $\mathrm{E}_{Y|X}(Y|X)$ stands for the expected value of Y with respect to the conditional density $f_{Y|X}(y|x)$, which is often called the *double expectation rule*.

■ **Example 6.26** Consider a stick of length ℓ. Jaden breaks it randomly into two pieces and keeps the piece, of length X, that contains the left end of the stick. Joanne then repeats the same process on the piece Jaden keeps, and let Y be the length of the remaining piece after breaking the second time. Find the expected value of Y.

- For simplicity, we assume that $\ell = 1$. Then, we have $X \sim \text{Uniform}(0,1)$, and $Y|X = x \sim \text{Uniform}(0,x)$ as in Example 6.4. Hence, by the law of total probability of expectations, $\mathrm{E}(Y) = \mathrm{E}(\mathrm{E}(Y|X)) = \mathrm{E}\left(\frac{X}{2}\right) = \frac{1}{4}$.
- Alternatively, since the marginal p.d.f. of Y is given $f_Y(y) = -\log(y)$ for $0 < y < 1$, we can directly evaluate $\mathrm{E}(Y) = \int_0^1 -y\log(y) = \frac{1}{4}$.

■

■ **Example 6.27** Suppose that X and Y have joint p.d.f.

$$f_{X,Y}(x,y) = \begin{cases} xe^{-y}, & \text{for } 0 < x < y < \infty, \\ 0, & \text{otherwise.} \end{cases} \quad \text{Find } \mathrm{E}\left(e^{-Y}\right).$$

- Note that the marginal p.d.f. of X is $f_X(x) = \int_x^\infty xe^{-y}dy = xe^{-x}$, for $x > 0$, which implies that $X \sim \text{Gamma}(2,1)$, and the conditional p.d.f. of Y given $X = x$ is

$$f_{Y|X}(y|x) = \frac{xe^{-y}}{xe^{-x}} = e^{-(y-x)}, \quad \text{for } y > x.$$

- Consider $g(Y) = e^{-Y}$. Then, by the law of unconscious statistician for the conditional expectation, we have $\mathrm{E}\left(e^{-Y}|X = x\right) = \int_x^\infty e^{-y}e^{-(y-x)}dy = \frac{e^{-x}}{2}$. Hence, by the double expectation rule, we obtain

$$\mathrm{E}\left(e^{-Y}\right) = \mathrm{E}\left(\mathrm{E}(e^{-Y}|X)\right) = \frac{1}{2}\mathrm{E}\left(e^{-X}\right) = \frac{1}{2}\int_0^\infty xe^{-2x}dx = \frac{1}{8}.$$

- Indeed, the marginal p.d.f. of Y is $f_Y(y) = \int_0^y xe^{-y}dx = \frac{y^2}{2}e^{-y}$, which implies that $Y \sim \text{Gamma}(3,1)$. Then, we can directly evaluate $\mathrm{E}\left(e^{-Y}\right) = \int_0^\infty \frac{y^2}{2}e^{-2y} = \frac{1}{8}$.

■

Theorem 6.2.4 Let X and Y be either continuous or discrete random variables with joint p.m.f. or p.d.f. $f_{X,Y}(x,y)$, and let $g: R \to R$ be any function. Then,

① $\mathrm{E}(g(X)\mathrm{E}(Y|X)) = \mathrm{E}(g(X)Y)$ and $\mathrm{E}(g(X)Y|X) = g(X)\mathrm{E}(Y|X)$.

② $\mathrm{E}(\mathrm{E}(Y|X)|X) = \mathrm{E}(Y|X)$.

Proof. For simplicity, we assume that X and Y are jointly continuous, but the discrete cases can be shown in a similar way. Then, we have

$$\mathrm{E}(g(X)\mathrm{E}(Y|X)) = \int_{-\infty}^{\infty} g(x)\mathrm{E}(Y|X=x)f_X(x)dx$$

$$= \int_{-\infty}^{\infty} g(x)\left(\int_{-\infty}^{\infty} yf_{Y|X}(y|x)dy\right)f_X(x)dx = \int_{-\infty}^{\infty}\int_{-\infty}^{\infty} g(x)yf_{X,Y}(x,y)dydx = \mathrm{E}(g(X)Y).$$

Also, it follows from Theorem 6.2.1 replacing $X=x$ with a random variable X that $\mathrm{E}(g(X)Y|X) = g(X)\mathrm{E}(Y|X)$, which completes the proof of ①. For ②, since $\mathrm{E}(Y|X)$ is a function of X, $h(X)$ it also follows from Theorem 6.2.1 by replacing Y with a constant 1. ■

Remark Theorem 6.2.4 ① is a generalized version of the double expectation rule with $g(X)=1$ and those results in Theorem 6.2.1 by conditioning on a random variable X. It also tells us that any function of X can be treated as a constant (factored out) when conditioning on X. Theorem 6.2.4 ② says that conditioning twice on X is the same as conditioning just once on X because it yields a function of X.

■ **Example 6.28** Let X and Y be two random variables with mean μ_X and μ_Y, and standard deviation σ_X and σ_Y, respectively. If the conditional expectation of Y given X is linear in X, then $\mathrm{E}(Y|X) = \mu_Y + \rho\,\dfrac{\sigma_Y}{\sigma_X}(X-\mu_X)$, where ρ denotes the correlation coefficient of X and Y.

- We are given that $\mathrm{E}(Y|X)$ is linear in X, that is $\mathrm{E}(Y|X) = \beta X + \alpha$, where α and β are two constants. Then, by the law of total probability of expectations and the linearity of expectations, it follows that $\mathrm{E}[\mathrm{E}(Y|X)] = \mathrm{E}(Y) = \beta\mathrm{E}(X) + \alpha$, which yields $\mu_Y = \beta\mu_X + \alpha$.

- Now, we multiply $\mathrm{E}(Y|X) = \beta X + \alpha$ with X and yield $\mathrm{E}(XY|X) = \beta X^2 + \alpha X$ by ② of Theorem 6.2.4 with $g(X) = X$. Once again, applying the law of total probability of expectations (or ③ of Theorem 6.2.4) and the linearity of expectations, we have $\mathrm{E}[\mathrm{E}(XY|X)] = \mathrm{E}(XY) = \beta\mathrm{E}(X^2) + \alpha\mathrm{E}(X)$, which yields $\mathrm{E}(XY) = \beta(\sigma_X^2 + \mu_X^2) + \alpha\mu_X$.

- Solving two equations for the unknown α and β, we get

$$\beta = \frac{\mathrm{E}(XY) - \mu_X\mu_Y}{\sigma_X^2} = \frac{\mathrm{Cov}(X,Y)}{\sigma_X^2} = \rho\,\frac{\sigma_Y}{\sigma_X}, \alpha = \mu_Y - \rho\,\frac{\sigma_Y}{\sigma_X}\mu_X.$$

- Hence, we have $\mathrm{E}(Y|X) = \rho\dfrac{\sigma_Y}{\sigma_X}X + \mu_Y - \rho\,\dfrac{\sigma_Y}{\sigma_X}\mu_X = \mu_Y + \rho\dfrac{\sigma_Y}{\sigma_X}(X-\mu_X)$.

■

■ **Example 6.29** Suppose X and Y are random variables with $\mathrm{E}(Y|X=x) = x-3$ and $\mathrm{E}(X|Y=y) = \frac{1}{4}y + 5$. What is the correlation coefficient of X and Y?

- From the result in Example 6.28, we have $\mathrm{E}(Y|X=x) = \mu_Y + \rho\,\dfrac{\sigma_Y}{\sigma_X}(x-\mu_X) = x-3$, which implies that $\rho\,\dfrac{\sigma_Y}{\sigma_X} = 1$. Similarly, since $\mu_X + \rho\,\frac{\sigma_X}{\sigma_Y}(y-\mu_Y) = \frac{1}{4}y+5$, we have $\rho\,\frac{\sigma_X}{\sigma_Y} = \frac{1}{4}$. By multiplying them, we thus get $\rho^2 = \frac{1}{4}$. Since $\rho\,\frac{\sigma_Y}{\sigma_X} = 1$, we get $\rho = \dfrac{1}{2}$. ■

Remark Since probabilities can be treated as expectations with indicator variables, that is, $P(A) = \mathrm{E}(I_A)$ for any event A, we also apply the law of total probability for expectations to $P(A)$,

$$P(A) = \mathrm{E}(I_A) = \mathrm{E}\left[\mathrm{E}(I_A|X)\right] = \int_{-\infty}^{\infty} \mathrm{E}(I_A|X=x) f_X(x)dx = \int_{-\infty}^{\infty} P(A|X=x) f_X(x)dx,$$

which is the continuous version of the law of total probability in Theorem 2.2.1.

■ **Example 6.30** Consider $P(X<Y)$, where X and Y are continuous. Let $A = \{X<Y\}$. Then, conditioning on either X or Y, we have

$$P(X<Y) = \begin{cases} \int_{-\infty}^{\infty} P(X<Y|Y=y) f_Y(y)dy = \int_{-\infty}^{\infty} P(X<y|Y=y) f_Y(y)dy \\ \int_{-\infty}^{\infty} P(X<Y|X=y) f_X(x)dx = \int_{-\infty}^{\infty} P(x<Y|Y=y) f_X(x)dx. \end{cases}$$

- Note that if X and Y are independent, then $P(X<y|Y=y) = P(X<y)$ and $P(x<Y|X=x) = P(x<Y)$. Thus, $P(X<Y) = \begin{cases} \int_{-\infty}^{\infty} F_X(y) f_Y(y)dy, \\ \int_{-\infty}^{\infty} (1-F_Y(x)) f_X(x)dx. \end{cases}$

- Suppose that X and Y are independent exponential random variables, $X \sim \mathrm{Exp}(\lambda_1)$ and $Y \sim \mathrm{Exp}(\lambda_2)$. Then,

$$\begin{aligned} P(X<Y) &= \int_0^{\infty} F_X(y) f_Y(y)dy = \int_0^{\infty} \left(1-e^{-\lambda_1 y}\right) \lambda_2 e^{-\lambda_2 y} dy \\ &= 1 - \frac{\lambda_2}{\lambda_1+\lambda_2} \int_0^{\infty} (\lambda_1+\lambda_2) e^{-(\lambda_1+\lambda_2)y} dy = \frac{\lambda_1}{\lambda_1+\lambda_2}, \end{aligned}$$

  which was also considered in Example 5.15.

- Suppose that X has a standard normal distribution, Y has a χ^2-distribution with parameter r, and that they are independent. Then, $\frac{X}{\sqrt{Y/r}}$ has a t-distribution with r degrees of freedom, as discussed in Example 5.44. Then, the c.d.f. of T can be derived as follows

$$F_T(t) = P\left(X \le t\sqrt{\frac{Y}{r}}\right) = \int_0^{\infty} F_X(t\sqrt{y/r}) f_Y(y)dy = \int_0^{\infty} \Phi(t\sqrt{y/r}) f_Y(y)dy,$$

  which was also shown in Example 6.10 in view of t-distribution as a scale mixture of normal distribution.

■

■ **Example 6.31 — Convolution (revisited).** Let $Y = X_1 + X_2$, where X_1 and X_2 are independent with c.d.f.'s F_{X_1} and F_{X_2} respectively. Then Y has c.d.f. F_Y given by

$$\begin{aligned} F_Y(y) &= P(X_1 + X_2 \leq y) = \mathrm{E}\left[\mathrm{E}(I_{\{X_1+X_2\leq y\}}|X_2)\right] = \mathrm{E}\left[\mathrm{E}(F_{X_1}(y-X_2)|X_2)\right] \\ &= \mathrm{E}\left[F_{X_1}(y-X_2)\right] = \mathrm{E}\left[F_{X_2}(y-X_1)\right], \end{aligned}$$

where the last equality holds since the roles of X_1 and X_2 can be switched.

- Suppose that X_1 and X_2 are continuous and independent with c.d.f.'s F_{X_1} and F_{X_2} respectively. Then $Y = X_1 + X_2$ has c.d.f. F_Y given by

$$\begin{aligned} F_Y(y) &= \mathrm{E}\left[F_{X_1}(y-X_2)\right] = \int_{-\infty}^{\infty} F_{X_1}(y-x_2) f_{X_2}(x_2)dx_2, \\ f_Y(y) &= \frac{d}{dy}F_Y(y) = \int_{-\infty}^{\infty} f_{X_1}(y-x_2) f_{X_2}(x_2)dx_2. \end{aligned}$$

- Let $X_1 \sim N(0,1)$ and $X_2 \sim \text{Poisson}(\lambda)$. Then, the c.d.f. of Y is given by

$$F_Y(y) = \mathrm{E}\left[F_{X_1}(y-X_2)\right] = \sum_{x_2=0}^{\infty} \left(\int_{-\infty}^{y-x_2} \frac{1}{\sqrt{2\pi}} e^{-\frac{1}{2}t^2} dt \right) \frac{e^{-\lambda}\lambda^{x_2}}{x_2!}.$$

■

6.3 Conditional variance

In addition to conditional expectation, we can define conditional variance as the conditional expectation of the squared distance between a random variable and its conditional mean.

Definition 6.3.1 The conditional variance of Y given $X = x$ is the variance with respect to a conditional distribution, that is,

$$\mathrm{Var}(Y|X=x) = \mathrm{Var}(Y|x) = \mathrm{E}([Y - \mathrm{E}(Y|x)]^2|x) = \mathrm{E}(Y^2|x) - [\mathrm{E}(Y|x)]^2.$$

- For a random variable Y with conditional p.d.f. $f_{Y|X}(y|x)$ given $X = x$, the conditional variance of Y given $X = x$ is

$$\mathrm{Var}(Y|x) = \int [y - \mathrm{E}(Y|x)]^2 f_{Y|X}(y|x)dy = \int y^2 f_{Y|X}(y|x)dy - [\mathrm{E}(Y|x)]^2.$$

- If Y is discrete with conditional p.m.f. $f_{Y|X}(y|x)$ given $X = x$, the conditional variance of Y given $X = x$ is

$$\mathrm{Var}(Y|x) = \sum_{\text{all } y} [y - \mathrm{E}(Y|x)]^2 f_{Y|X}(y|x) = \sum_{\text{all } y} y^2 f_{Y|X}(y|x) - [\mathrm{E}(Y|x)]^2.$$

Remark By the linearity property and the law of unconscious statistician for conditional expectations in Theorem 6.2.1, we have

$$\begin{aligned}\mathrm{Var}(Y|x) &= \mathrm{E}([Y-\mathrm{E}(Y|x)]^2|x) = \mathrm{E}\left(\left[Y^2 - 2Y\mathrm{E}(Y|x) + \mathrm{E}(Y|x)^2\right]|x\right)\\ &= \mathrm{E}(Y^2|x) - 2\mathrm{E}(Y|x)^2 + [\mathrm{E}(Y|x)]^2 = \mathrm{E}(Y^2|x) - \mathrm{E}(Y|x)^2.\end{aligned}$$

The conditional variance of Y given $X = x$, $\mathrm{Var}(Y|x)$ is a function of x, and we can define $\mathrm{Var}(Y|X)$ as a random variable by replacing x with X. Hence,

$$\mathrm{Var}(Y|X) = \mathrm{E}([Y-\mathrm{E}(Y|X)]^2|X) = \mathrm{E}(Y^2|X) - [\mathrm{E}(Y|X)]^2.$$

■ **Example 6.32** Consider X and Y with joint p.d.f. as given in Example 6.27

$$f_{X,Y}(x,y) = \begin{cases} xe^{-y}, & \text{for } 0 < x < y < \infty, \\ 0, & \text{otherwise.} \end{cases} \quad \text{Find } \mathrm{Var}(Y|X=x).$$

- Note that the conditional distribution of Y given $X = x$ is a shifted exponential distribution, an exponential distribution with parameter $\lambda = 1$ and location parameter x, which has a conditional p.d.f. $f_{Y|X}(y|x) = \dfrac{xe^{-y}}{xe^{-x}} = e^{-(y-x)}$, for $y > x$.
- Thus, the conditional expectation $\mathrm{E}(Y|X=x) = \mathrm{E}(Y-x|X=x) + x = 1 + x$ and

$$\begin{aligned}\mathrm{E}(Y^2|X=x) &= \int_x^{\infty} y^2 f_{Y|X}(y|x) = \mathrm{E}((Y-x+x)^2|X=x)\\ &= \mathrm{E}((Y-x)^2|X=x) + 2x\mathrm{E}(Y-x|X=x) + x^2 = 2 + 2x + x^2,\end{aligned}$$

which follows from $\mathrm{E}((Y-x)^2|X=x) = 2$ and $\mathrm{E}(Y-x|X=x) = 1$ since $Y - x|X = x \sim \mathrm{Exp}(1)$. Hence, $\mathrm{Var}(Y|X=x) = \mathrm{E}(Y^2|X=x) - (\mathrm{E}(Y|X=x))^2 = 1$.

■

■ **Example 6.33** We return to the uniform density on the unit disk in Example 6.2.

- As given in Example 6.2, the conditional distribution of Y given $X = x$ is the uniform distribution, $Y|X = x \sim \mathrm{Uniform}[-\sqrt{1-x^2}, \sqrt{1-x^2}]$. Note that if W is uniformly distributed on (a,b), then $\mathrm{E}(W) = (b+a)/2$ and $\mathrm{Var}(W) = (b-a)^2/12$.
- Hence, we have $\mathrm{E}(Y|X=x) = 0$ and $\mathrm{Var}(Y|X=x) = \frac{4(1-x^2)}{12}$, which implies that $\mathrm{E}(Y|X) = 0$ and $\mathrm{Var}(Y|X) = \dfrac{1}{3}(1-X^2)$.

■

Theorem 6.3.1 — Law of total probability for variances. The expected value of conditional variance $\mathrm{Var}(Y|X)$ is generally smaller than $\mathrm{Var}(Y)$. In fact,

$$\mathrm{E}[\mathrm{Var}(Y|X)] + \mathrm{Var}[\mathrm{E}(Y|X)] = \mathrm{Var}(Y).$$

Proof. Let $Z = \mathrm{E}(Y|X)$. Then, $\mathrm{E}(Z) = \mathrm{E}(\mathrm{E}(Y|X))$. Thus, we have

$$\mathrm{E}(\mathrm{Var}(Y|X)) = \mathrm{E}(\mathrm{E}(Y^2|X) - Z^2) = \mathrm{E}(Y^2) - \mathrm{E}(Z^2),$$
$$\mathrm{Var}(\mathrm{E}(Y|X)) = \mathrm{E}(Z^2) - (\mathrm{E}[Z])^2 = \mathrm{E}(Z^2) - (\mathrm{E}[Y])^2.$$

Hence, $\mathrm{E}[\mathrm{Var}(Y|X)] + \mathrm{Var}[\mathrm{E}(Y|X)] = \mathrm{Var}(Y)$.

- Alternatively, we have that $\mathrm{Var}(Y) = \mathrm{E}\left[(Y-\mathrm{E}(Y))^2\right] = \mathrm{E}\left\{\mathrm{E}\left[(Y-\mathrm{E}(Y))^2|Y\right]\right\}$ by the law of total probability for expectations. Then, we have

$$(Y-\mathrm{E}(Y))^2 = (Y-\mathrm{E}(Y|X)+\mathrm{E}(Y|X)-\mathrm{E}(Y))^2$$
$$= (Y-\mathrm{E}(Y|X))^2 + (\mathrm{E}(Y|X)-\mathrm{E}(Y))^2 + 2(Y-\mathrm{E}(Y|X))(\mathrm{E}(Y|X)-\mathrm{E}(Y)).$$

- Notice that $\mathrm{E}\left[(Y-\mathrm{E}(Y|X))^2|X\right] = \mathrm{Var}(Y|X)$ by the definition of conditional variance, $\mathrm{E}\left[(\mathrm{E}(Y|X)-\mathrm{E}(Y))^2|X\right] = (\mathrm{E}(Y|X)-\mathrm{E}(Y))^2$ by Theorem 6.2.1 with $\mathrm{E}(g(X)|X) = g(X)$, where $g(x) = (\mathrm{E}(Y|x)-\mathrm{E}(Y))^2$, and that $\mathrm{E}((Y-\mathrm{E}(Y|X))(\mathrm{E}(Y|X)-\mathrm{E}(Y))|X) = 0$ by Theorem 6.2.4 with $g(X) = \mathrm{E}(Y|X) - \mathrm{E}(Y)$.

- Hence, using the linearity of conditional expectations, we see that

$$\begin{aligned}\mathrm{Var}(Y) &= \mathrm{E}\left\{\mathrm{E}\left[(Y-\mathrm{E}(Y))^2|Y\right]\right\} = \mathrm{E}(\mathrm{Var}(Y|X)) + \mathrm{E}((\mathrm{E}(Y|X)-\mathrm{E}(Y))^2) \\ &= \mathrm{E}(\mathrm{Var}(Y|X)) + \mathrm{Var}(\mathrm{E}(Y|X)).\end{aligned}$$

■

Remark Define a random variable Z as $Z = \mathrm{E}(Y|X)$. Then,

- $0 \le \mathrm{Var}(Y|X) = \mathrm{E}[(Y-Z)^2|X] = \mathrm{E}[Y^2|X] - Z^2$.
- $0 \le \mathrm{E}[\mathrm{Var}(Y|X)] = \mathrm{E}[Y^2] - \mathrm{E}[Z^2]$, which is equivalent to $\mathrm{E}(Z^2) \le \mathrm{E}(Y^2)$.
- $\mathrm{Var}(\mathrm{E}(Y|X)) = \mathrm{E}(Z^2) - [\mathrm{E}(Y)]^2 \le \mathrm{E}(Y^2) - [\mathrm{E}(Y)]^2 = \mathrm{Var}(Y)$.

■ **Example 6.34** Suppose that X and Y have joint p.d.f.

$$f_{X,Y}(x,y) = \begin{cases} \sqrt{\frac{2}{\pi}}\frac{e^{-y/x}}{x}e^{-x^2/2}, & \text{for } x>0,\ y>0, \\ 0, & \text{otherwise.}\end{cases} \quad \text{Find Var}(Y).$$

- Note that $Y|X = x \sim \text{Exponential}(1/x)$ with $f_{Y|X}(y|x) = \dfrac{e^{-y/x}}{x}$ and that $\mathrm{E}(Y|X) = X$ and $\mathrm{Var}(Y|X) = X^2$. Also, note that the marginal distribution of X is $f_X(x) = \dfrac{f_{X,Y}(x,y)}{f_{Y|X}(y|x)}$, which is given by $f_X(x) = \sqrt{\dfrac{2}{\pi}}e^{-\frac{x^2}{2}}$, for $x>0$. Indeed, X has the same distribution with $|Z|$, $Z \sim N(0,1)$.

- By Theorem 6.3.1, it follows that $\mathrm{Var}(Y) = \mathrm{E}(\mathrm{Var}(Y|X)) + \mathrm{Var}(\mathrm{E}(Y|X))$. Thus, using the change of variable with $u = x^2/2$ and $du = \sqrt{2u}dx$, we have

$$\mathrm{E}(\mathrm{Var}(Y|X)) = \mathrm{E}(X^2) = \int_0^\infty x^2 \sqrt{\frac{2}{\pi}} e^{-\frac{x^2}{2}} dx$$
$$= \sqrt{\frac{2}{\pi}} \int_0^\infty \sqrt{2} u^{\frac{3}{2}-1} e^{-u} \frac{1}{2\sqrt{u}} du = \frac{2}{\sqrt{\pi}} \Gamma\left(\frac{3}{2}\right) = 1 = \mathrm{E}(|Z|^2),$$
$$\mathrm{Var}(\mathrm{E}(Y|X)) = \mathrm{Var}(X) = \mathrm{E}(X^2) - \mathrm{E}(X)^2 = 1 - \left(\int_0^\infty x \sqrt{\frac{2}{\pi}} e^{-\frac{x^2}{2}} dx\right)^2 = 1 - \frac{2}{\pi},$$
$$\mathrm{Var}(Y) = \mathrm{E}(\mathrm{Var}(Y|X)) + \mathrm{Var}(\mathrm{E}(Y|X)) = 2 - \frac{2}{\pi}.$$

■

■ **Example 6.35 — Scale mixture of normal : Laplace distribution.** Suppose that the conditional distribution of Y given $X = x$ is a normal distribution with mean 0 and variance x, $Y|X = x \sim N(0,x)$ and that the marginal distribution of X is a chi-squared distribution with two degrees of freedom, $X \sim \chi^2(2)$, which is exponentially distributed with parameter $1/2$. Find $\mathrm{E}(Y)$, $\mathrm{Var}(Y)$ and $M_Y(t) = \mathrm{E}(e^{tY})$.

- By the law of total probability for expectations and variances, we have $\mathrm{E}(Y) = \mathrm{E}(\mathrm{E}(Y|X)) = 0$, and $\mathrm{Var}(Y) = \mathrm{E}(\mathrm{Var}(Y|X)) + \mathrm{Var}(\mathrm{E}(Y|X)) = \mathrm{E}(X) = 2$.

- For the moment generating function $M_Y(t)$, by the law of total probability for expectations and the law of unconscious statistician for the conditional expectation, we have

$$M_Y(t) = \mathrm{E}\left(\mathrm{E}\left(e^{tY}|X\right)\right) = \mathrm{E}\left(e^{\frac{1}{2}t^2 X}\right) = \int_0^\infty e^{\frac{1}{2}t^2 x} \frac{1}{2} e^{-\frac{1}{2}x} dx = \frac{1}{1-t^2}, \quad \text{for } t^2 < 1.$$

- Note that $M_Y(t)$ is the same as the m.g.f. of Laplace (double exponential) distribution as shown in Example 4.32. Hence, we conclude that Y has Laplace (double exponential) distribution, $Y \sim \mathrm{DE}(0,1)$, by the uniqueness of m.g.f. In other words, Laplace distribution is a scale mixture of normal distribution.

■

■ **Example 6.36** Let X and Y be two random variables with means μ_X and μ_Y, and standard deviations σ_X and σ_Y, respectively. If the conditional expectation of Y given X is linear in X, then $\mathrm{E}(Y|X) = \mu_Y + \rho \frac{\sigma_Y}{\sigma_X}(X - \mu_X)$ as shown in Example 6.28, where ρ denotes the correlation coefficient of X and Y.

- Thus, by the law of total probability for variances we have

$$\begin{aligned} \mathrm{E}[\mathrm{Var}(Y|X)] &= \mathrm{Var}(Y) - \mathrm{Var}[\mathrm{E}(Y|X)] = \mathrm{Var}(Y) - \rho^2 \frac{\sigma_Y^2}{\sigma_X^2} \mathrm{Var}(X) \\ &= (1-\rho^2)\,\mathrm{Var}(Y). \end{aligned}$$

■

A common use for conditional expectation (mean) is prediction, and we can view $\mathrm{E}(Y|X)$ as the best predictor for Y given X with the mean squared error loss $\mathrm{E}(Y-d)^2$. Suppose that we want to predict a random variable Y using a prediction d, subject to squared error loss $(Y-d)^2$. Suppose further that we get to observe another random variable X before we have to choose d. Then d can be a function of X, $d(X)$. Then, the best predictor for Y is $d(X)=\mathrm{E}(Y|X)$ as in the following theorem.

Theorem 6.3.2 — Best predictor : conditional mean. Let X and Y be jointly distributed random variables with $\mathrm{E}(Y^2)<\infty$. Then, the random variable $d(X)$ that minimizes $\mathrm{E}(Y-d(X))^2$ is $d(X)=\mathrm{E}(Y|X)$, and the value of the minimum is $\mathrm{E}(\mathrm{Var}(Y|X))$.

Proof. Note that

$$\begin{aligned} & \mathrm{E}(Y-d(X))^2 = \mathrm{E}(Y-\mathrm{E}(Y|X)+\mathrm{E}(Y|X)-d(X))^2 \\ = \ & \mathrm{E}(Y-\mathrm{E}(Y|X))^2+\mathrm{E}(E(Y|X)-d(X))^2+2\mathrm{E}[(Y-\mathrm{E}(Y|X))(\mathrm{E}(Y|X)-d(X))] \\ = \ & \mathrm{E}(Y-\mathrm{E}(Y|X))^2+\mathrm{E}(\mathrm{E}(Y|X)-d(X))^2, \end{aligned}$$

where the last equality follows from

$$\mathrm{E}[(Y-\mathrm{E}(Y|X))(\mathrm{E}(Y|X)-d(X))] = \mathrm{E}\{\mathrm{E}[(Y-\mathrm{E}(Y|X))(\mathrm{E}(Y|X)-d(X))|X]\} = 0.$$

Hence, $\mathrm{E}(Y-d(X))^2$ is minimized by taking $d(X)=\mathrm{E}(Y|X)$. Then, the value of the minimum is $\mathrm{E}(Y-\mathrm{E}(Y|X))^2 = \mathrm{E}\left(\mathrm{E}\left[(Y-\mathrm{E}(Y|X))^2|X\right]\right) = \mathrm{E}(\mathrm{Var}(Y|X))$. ■

Remark Notice that $\mathrm{E}[\mathrm{Var}(Y|X)]$ is the mean squared error (M.S.E.) $\mathrm{E}(Y-d(X))^2$ for the best predictor for Y based on X.

- $\mathrm{Var}(Y|X=x)$ is the conditional M.S.E. of that same predictor given $X=x$.
- $\mathrm{E}(Y)$ is not as good a predictor for Y as $\mathrm{E}(Y|X)$ since

$$\begin{aligned} & \mathrm{E}(Y-\mathrm{E}(Y|X))^2 = \mathrm{E}(\mathrm{Var}(Y|X)] \\ \leq \ & \mathrm{E}(\mathrm{Var}(Y|X))+\mathrm{Var}(\mathrm{E}(Y|X)) = \mathrm{Var}(Y) = \mathrm{E}(Y-\mathrm{E}(Y))^2. \end{aligned}$$

■ **Example 6.37** Let X and Y be jointly distributed random variables with $\mathrm{E}(Y^2)<\infty$. Define Z to be the difference between Y and $\mathrm{E}(Y|X)$, $Z=Y-\mathrm{E}(Y|X)$, which is often called the *residual*.

- It is obvious that $\mathrm{E}(Z)=0$ and $\mathrm{Var}(Z)=\mathrm{E}(\mathrm{Var}(Y|X))$ by the law of total probability of expectations and variances as discussed in Theorem 6.2.4 and 6.3.1.

- Further, Z and $\mathrm{E}(Y|X)$ are uncorrelated :

$$\begin{aligned} & \mathrm{Cov}(Z,\mathrm{E}(Y|X)) = \mathrm{Cov}(Y-\mathrm{E}(Y|X),\mathrm{E}(Y|X)) \\ = \;& \mathrm{E}\{[Y-\mathrm{E}(Y|X)]\mathrm{E}(Y|X)\} = \mathrm{E}(\mathrm{E}\{[Y-\mathrm{E}(Y|X)]\mathrm{E}(Y|X)|X\}) = 0. \end{aligned}$$

- More generally, for any function $h(X)$ with $\mathrm{E}(h(X)) < \infty$, Z and $h(X)$ are uncorrelated, $\mathrm{Cov}(Z,h(X)) = \mathrm{E}(Zh(X)) = \mathrm{E}[\mathrm{E}(h(X)(Y-\mathrm{E}(Y|X)|X)] = 0$.

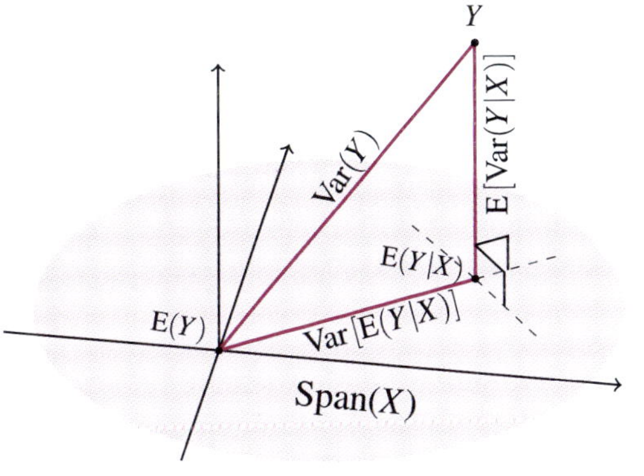

Figure 6.4: Orthogonality of $Z = Y - \mathrm{E}(X|Y)$ to $\mathbf{Span}(X)$

- The conditional expectation $\mathrm{E}(Y|X)$ can be viewed as a projection. Consider the space of all random variables with finite second moments. Then, the inner product is $< Y_1,Y_2 >\equiv \mathrm{E}(Y_1Y_2)$ or $\mathrm{Cov}(Y_1,Y_2)$ if Y_1 and Y_2 are centered (with zero mean) as discussed in Theorem 5.4.4.

- Then $\mathrm{E}(Y|X)$ is the *projection* of Y onto the subspace consisting of functions of X, i.e., the closest(optimal) point (best prediction) to Y within the space of functions of X.

- Geometrically, the residual $Z = Y - E(Y|X)$ should be uncorrelated with any function $h(X)$ in the space spanned by X, $\mathbf{Span}(X)$, or orthogonal (perpendicular) to the plane $\mathbf{Span}(X)$, as shown in Figure 6.4. Indeed, we have shown Z and $h(X)$ are uncorrelated.

■

Remark For two-dimensional nonzero vectors $\mathbf{v}$ and $\mathbf{w}$, the *projection* of $\mathbf{v}$ onto $\mathbf{w}$ is the vector $\mathbf{u}$ along the same line L as $\mathbf{w}$ whose terminal point is obtained by dropping a perpendicular line from the terminal point of $\mathbf{v}$ to L as shown in Figure 6.5. In other words, the projection of the point $\mathbf{v}$ in the plane onto a line is the point in the set that is closest to the point.

- More generally, let $\mathscr{V}$ be a subspace of R^n. A projection from R^n into $\mathscr{V}$ is a linear transformation $\mathbf{P}$ such that ① $\mathbf{Py} \in \mathscr{V}$, for all $\mathbf{y} \in R^n$ and ② $\mathbf{Pv} = \mathbf{v}$ for all $\mathbf{v} \in \mathscr{V}$.

- For all $\mathbf{y} \in R^n$, there exists a vector $\widehat{\mathbf{y}} \in \mathscr{V}$ such that $\|\mathbf{y} - \widehat{\mathbf{y}}\| \le \|\mathbf{y} - \mathbf{v}\|$ for all $\mathbf{v} \in \mathscr{V}$, and the vector $\widehat{\mathbf{y}}$ is called the *orthogonal projection* of $\mathbf{y}$ onto $\mathscr{V}$.

- Let $\mathscr{L}_2$ denote the set of real-valued random variables with finite second moments and let $\mathbf{Span}(X)$ denote the elements of $\mathscr{L}_2$ that are (spanned by) functions of X. For a given random variable $Y \in \mathscr{L}_2$, let $\mathbf{P}Y$ denote the projection of Y onto $\mathbf{Span}(X)$, define to be the element of $\mathbf{Span}(X)$ such that $Y - \mathbf{P}Y$ is orthogonal to all elements of $\mathbf{Span}(X)$. Then, $\mathbf{P}Y = \mathrm{E}(Y|X)$ as shown in Example 6.37.

- Also, note that $\mathrm{E}(\cdot|X)$ is a linear operator, $\mathrm{E}(Y|X) \in \mathbf{Span}(X)$, and $\mathrm{E}(h(X)|X)$ for all $h(X) \in \mathbf{Span}(X)$, which satisfies two conditions ① and ② for projection.

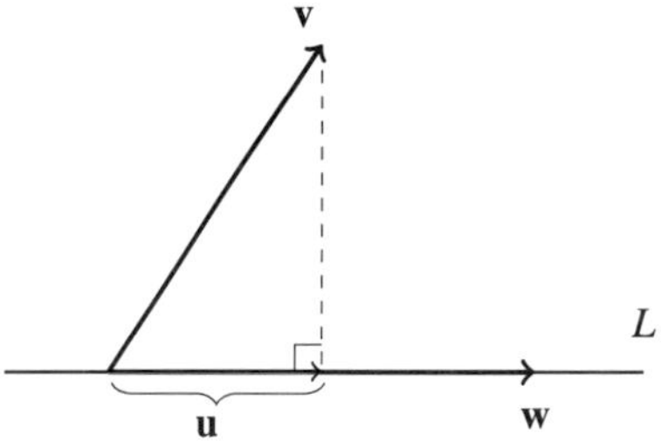

Figure 6.5: Projection of **v** to the line L

■ **Example 6.38 — Best linear predictor (Simple linear regression).** Let X and Y be random variables with means μ_X and μ_Y, variances σ_X^2 and σ_Y^2 and correlation ρ. Suppose that we observe X and then predict Y by a linear function $aX + b$. Consider the squared error, $(Y - [\beta X + \alpha])^2$. How should we choose β and α to minimize $\mathrm{E}\{(Y - [\beta X + \alpha])^2\}$?

- The mean squared error (M.S.E.) is

$$\begin{aligned} & \mathrm{E}\{(Y - [\beta X + \alpha])^2\} = \mathrm{Var}(Y - \beta X - \alpha) + [\mathrm{E}(Y - \beta X - \alpha)]^2 \\ = \; & \mathrm{Var}(Y) + \beta^2 \mathrm{Var}(X) - 2\beta \mathrm{Cov}(X,Y) + (\mu_Y - [\beta \mu_X + \alpha])^2. \end{aligned}$$

- Since α only appears in one place, we must choose $\alpha = \mu_Y - \beta\mu_X$ to make the last term zero. The value of α depends on what we choose for β, but it must be $\mu_Y - \beta\mu_X$ no matter what we choose for β. The rest of the M.S.E. is a quadratic in β that is minimized by $\beta = \dfrac{\mathrm{Cov}(X,Y)}{\mathrm{Var}(X)} = \dfrac{\rho\sigma_Y}{\sigma_X}$.

- Alternatively, since we would like to minimize $S(\alpha,\beta) = \mathrm{E}((Y - \alpha - \beta X)^2)$ with respect to the unknowns α and β, we differentiate $S(\alpha,\beta)$ with respect to α and β

$$\frac{\partial S}{\partial \alpha} = -2\mathrm{E}(Y - \alpha - \beta X), \text{ and } \frac{\partial S}{\partial \beta} = -2\mathrm{E}(X(Y - \alpha - \beta X)).$$

Setting $\frac{\partial S}{\partial \alpha}$ and $\frac{\partial S}{\partial \beta}$ equal to zero gives the solutions α and β as

$$\alpha = \mathrm{E}(Y) - \beta \mathrm{E}(X), \text{ and } \beta = \frac{\mathrm{Cov}(X,Y)}{\mathrm{Var}(Y)} = \frac{\rho \sigma_Y}{\sigma_X}.$$

- If we plug the values $\beta = \rho\sigma_Y/\sigma_X$ and $\alpha = \mu_Y - \beta\mu_X$ into the formula for M.S.E.,

$$\mathrm{E}(Y - (\beta X + \alpha))^2 = \sigma_Y^2 + \rho^2 \frac{\sigma_Y^2}{\sigma_X^2}\sigma_X^2 - 2\rho\frac{\sigma_Y}{\sigma_X}\rho\sigma_Y\sigma_X = \sigma_Y^2(1-\rho^2) \le \sigma_Y^2.$$

That is, if we did not get to observe X first, then we would use a constant μ_Y to predict Y, and the M.S.E. would be σ_Y^2. By using $\beta X + \alpha$ to predict Y, we reduce M.S.E. by a fraction of ρ^2. The larger ρ^2 is, the more we can reduce M.S.E. by using $\beta X + \alpha$ to predict Y.

- Since the best predictor for Y given X is $\mathrm{E}(Y|X)$, when $\mathrm{E}(Y|X)$ is assumed to be linear in X, $\mathrm{E}(Y|X) = \mu_Y + \rho\frac{\sigma_Y}{\sigma_X}(X - \mu_X)$ in Example 6.28 is the best linear predictor Y given X, which is exactly the same as $\beta X + \alpha$.

■

Remark Let $(x_1, y_1), \ldots, (x_n, y_n)$ be points in the plane. The *least squares* problem is to find the line $y = \beta x + \alpha$ that minimizes $\sum_{i=1}^n [y_i - (\beta x_i + \alpha)]^2$.

- Suppose that the joint p.m.f. of X and Y is $P\{X = x_i,\ Y = y_i\} = 1/n$ for $i = 1, \ldots, n$. Then, we have $\mathrm{E}[(Y - (\beta X + \alpha))^2] = \frac{1}{n}\sum_{i=1}^n [y_i - (\beta x_i + \alpha)]^2$, and the least squares problem is equivalent to finding the best linear predictor for Y given $X = x$ of the form $\beta x + \alpha$, which is given $\beta x + \alpha = \mu_y + \rho\frac{\sigma_Y}{\sigma_X}(x - \mu_X)$.
- Note that μ_X, μ_Y, σ_X^2, σ_Y^2, and $\rho\sigma_Y/\sigma_x$ are given by

$$\mu_X = \frac{1}{n}\sum_{i=1}^n x_i = \bar{x},\ \mu_Y = \frac{1}{n}\sum_{i=1}^n y_i = \bar{y},$$
$$\sigma_X^2 = \frac{1}{n}\sum_{i=1}^n (x_i - \bar{x})^2 = s_x^2,\ \sigma_Y^2 = \frac{1}{n}\sum_{i=1}^n (y_i - \bar{y})^2 = s_y^2,$$
$$\rho\frac{\sigma_Y}{\sigma_X} = \frac{\mathrm{Cov}(X,Y)}{\mathrm{Var}(X)} = \frac{\sum_{i=1}^n (x_i - \bar{x})(y_i - \bar{y})}{\sum_{i=1}^n (x_i - \bar{x})^2}.$$

- Note that $\beta = \rho\frac{\sigma_Y}{\sigma_X}$ is the slope of the least squares lines. If $\rho > 0$, then the least squares line has positive slope, and y tends to increase with x. If $\rho < 0$, then the least squares line has negative slope and y tends to decrease as x increases.

■ **Example 6.39 — Beta-Binomial distribution : mixed case.** Suppose that the joint p.m.f./p.d.f. of X and Y is

$$f_{X,Y}(x,y) = \frac{\Gamma(\alpha+\beta)}{\Gamma(\alpha)\Gamma(\beta)}\binom{n}{x} y^{x+\alpha-1}(1-y)^{n-x+\beta-1}, \text{ for } x = 0, \ldots, n \text{ and } 0 < y < 1.$$

- First, note that $X|Y = y \sim \text{Binomial}(n, y)$ and $Y \sim \text{Beta}(\alpha, \beta)$. Then, the marginal distribution of X is given by

$$\begin{aligned} f_X(x) &= \frac{\Gamma(\alpha+\beta)}{\Gamma(\alpha)\Gamma(\beta)}\binom{n}{x}\int_0^1 y^{x+\alpha-1}(1-y)^{n-x+\beta-1}dy \\ &= \frac{\Gamma(\alpha+\beta)}{\Gamma(\alpha)\Gamma(\beta)}\binom{n}{x}\frac{\Gamma(x+\alpha)\Gamma(n-x+\beta)}{\Gamma(\alpha+\beta+n)}, \ x = 0, 1, \ldots, n, \end{aligned}$$

  which is known as the *Beta-Binomial* distribution.

- Next, we get $f_Y(y|x) = \dfrac{f_{X,Y}(x,y)}{f_X(x)} = \dfrac{\Gamma(\alpha+\beta+n)}{\Gamma(x+\alpha)\Gamma(n-x+\beta)} y^{x+\alpha-1}(1-y)^{n-x+\beta-1}$ for $0 < y < 1$, which implies that $Y|X = x \sim \text{Beta}(x+\alpha, n-x+\beta)$. Thus, we have $\text{E}(Y|X) = \dfrac{X+\alpha}{n+\alpha+\beta}$, the best predictor for Y given X.

- Note that $\text{Var}(Y) = \frac{\alpha\beta}{(\alpha+\beta)^2(\alpha+\beta+1)}$ and $\text{Var}[\text{E}(Y|X)] = \frac{1}{(n+\alpha+\beta)^2}\text{Var}(X)$. Then, by applying the law of total probability for variances, we have

$$\text{Var}(X) = \text{E}[\text{Var}(X|Y)] + \text{Var}[\text{E}(X|Y)] = \text{E}(nY[1-Y]) + n^2\text{Var}(Y).$$

  Thus, we can compute the mean squared error for the best predictor for Y given X, which is $\text{E}(\text{Var}(Y|X)) = \text{Var}(Y) - \text{Var}(\text{E}(Y|X))$. Hence, we have

$$\text{E}(\text{Var}(Y|X)) = \frac{\alpha\beta}{(\alpha+\beta)^2(\alpha+\beta+1)} - \frac{1}{(n+\alpha+\beta)^2}\left\{\text{E}(nY[1-Y]) + n^2\text{Var}(Y)\right\}.$$

- For example, assuming that $\alpha = \beta = 1$, we have

$$\begin{aligned} \text{E}(nY[1-Y]) &= n\int_0^1 y(1-y)dy = \frac{n}{6}, \ n^2\text{Var}(Y) = \frac{n^2}{12}, \\ \text{E}(\text{Var}(Y|X)) &= \frac{1}{12} - \frac{1}{(n+2)^2}\left[\frac{n}{6} + \frac{n^2}{12}\right] = \frac{1}{6(n+2)}. \end{aligned}$$

■

Remark Suppose that there are $N+1$ urns, each containing a total of N red and blue balls; the urn number k contains k red and $N-k$ blue balls ($k = 0, 1, 2, \ldots, N$). An urn is chosen at random and n balls are drawn at random from it (the same urn for the n draws), the ball drawn being replaced each time. Let X denote the number of red balls out of the n balls drawn.

- Let Θ be the random variable to denote the proportion of red balls in the urn chosen. Let R_N denote the set of $N+1$ possible values of the proportion(θ) of Θ, $\theta \in R_N = \left\{0, \dfrac{1}{N}, \dfrac{2}{N}, \ldots, \dfrac{N-1}{N}, 1\right\}$, which are all equally likely, $P(\Theta = \theta) = \frac{1}{N+1}$. Given the proportion of red balls, $\theta \in R_N$, that is, an urn is chosen, then X, the number of red balls out of the chosen urn, has a binomial distribution with parameters n and θ, $X|\Theta = \theta \sim \text{Binomial}(n, \theta)$, $P(X = x|\Theta = \theta) = \dbinom{n}{x}\theta^x(1-\theta)^{n-x}$. In addition, the marginal distribution of X is obtained by averaging with respect to the discrete uniform distribution for Θ over R_N, $P(\Theta=\theta)=\frac{1}{N+1}$, $\theta\in R_N$, which is the *prior* probability for Θ. Then, we have

$$P(X=x)=\sum_{\theta\in R_N}\binom{n}{x}\theta^x(1-\theta)^{n-x}\frac{1}{N+1}=\binom{n}{x}\frac{1}{(N+1)}\sum_{k=0}^{N}\frac{k}{N}^x\left(1-\frac{k}{N}\right)^{n-x}.$$

- For a fixed value of n, as the number of urns N tends to ∞, we have the approximation

$$\frac{1}{(N+1)}\sum_{k=0}^{N}\frac{k}{N}^x\left(1-\frac{k}{N}\right)^{n-x}\approx\int_0^1\theta^x(1-\theta)^{n-x}d\theta=B(x+1,n-x+1),$$

which is obtained by taking the average value of $\theta^x(1-\theta)^{n-x}$ at $N+1$ equally spaced points θ between 0 and 1 that converges to the beta integral as $N\to\infty$. Thus, we have

$$P(X=x)\overset{N\to\infty}{\to}\int_0^1\binom{n}{x}\theta^x(1-\theta)^{n-x}d\theta=\binom{n}{x}B(x+1,n-x+1)=\frac{1}{n+1}.$$

- Suppose that there are $n+1$ balls, n white balls and 1 black ball and that they are randomly located on the unit interval $[0,1]$ such that the positions of the balls are i.i.d. with Uniform$(0,1)$. Let X be the number of white balls to the left of the black ball, which has the Binomial distribution with parameters n and θ, given the position $(\Theta=\theta)$ of the black ball, $X|\Theta=\theta\sim$ Binomial(n,θ), $P(X=x|\Theta=\theta)=\binom{n}{x}\theta^x(1-\theta)^{n-x}$, $x=0,1,\ldots,n$, where the position of the black ball can be at random on $(0,1)$, which implies that $f_\Theta(\theta)=1$ for $0<\theta<1$. Thus,

$$P(X=x)=\int_0^1 P(X=x|\Theta=\theta)f_\Theta(\theta)d\theta=\int_0^1\binom{n}{x}\theta^x(1-p)^{n-x}d\theta=\frac{1}{n+1}.$$

Alternatively, suppose there are $n+1$ balls randomly located on the unit interval. Among them, we choose one ball at random and count the number of balls to the left of the ball chosen, denoted by X. Then, the chosen ball can be any one of the $n+1$ balls, so $X=x$ are all equally likely, $P(X=x)=\frac{1}{n+1}$ for $x=0,1,\ldots,n$.

- Using Bayes' theorem, the conditional distribution of Θ given that $X=x$ is

$$f_{\Theta|X}(\theta|x)=\frac{f_{X|\theta}f_\Theta(\theta)}{f_X(x)}=(n+1)\binom{n}{x}\theta^x(1-\theta)^{n-x},\ \text{for } 0<\theta<1,$$

which implies that $\Theta|X=x\sim$ Beta$(x+1,n-x+1)$.

- The underlying idea is to assume a *prior* distribution for Θ, a distribution representing the plausibility of each possible value of θ *before* the data X is observed. Then, to make inferences about θ, one simply considers the conditional distribution of Θ given the observed data X. This is referred to as the *posterior* distribution, since it represents the plausibility of each possible value of θ *after* seeing the data. Here, the prior distribution for Θ is a beta distribution distribution with parameters 1 and 1, Beta$(1,1)$, that is, Uniform$(0,1)$, and the posterior distribution for Θ is also a beta distribution with parameters $x+1$ and $n-x+1$. This is known as the *Beta-binomial conjugacy*, in that if the prior distribution for Θ is a beta distribution and the conditional distribution of X given $\Theta=\theta$ is a binomial distribution with parameters n and θ, then, the posterior distribution is also a beta distribution, in other words, belongs to the family of beta distribution.

■ **Example 6.40 — Noncentral χ^2-distribution.** Suppose that X has a Poisson distribution with parameter λ, $X \sim \text{Poisson}(\lambda)$ and that the conditional distribution of Y given $X = x$ has χ^2-distribution with $1+2x$ degrees of freedom $Y|X = x \sim \chi^2(1+2x)$. Find the distribution of Y.

- Note that the joint distribution is the mixed one as X is discrete but Y is continuous. Thus, the marginal p.d.f. of Y is obtained by

$$f_Y(y) = \sum_{x=0}^{\infty} f_{X,Y}(x,y) = \sum_{x=0}^{\infty} f_{Y|X}(y|x) f_X(x) = \sum_{x=0}^{\infty} \frac{y^{\frac{1}{2}+x-1}}{\Gamma(\frac{1}{2}+x)} e^{-y/2} \cdot \frac{e^{-\lambda}\lambda^x}{x!}.$$

Also, the mean and variance of Y are computed from the law of total probability for expectations and variances,

$$\begin{aligned} \mathrm{E}(Y) &= \mathrm{E}\left[\mathrm{E}(Y|X)\right] = \mathrm{E}(1+2X) = 1+2\lambda, \\ \mathrm{Var}(Y) &= \mathrm{Var}\left[\mathrm{E}(Y|X)\right] + \mathrm{E}\left[\mathrm{Var}(Y|X)\right] = \mathrm{Var}(1+2X) + \mathrm{E}(2+4X) = 8\lambda. \end{aligned}$$

- Let $W = (Z+\theta)^2$, where $Z \sim N(0,1)$. The p.d.f. of W can be found by differentiating the c.d.f of W,

$$\begin{aligned} f_W(w) &= \frac{d}{dw} F_W(w) = \frac{d}{dw} P(\theta - \sqrt{w} \le Z \le \theta + \sqrt{w}) \\ &= \frac{1}{2\sqrt{w}\sqrt{2\pi}} \exp\left(-\frac{1}{2}[\sqrt{w}+\theta]^2\right) + \frac{1}{2\sqrt{w}\sqrt{2\pi}} \exp\left(-\frac{1}{2}[\sqrt{w}-\theta]^2\right) \\ &= \frac{1}{2\sqrt{w}\sqrt{2\pi}} \exp\left(-\frac{1}{2}[w+\theta^2]\right) \left[\exp(\sqrt{w}\theta) + \exp(-\sqrt{w}\theta)\right]. \end{aligned}$$

Note that the last two exponentials can be expanded and that all of the odd power terms cancel so that $\exp(\sqrt{w}\theta) + \exp(-\sqrt{w}\theta) = 2\sum_{x=0}^{\infty} \frac{w^x \theta^{2x}}{(2x)!}$. Thus, we have

$$\begin{aligned} f_w(w) &= \frac{1}{\sqrt{w}\sqrt{2\pi}} \exp\left(-\frac{1}{2}[w+\theta^2]\right) \sum_{x=0}^{\infty} \frac{w^x\theta^{2x}}{(2x)!} \\ &= \sum_{x=0}^{\infty} \frac{\exp(-\lambda)\lambda^x}{x!} \frac{(1/2)^{(1+2x)/2}}{\Gamma((1+2x)/2)} w^{(1+2x)/2-1} \exp(-w/2), \end{aligned}$$

where $\lambda = \theta^2/2$ and we have used the fact that $\Gamma\left(x + \frac{1}{2}\right) = \frac{\sqrt{\pi}}{2^x} 1 \cdot 3 \cdots (2x-1)$.

- Note that $f_W(w)$ has the same form of $f_Y(w)$, a Poisson weighted average of χ^2_{1+2x} densities. That is, suppose that X has a Poisson distribution with parameter $\lambda = \theta^2/2$ and that Y conditional on $X = x$ has $\chi^2(1+2x)$ distribution. Then, the marginal distribution of Y has the same distribution with W. This is called the *noncentral χ^2-distribution* with 1 degree of freedom and noncentral parameter λ denoted $\chi^2(1,\lambda)$. Then, the m.g.f of Y, the $\chi^2(1,\lambda)$ distribution, can also be computed by conditioning

$$M_Y(t) = \mathrm{E}(\mathrm{E}[\exp(tY)|X]) = \mathrm{E}[(1-2t)^{1+2X}] = (1-2t)^{-1/2} \exp\left[\frac{2t\lambda}{1-2t}\right], \; t < 1/2.$$

■

6.4 Bivariate normal distributions

In most experiments several continuous random variables are observed on the same individual (e.g., height and weight, or midterm and final scores, father's height and son's height, etc.) A popular model for the joint distribution of continuous random variables is the family of bivariate normal distributions.

Definition 6.4.1 Let X_1 and X_2 be random variables with the following joint p.d.f.:

$$\begin{aligned} f_{X_1,X_2}(x_1,x_2) &= \frac{1}{2\pi\sigma_1\sigma_2\sqrt{1-\rho^2}}\exp\left(-\frac{1}{2(1-\rho^2)}\left[\frac{(x_1-\mu_1)^2}{\sigma_1^2}+\frac{(x_2-\mu_2)^2}{\sigma_2^2}\right.\right. \\ &\qquad \left.\left.-\frac{2\rho(x_1-\mu_1)(x_2-\mu_2)}{\sigma_1\sigma_2}\right]\right). \end{aligned}$$

We say that X_1 and X_2 have a *bivariate normal distribution with means μ_1 and μ_2, variances σ_1^2 and σ_2^2 and correlation ρ*, denoted as $N_2(\mu_1,\mu_2,\sigma_1^2,\sigma_2^2,\rho)$.

■ **Example 6.41** When $\sigma,\tau>0$ and $|\rho|<1$, consider a bivariate function $f(x,y)$

$$f_{X,Y}(x,y)=\frac{1}{2\pi\sigma\tau(1-\rho^2)^{\frac{1}{2}}}\exp\left[-\frac{1}{2(1-\rho^2)}\left(\frac{x^2}{\sigma^2}-\frac{2\rho xy}{\sigma\tau}+\frac{y^2}{\tau^2}\right)\right].$$

- If $f(x,y)$ is a joint p.d.f. of two random variables X and Y, then we have the marginal p.d.f. of X, $f_X(x)=\int_{-\infty}^{\infty}f(x,y)dy$

$$f_X(x)=\frac{1}{2\pi\sigma\tau(1-\rho^2)^{\frac{1}{2}}}\int_{-\infty}^{\infty}\exp\left[-\frac{1}{2(1-\rho^2)}\left(\left(\frac{y}{\tau}-\frac{\rho x}{\sigma}\right)^2+\frac{x^2}{\sigma^2}-\frac{\rho^2x^2}{\sigma^2}\right)\right]dy$$

  Now setting $\frac{y}{\tau}-\frac{\rho x}{\sigma}=u$, and recalling $\int_{-\infty}^{\infty}\exp\left(-\frac{u^2}{2(1-\rho^2)}\right)\tau du=\left(2\pi\left(1-\rho^2\right)\right)^{\frac{1}{2}}\tau$ yields $f_X(x)=\frac{1}{\sqrt{2\pi\sigma^2}}\exp\left(-\frac{x^2}{2\sigma^2}\right)$, which is the $N\left(0,\sigma^2\right)$ density.
- Therefore $f(x,y)$ is a joint p.d.f. Similarly, we have $f_Y(y)$ is the $N\left(0,\tau^2\right)$ density.

■

Theorem 6.4.1 Let Z_1 and Z_2 be independent random variables having the standard normal distribution. Let $X=Z_1$ and $Y=\rho Z_1+\sqrt{1-\rho^2}Z_2$, $|\rho|<1$. Then, the pair of X and Y has the bivariate normal distribution, $N_2(0,0,1,1,\rho)$, which is often called the standard bivariate normal distribution.

Proof. The joint p.d.f. of Z_1 and Z_2 is $f_{Z_1,Z_2}(z_1,z_2)=\frac{1}{2\pi}\exp\left\{-\frac{1}{2}(z_1^2+z_2^2)\right\}$, and the inverse transformation is $Z_1=X$ and $Z_2=(Y-\rho X)/\sqrt{1-\rho^2}$. Thus, the joint p.d.f. of X

and Y is given by

$$f_{X,Y}(x,y) = f_{Z_1,Z_2}\left(x, \frac{y-\rho x}{\sqrt{1-\rho^2}}\right)|J| = \frac{1}{2\pi\sqrt{1-\rho^2}}\exp\left\{-\frac{1}{2(1-\rho^2)}(x^2-2\rho xy+y^2)\right\},$$

which is the bivariate normal density. ∎

Example 6.42 Let X and Y have the standard bivariate normal joint density

$$f_{X,Y}(x,y) = \frac{1}{2\pi(1-\rho^2)^{\frac{1}{2}}}\exp\left(-\frac{x^2-2\rho xy+y^2}{2(1-\rho^2)}\right).$$

- The joint m.g.f. of X and Y is $M_{X,Y}(t_1,t_2) = \mathrm{E}\left(e^{t_1X+t_2Y}\right)$. Thus,

$$\begin{aligned} M_{X,Y}(t_1,t_2) &= \int_{-\infty}^{\infty}\int_{-\infty}^{\infty} f_{X,Y}(x,y)e^{t_1x+t_2y}dxdy \\ &= \frac{1}{2\pi}\int_{-\infty}^{-\infty}\int_{-\infty}^{-\infty}\exp\left(-\frac{1}{2}u^2+(t_1+t_2\rho)u-\frac{1}{2}v^2+t_2\left(1-\rho^2\right)^{\frac{1}{2}}v\right)dudv \\ &= \mathrm{E}\left(e^{(t_1+t_2\rho)U}\right)\mathrm{E}\left(e^{t_2(1-\rho^2)^{\frac{1}{2}}V}\right) = e^{\frac{1}{2}(t_1+t_2\rho)^2}e^{\frac{1}{2}t_2^2(1-\rho^2)}, \end{aligned}$$

by the change of variables with $u=x$ and $v=(y-\rho x)/\sqrt{1-\rho^2}$. Note that since U and V are standard normal random variables, the last equality follows from the m.g.f. of a standard normal random variable. Hence, we have

$$M_{X,Y}(t_1,t_2) = \exp\left(\frac{1}{2}(t_1^2+2\rho t_1t_2+t_2^2)\right).$$

Alternatively, we can represent X and Y with linear combinations of two independent standard normal random variables Z_1 and Z_2 as given in Theorem 6.4.1. Then, the joint m.g.f. of X and Y, $M_{X,Y}(t_1,t_2)$ is given by

$$\begin{aligned} M_{X,Y}(t_1,t_2) &= \mathrm{E}\left(e^{t_1X+t_2Y}\right) = \mathrm{E}\left(e^{t_1Z_1+t_2(\rho Z_1+\sqrt{1-\rho^2}Z_2)}\right) \\ &= \mathrm{E}\left(e^{(t_1+t_2\rho)Z_1}\right)\mathrm{E}\left(e^{t_2\sqrt{1-\rho^2}Z_2}\right) = e^{\frac{1}{2}(t_1+t_2\rho)^2}e^{\frac{1}{2}t_2^2(1-\rho^2)}. \end{aligned}$$

- Let $U=aX$, $V=bY$ and $W=U+V$ for constants a and b. The joint p.d.f. of U and V is $f_{U,V}(u,v) = \frac{1}{ab}f_{X,Y}\left(\frac{u}{a},\frac{v}{b}\right)$. Thus, the p.d.f. of W is given by $f_W(w) = \int_{-\infty}^{\infty}\frac{1}{ab}f_{X,Y}\left(\frac{u}{a},\frac{w-u}{b}\right)du = \frac{1}{(2\pi\xi^2)^{\frac{1}{2}}}\exp\left(-\frac{w^2}{2\xi^2}\right)$, where $\xi^2 = a^2+2\rho ab+b^2$ from some tedious rearrangements of the exponent in the integrand. Hence, $W = aX+bY \sim N\left(0,a^2+2\rho ab+b^2\right)$. On the other hand, we compute the m.g.f. of W

$$\begin{aligned} M_W(t) &= \mathrm{E}\left(e^{tW}\right) = \mathrm{E}\left(e^{taX+tbY}\right) = M_{X,Y}(ta,tb) \\ &= \exp\left(\frac{1}{2}(a^2t^2+2\rho abt+b^2t^2)\right) = \exp\left(\rho abt+\frac{1}{2}(a^2+b^2)t^2\right), \end{aligned}$$

which is the m.g.f. of a normal distribution with mean ρab and variance (a^2+b^2). Therefore, by the uniqueness of m.g.f., we also conclude that $W = aX + bY \sim N\left(0, a^2+2\rho ab+b^2\right)$.

■

Theorem 6.4.2 Let X and Y have the bivariate normal distribution with a joint p.d.f.

$$f_{X,Y}(x,y) = \frac{1}{2\pi\sigma\tau(1-\rho^2)^{\frac{1}{2}}} \exp\left(-\frac{1}{2(1-\rho^2)}\left(\frac{x^2}{\sigma^2} - \frac{2\rho xy}{\sigma\tau} + \frac{y^2}{\tau^2}\right)\right).$$

Then, we have

- $X \sim N(0,\sigma^2)$ and $Y \sim N(0,\tau^2)$. In addition, X and Y are independent if and only if they are uncorrelated, i.e. $\rho = 0$.
- The conditional distribution of Y given $X = x$ is the normal distribution with mean and variance given by $\mathrm{E}(Y|X=x) = \rho\frac{\tau}{\sigma}x$, $\mathrm{Var}(Y|X=x) = \tau^2(1-\rho^2)$.

Proof. The proof relies on the results in Examples 6.36 and 6.37. Here, we assumed $\mathrm{E}(X) = \mathrm{E}(Y) = 0$, but they do not have to be zero.

- It follows from Example 6.36 that the marginal distributions of X and Y are normal distributions, $X \sim N(0,\sigma^2)$ and $Y \sim N(0,\tau^2)$. If X and Y are independent, it is obvious that $\rho = 0$. Assume that X and Y are uncorrelated, that is $\rho = 0$. Since the joint p.d.f. $f_{X,Y}(x,y)$ is written as

$$f_{X,Y}(x,y) = \frac{1}{\sqrt{2\pi\sigma^2}} \exp\left(-\frac{x^2}{2\sigma^2}\right) \times \frac{1}{\sqrt{2\pi(1-\rho^2)\tau}} \exp\left(-\frac{(y-[\rho\tau x/\sigma])^2}{2(1-\rho^2)\tau^2}\right),$$

  we see that the joint p.d.f. factors into the product of the marginal p.d.f. of X and the marginal p.d.f. of Y. Hence, X and Y are independent.

- The conditional p.d.f. of Y given $X = x$ is

$$f_{Y|X}(y|x) = \frac{f_{X,Y}(x,y)}{f_X(x)} = \frac{1}{\sqrt{2\pi(1-\rho^2)\tau}} \exp\left(-\frac{(y-[\rho\tau x/\sigma])^2}{2(1-\rho^2)\tau^2}\right).$$

  That is, the conditional distribution of Y given $X = x$ is the normal distribution with mean $\rho\tau x/\sigma$ and variance $\tau^2(1 \quad \rho^2)$, $N\left(\rho\tau x/\sigma, \tau^2\left(1 \quad \rho^2\right)\right)$. Further when $\mathrm{E}(X) = \mu_X \neq 0$ and $\mathrm{E}(Y) = \mu_Y \neq 0$, define $U = X - \mu_X$ and $V = Y - \mu_Y$. Then, using the change of variables formula, we see that $f_{U,V}(u,v) = f_{X,Y}(u+\mu_X, v+\mu_Y)|J(u,v)|$ and that

$$\mathrm{E}(V|U=u) = \rho\frac{\tau}{\sigma}u \Rightarrow \mathrm{E}(Y|X=x) = \mathrm{E}(Y) + \frac{\mathrm{Cov}(X,Y)}{\mathrm{Var}(X)}(x-\mu_X),$$

$$\mathrm{Var}(V|U=u) = \mathrm{Var}(Y|X=x) = \mathrm{Var}(Y)(1-\rho^2) = \tau^2(1-\rho^2).$$

- By the symmetry of the joint p.d.f., we easily see that the conditional distribution of X given $Y = y$ is the normal distribution with mean $\rho\frac{\sigma}{\tau}y$ and variance $\sigma^2(1-\rho^2)$. ■

Example 6.43 The bivariate normal distribution is the distribution of a linear transformation of two independent standard normal random variables. Let Z_1 and Z_2 be independent standard normal random variables. Define $X = \mu_X + \sigma_X Z_1$ and $Y = \mu_Y + \rho\sigma_Y Z_1 + \sqrt{1-\rho^2}\sigma_Y Z_2$ for $\sigma_X, \sigma_Y > 0$ and $|\rho| < 1$. Then,

- It is obvious that the marginal distributions of X and Y are normal distributions with

$$\mathrm{E}(X) = \mu_X + \sigma_X \mathrm{E}(Z_1) = \mu_X, \;\; \mathrm{E}(Y) = \mu_Y + \rho\sigma_Y \mathrm{E}(Z_1) + \sqrt{1-\rho^2}\sigma_Y \mathrm{E}(Z_2) = \mu_Y,$$
$$\mathrm{Var}(X) = \sigma_X^2 \mathrm{Var}(Z_1) = \sigma_X^2, \;\; \mathrm{Var}(Y) = \rho^2\sigma_Y^2 \mathrm{Var}(Z_1) + (1-\rho^2)\sigma_Y^2 \mathrm{Var}(Z_2) = \sigma_Y^2.$$

- Also, the covariance and the correlation coefficient between X and Y are given

$$\mathrm{Cov}(X_1, X_2) = \rho\sigma_X\sigma_Y \mathrm{Var}(Z_1) = \rho\sigma_X\sigma_Y, \;\; \rho(X_1, X_2) = \frac{\mathrm{Cov}(X_1, X_2)}{\sqrt{\mathrm{Var}(X_1)\mathrm{Var}(X_2)}} = \rho.$$

- In addition, the joint distribution of X and Y is bivariate normal, which is explained as follows: Let $U = \rho Z_1 + \sqrt{1-\rho^2}Z_2$. The conditional distribution of U given $Z_1 = z_1$ should be a normal distribution with mean ρz_1 and variance $1-\rho^2$ because Z_1 and Z_2 are independent. The joint p.d.f. of Z_1 and U is then

$$f_{Z_1,U}(z_1, u) = f_{Z_1}(z_1) f_{U|Z_1}(u|z_1) = \frac{1}{\sqrt{2\pi}}\exp\left(-\frac{z_1^2}{2}\right)\frac{1}{\sqrt{2\pi(1-\rho^2)}}\exp\left(-\frac{(u-\rho z_1)^2}{2(1-\rho^2)}\right).$$

Next, transform to $X = \mu_X + \sigma_X Z_1$ and $Y = \mu_Y + \sigma_Y U$. Then, the joint p.d.f. of X and Y is given by

$$\frac{1}{\sigma_X\sigma_Y} f_{Z_1,U}\left(\frac{x-\mu_X}{\sigma_X}, \frac{y-\mu_Y}{\sigma_Y}\right)|J(x,y)|$$
$$= \frac{1}{\sqrt{2\pi}\sigma_X}\exp\left(-\frac{(x-\mu_X)^2}{2\sigma_X^2}\right)\frac{1}{\sqrt{2\pi(1-\rho^2)}\sigma_Y}\exp\left(-\frac{\left(y-\left[\mu_Y+\rho\frac{(x-\mu_X)}{\sigma_X}\right]\right)^2}{2(1-\rho^2)\sigma_Y^2}\right),$$

which simplifies to the bivariate normal p.d.f. given in Definition 6.4.1. ■

The bivariate normal distribution is the joint distribution of two particular linear combinations of two independent standard normal random variables as discussed in Theorem 6.4.1 and Example 6.38. Thus, in general, if Y_1 and Y_2 are independent normal random variables, then $X_1 = a_1Y_1 + a_2Y_2 + a_3$ and $X_2 = b_1Y_1 + b_2Y_2 + b_3$ have a bivariate normal distribution with the appropriate means, variances and correlation.

Example 6.44 Let Y_i have a normal distribution with mean μ_i and variance σ_i^2 for $i = 1, 2$. Suppose that Y_1 and Y_2 are independent. Let $X_1 = (Y_1 + Y_2)/2$ and $X_2 = Y_2 - Y_1$.

- The joint distribution is bivariate normal. The mean of X_1 is $(\mu_1+\mu_2)/2$ and the variance is $(\sigma_1^2+\sigma_2^2)/4$. The mean of X_2 is $\mu_2-\mu_1$ and its variance is $\sigma_1^2+\sigma_2^2$. The covariance between X_1 and X_2 is

$$\mathrm{Cov}(X_1,X_2) = -\frac{1}{2}\mathrm{Var}(Y_1)+\frac{1}{2}\mathrm{Var}(Y_2) = \frac{\sigma_2^2-\sigma_1^2}{2},\ \rho = \frac{\sigma_2^2-\sigma_1^2}{\sigma_1^2+\sigma_2^2}.$$

- Consider two i.i.d. normal random variables Y_1 and Y_2 with common mean μ and common variance σ^2. The joint distribution in this case is bivariate normal with means μ and 0, variances $\sigma^2/2$ and $2\sigma^2$ and covariance 0. Hence X_1 and X_2 are independent, even though they are functions of the same two random variables.

■

■ **Example 6.45** Two linear combinations of bivariate normals also have a bivariate normal distribution. That is, if X_1 and X_2 have the standard bivariate normal joint distribution, $N_2(0,0,1,1,\rho)$, then each of them is a linear combination of independent normals Z_1 and Z_2. Thus, every linear combination of X_1 and X_2 is also a linear combination of Z_1 and Z_2. For example, $aX_1+bX_2+c = [a\sigma_1+b\rho\sigma_2]Z_1+b\sqrt{1-\rho^2}\sigma_2 Z_2+a\mu_1+b\mu_2+c$. ■

■ **Example 6.46** Let X and Y have the standard bivariate normal distribution with a joint p.d.f. given by $f_{X,Y}(x,y) = \dfrac{1}{2\pi(1-\rho^2)^{\frac{1}{2}}}\exp\left(-\dfrac{1}{2(1-\rho^2)}(x^2-2\rho xy+y^2)\right)$. Define $W=XY$. Then, we can compute $\mathrm{E}(W)$, $\mathrm{Var}(W)$ and $M_W(t)=\mathrm{E}(e^{tW})$ using the law of total probability for expectations and variances.

- Note that $X\sim N(0,1)$, $Y\sim N(0,1)$, and $X|Y=y\sim N\left(\rho y,1-\rho^2\right)$. Thus, we have

$$\mathrm{E}(X|Y)=\rho Y,\ \mathrm{E}(X^2|Y)=\mathrm{Var}(X|Y)+(\mathrm{E}[X|Y])^2 = 1-\rho^2+\rho^2Y^2,$$
$$\mathrm{E}\left(e^{tX}|Y\right) = \exp\left(\rho Yt+\frac{1}{2}(1-\rho^2)t^2\right).$$

- Now, we evaluate

$$\begin{aligned}
\mathrm{E}(W) &= \mathrm{E}(\mathrm{E}(XY|Y)) = \mathrm{E}(\rho Y^2) = \rho,\\
\mathrm{Var}(W) &= \mathrm{Var}(\mathrm{E}(XY|Y))+\mathrm{E}(\mathrm{Var}(XY|Y)) = \mathrm{Var}(\rho Y^2)+\mathrm{E}(Y^2) = 2\rho^2+1,\\
\mathrm{E}\left(e^{tW}\right) &= \mathrm{E}\left[\mathrm{E}\left(e^{tXY}|Y\right)\right] = \mathrm{E}\left[\exp\left(\rho Y^2t+\frac{1}{2}(1-\rho^2)t^2Y^2\right)\right].\\
&= \frac{1}{(1-2\rho t-(1-\rho^2)t^2)^{1/2}},
\end{aligned}$$

which follows from $Y^2\sim\chi^2(1)$, $\mathrm{E}(Y^2)=1$, $\mathrm{Var}(Y^2)=2$, and $\mathrm{E}(e^{tY^2}) = \left(\dfrac{1}{1-2t}\right)^{1/2}$ for $t<1/2$.

■

■ **Example 6.47** Let X and Y have the standard bivariate normal distribution with a joint p.d.f. given by $f_{X,Y}(x,y) = \dfrac{1}{2\pi(1-\rho^2)^{\frac{1}{2}}} \exp\left(-\dfrac{1}{2(1-\rho^2)}(x^2-2\rho xy+y^2)\right)$. Then,

$$P(X \le 0,\ Y \le 0) = \frac{1}{4} + \frac{1}{2\pi}\tan^{-1}\left(\frac{\rho}{\sqrt{1-\rho^2}}\right), \text{ and } \mathrm{E}(Y|X \ge 0) = \frac{\sqrt{2}\rho}{\sqrt{\pi}}.$$

When $\rho = 0$, i.e., X and Y are independent, it is obvious that $P(X \le 0,\ Y \le 0) = \frac{1}{4}$ and $\mathrm{E}(Y|X \ge 0) = 0$.

- Since $f_{X,Y} = f_X(x) f_{Y|X}(y|x)$, $X \sim N(0,1)$ and $Y|X = x \sim N(\rho x, 1-\rho^2)$, Then, we have $P(X \le 0,\ Y \le 0) = \int_{-\infty}^{0} \frac{1}{\sqrt{2\pi}} e^{-\frac{x^2}{2}} dx \times \int_{-\infty}^{0} \frac{1}{\sqrt{2\pi(1-\rho^2)}} e^{-\frac{(y-\rho x)^2}{2(1-\rho^2)}} dy$.
- Thus, using the change of variable formula with the polar coordinates, we evaluate $P(X \le 0,\ Y \le 0)$, the probability that X and Y belong to negative quadrant as follows.

$$\begin{aligned}
&\int_{-\infty}^{0} \frac{1}{\sqrt{2\pi}} e^{-\frac{x^2}{2}} dx \times \int_{-\infty}^{0} \frac{1}{\sqrt{2\pi(1-\rho^2)}} e^{-\frac{(y-\rho x)^2}{2(1-\rho^2)}} dy \\
&= \int_{-\infty}^{0} \frac{1}{\sqrt{2\pi}} e^{-\frac{x^2}{2}} dx \times \int_{-\infty}^{-\frac{\rho x}{\sqrt{1-\rho^2}}} \frac{1}{\sqrt{2\pi}} e^{-\frac{y^2}{2}} dy = \frac{1}{2\pi} \int_{-\infty}^{0} \int_{-\infty}^{-\frac{\rho x}{\sqrt{1-\rho^2}}} e^{-\frac{1}{2}(x^2+y^2)} dy dx \\
&= \frac{1}{2\pi} \int_{\pi-\tan^{-1}\left(\rho/\sqrt{1-\rho^2}\right)}^{\frac{3}{2}\pi} \int_{0}^{\infty} e^{-\frac{1}{2}r^2} r dr d\theta = \frac{1}{4} + \frac{1}{2\pi}\tan^{-1}\left(\frac{\rho}{\sqrt{1-\rho^2}}\right).
\end{aligned}$$

- Note that $Y|X = x \sim N(\rho x, 1-\rho^2)$, that is, $\mathrm{E}(Y|X = x) = \rho x$. Thus,

$$\mathrm{E}(Y|X \ge 0) = \frac{\mathrm{E}(Y I_{X \ge 0})}{P(X \ge 0)} = 2\mathrm{E}(\mathrm{E}(Y I_{X \ge 0}|X)) = 2\int_{0}^{\infty} \rho x \frac{1}{\sqrt{2\pi}} e^{-\frac{1}{2}x^2} dx = \frac{\sqrt{2}\rho}{\sqrt{\pi}}.$$

■

Remark Let Z_1 and Z_2 be two independent standard normal random variables. Then, a pair of the random variables X and Y with the standard bivariate normal distribution can be constructed as $X = Z_1$ and $Y = X\cos\theta + Z_2\sin\theta$, where θ be an angle between X and Y.

- Since X and Y are linear combinations of two independent standard normal random vairiables, X and Y have the bivariate normal distribution with

$$\begin{aligned}
&\mathrm{E}(X) = \mathrm{E}(Z_1) = 0,\ \mathrm{E}(Y) = \cos\theta \mathrm{E}(X) + \sin\theta \mathrm{E}(Z_2) = 0, \\
&\mathrm{Var}(X) = \mathrm{Var}(Z_1) = 1,\ \mathrm{Var}(Y) = \cos^2\theta \mathrm{Var}(X) + \sin^2\theta \mathrm{Var}(Z_2) = 1, \\
&\rho = \frac{\mathrm{Cov}(X,Y)}{\sqrt{\mathrm{Var}(X)\mathrm{Var}(Y)}} = \mathrm{E}(XY) = \cos\theta \mathrm{E}(X^2) + \mathrm{E}(Z_1)\mathrm{E}(Z_2)\sin\theta = \cos\theta.
\end{aligned}$$

- Note that the correlation ρ is $\cos\theta$, where θ is an angle between two random variables defined in L^2 space as remarked in Section 5.4. Thus, $\cos\theta = \rho$ and $\sin\theta = \sqrt{1-\cos^2\theta} = \sqrt{1-\rho^2}$, which implies $Y = \rho X + \sqrt{1-\rho^2}Z_2$ and $X = Z_1$ as given in Theorem 6.4.1.

- Francis Galton (1822-1911) measured the heights of 1,078 fathers and sons and calculated the expected value of the height of a son given the height of the father. Galton observed that tall fathers tend to have tall sons but the sons are not, on average, as tall as the fathers. Also, short fathers tend to have short sons, but the sons are not as short on average as their fathers. Galton used this to show that tall fathers can be expected to have sons who are less tall on the average. For example, students who do very well on the midterm can be expected to do less well on the final. This phenomenon is called *regression to the mean*.

- Regression, as we know it today, was invented in Galton's 1877 paper "Typical laws of heredity" on his analysis about the question "why do the characteristics (mean and variance) of a hereditary attribute (such as height) from an isolated human population remain constant from generation to generation?"

- Galton's analysis can be explained as the bivariate normal distribution, $(X,Y)^\top \sim N_2(\mu_X, \mu_Y, \sigma_X^2, \sigma_Y^2, \rho)$, $\mathrm{E}(Y|X=x) = \mu_Y + \rho\sigma_Y/\sigma_X(x-\mu_X)$ and $\mathrm{Var}(Y|X=x) = \sigma_Y^2(1-\rho^2)$. For the standard bivariate normal distribution, $\mathrm{E}(Y|X=x) = \rho x$. Thus, if X and Y are positively correlated $(0 < \rho < 1)$, and if $x > \mathrm{E}(X)$, then $\mathrm{E}(Y|X=x) = \rho x$ will be less than x. (i.e., we have regression to the mean).

- Suppose that X and Y denote the midterm and the final scores in STAT 221, which are designed following the bivariate normal distributions $(X,Y)^\top \sim N_2(75, 75, 10^2, 10^2, 0.6)$. Then, the prediction for Y given $X = x$ is $\mathrm{E}(Y|X=x) = 75 + 0.6(x-75)$, and the regression line $y = 30 + 0.6x$. For example, a student who got 90 points in the midterm is predicted to be 84, less than the midterm score 90, which is "regression to the mean" as show in Figure 6.6

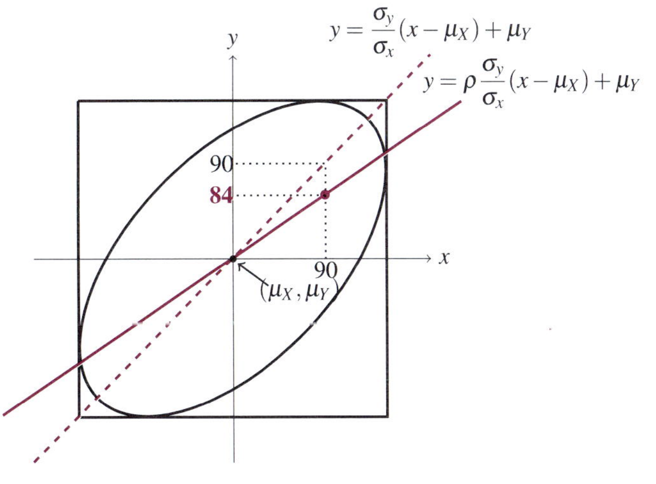

Figure 6.6: Regression to the mean

Also, note that the event that midterm score is above average can be interpreted as the event $X > \mathrm{E}(X)$. Then, the average final score of students who scored above average ($\mu_X = 75$ points) on the midterm is, $\mathrm{E}\left(\frac{Y-\mu_Y}{\sigma_Y}|X>75\right) = \frac{\sqrt{2}\rho}{\sqrt{\pi}}$, as in Example 6.47. Thus, the expected score is $75 + 10 \times 0.6\sqrt{\frac{2}{\pi}} \doteq 79.787$.

■ **Example 6.48 — Normal-normal model.** Suppose that $X|Y = y \sim N(y, \sigma^2)$ and $Y \sim N(\mu, \tau^2)$. Then, the joint p.d.f. of X and Y is written as

$$f_{X,Y}(x,y) = f_{X|Y}(x|y)f_Y(y) = \frac{\exp\left(-(x-y)^2/(2\sigma^2)\right)}{\sqrt{2\pi\sigma^2}} \times \frac{\exp\left(-(y-\mu)^2/(2\tau^2)\right)}{\sqrt{2\pi\tau^2}}.$$

- The conditional p.d.f. of Y given $X = x$ is proportion to the joint p.d.f., that is, the product of $f_{X|Y}$ and f_Y,

$$f_{Y|X}(y|x) = \frac{f_{X,Y}(x,y)}{f_X(x)} \propto f_{X|Y}(x|y)f_Y(y) \propto \exp\left(-\frac{y^2-2\mu y}{2\tau^2} - \frac{y^2-2xy}{2\sigma^2}\right)$$

since multiplicative terms such as $\exp\left(-\mu^2/(2\tau^2)\right)$ do not contain y. Then, letting $\frac{1}{\sigma_1^2} = \frac{1}{\sigma^2} + \frac{1}{\tau^2}$ and $\frac{\mu_1}{\sigma_1^2} = \frac{\mu}{\tau^2} + \frac{x}{\sigma^2}$, it is simplified to $f_{Y|X}(y|x) \propto \exp\left(-\frac{y^2-2\mu_1 y}{2\sigma_1^2}\right)$, which implies that $Y|X = x \sim N(\mu_1, \sigma_1^2)$.

- Note that X and Y have a bivariate normal distribution $(X,Y) \sim N_2(\mu_X, \mu_Y, \sigma_X^2, \sigma_Y^2, \rho)$,

$$\begin{aligned}
&\mu_Y = \mathrm{E}(Y) = \mu,\ \sigma_Y^2 = \mathrm{Var}(Y) = \tau^2,\ \mu_X = \mathrm{E}(X) = \mathrm{E}(\mathrm{E}(X|Y)) = \mu, \\
&\sigma_X^2 = \mathrm{Var}(X) = \mathrm{E}(\mathrm{Var}(X|Y)) + \mathrm{Var}(\mathrm{E}(X|Y)) = \sigma^2 + \tau^2, \\
&\rho = \frac{\mathrm{Cov}(X,Y)}{\sqrt{\mathrm{Var}(X)\mathrm{Var}(Y)}} = \frac{\mathrm{E}(\mathrm{E}(XY|Y)) - \mathrm{E}(X)\mathrm{E}(Y)}{\sqrt{(\sigma^2+\tau^2)\tau^2}} \\
&\quad = \frac{\mathrm{E}(Y^2) - \mathrm{E}(X)\mathrm{E}(Y)}{\sqrt{(\sigma^2+\tau^2)\tau^2}} = \frac{\tau^2+\mu^2-\mu^2}{\sqrt{(\sigma^2+\tau^2)\tau^2}} = \sqrt{\frac{\tau^2}{\sigma^2+\tau^2}}.
\end{aligned}$$

Hence, by Theorem 6.4.2, we also have

$$\begin{aligned}
\mathrm{E}(Y|x) &= \mu_Y + \rho\frac{\sigma_Y}{\sigma_X}(x-\mu_X) = \mu + \frac{\tau^2}{\tau^2+\sigma^2}(x-\mu) \\
&= \frac{\tau^2}{\tau^2+\sigma^2}x + \frac{\sigma^2}{\sigma^2+\tau^2}\mu = \frac{\mu/\tau^2 + x/\sigma^2}{1/\tau^2+1/\sigma^2} = \mu_1, \\
\mathrm{Var}(Y|x) &= \sigma_Y^2(1-\rho^2) = \frac{\sigma^2\tau^2}{\sigma^2+\tau^2} = \sigma_1^2.
\end{aligned}$$

■

Ⓡ Consider computing bivariate normal probabilities

$$P(X \le a, Y \le b) = \int_{-\infty}^{b}\int_{-\infty}^{a} f_{X,Y}(x,y)dxdy,$$

where $f_{X,Y}$ is the bivariate normal p.d.f. The most naïve approach to approximating this would be the simulate a lot of bivariate normals and compute the proportion with $X \le a$ and $Y \le b$.

- A simple Monte Carlo method for approximating this would be the simulate a lot of bivariate normals and compute the proportion with $X \le a$ and $Y \le b$.

$$P(X \le a, Y \le b) = \int_{-\infty}^{a}\int_{-\infty}^{b} f_{X,Y}(x,y)dydx \approx \frac{1}{n}\sum_{i=1}^{n} I(X_i \le a, Y_i \le b).$$

- For bivariate normal random variables X and Y, $(X,Y)^\top \sim N_2(0,0,1,1,\rho)$, we generate two independent standard normal random variables Z_1 and Z_2, and transform them $X = Z_1$ and $Y = \rho Z_1 + \sqrt{1-\rho^2}Z_2$. Alternatively, we can rewrite the joint p.d.f. $f_{X,Y}(x,y) = f_X(x)f_{Y|X}(y|x)$ as a mixture using the multiplication theorem. Then, we can generate $X \sim N(0,1)$ and then generate $Y|X = x \sim N(\rho x, 1-\rho^2)$.

- Consider the case with $\mu_1 = \mu_2 = 0$, $\sigma_1^2 = \sigma_2^2 = 1$, and $\rho = 1/2$. Also, let $a = 0$ and $b = 0$. Indeed, it is the probability that X and Y belong to negative quadrant that we have analytically evaluated in Example 6.47 as $P(X \le 0,\ Y \le 0) = \frac{1}{4} + \frac{\tan^{-1}(\rho/\sqrt{1-\rho^2})}{2\pi}$. Thus, the exact value is $\frac{1}{4} + \frac{\tan^{-1}(1/\sqrt{3})}{2\pi} = 1/3 \doteq 0.3333$.

```
x<-rnorm(10000,0,1)
y<-rnorm(10000,0.5*x, sqrt(1-.5^2))
mean(x<=0&y<=0)
1] 0.3382
```

- A more efficient way is to consider the mixture representation for the integral by rewriting $f_{X,Y}(x,y) = f_X(x)f_{Y|X}(y|x)$, and doing the inner integral first

$$\begin{aligned} P(X \le a, Y \le b) &= \int_{-\infty}^{a} f_X(x)\Phi\left(\frac{b-\rho x}{\sqrt{1-\rho^2}}\right)dx \\ &= \int f_X(x)I_{(-\infty,a]}(x)\Phi\left(\frac{b-\rho x}{\sqrt{1-\rho^2}}\right)dx. \end{aligned}$$

- We could now simulate a lot of X_i's from its marginal distribution and average the values of $I_{(-\infty,a]}(X)\Phi\left(\frac{b-\rho X}{\sqrt{1-\rho^2}}\right)$. Thus,

$$P(X \le a, Y \le b) \approx \frac{1}{n}\sum_{i=1}^{n} I_{(-\infty,a]}(X_i)\Phi\left(\frac{b-\rho X_i}{\sqrt{1-\rho^2}}\right).$$

```
x<-rnorm(10000,0,1)
gx <-function(x){pnorm((0-.5*x)/sqrt(1-.5^2))*(x<=0)}
mean(gx(x))
[1] 0.3312297
```

6.5 Multivariate distributions

Everything that we have developed so far can be extended in a straightforward way to joint distributions of three or more random variables. Many of the results have the same form regardless of how many of the random variables under consideration are discrete or continuous.

Definition 6.5.1 — Multivariate Random vector. An n-dimensional *multivariate random vector* $\mathbf{X}$ is a vector whose ith coordinate X_i is a random variable for each $i = 1, \ldots, n$, $\mathbf{X} = (X_1, \ldots, X_n)^\top$.

- If $X_1, \ldots, X_n$ are all discrete, the *joint p.m.f.* is

$$f_{\mathbf{X}}(x_1, \ldots, x_n) = P(X_1 = x_1, \ldots, X_n = x_n), \ f_{\mathbf{X}} \geq 0, \ \sum_{\text{all } x_1, \ldots, x_n} f_{\mathbf{X}}(x_1, \ldots, x_n) = 1.$$

- We say that $X_1, \ldots, X_n$ have a *joint continuous distribution with joint p.d.f.* f if, for each n-dimensional set A,

$$P(\mathbf{X} \in A) = P[(X_1, \ldots, X_n) \in A] = \int_A \cdots \int f_{\mathbf{X}}(x_1, \ldots, x_n) dx_1 \cdots dx_n.$$

  For f to be a joint p.d.f. we need $f \geq 0$ and $\int \cdots \int f_{\mathbf{X}}(x_1, \ldots, x_n) dx_1 \cdots dx_n = 1$.

- The marginal p.m.f. or p.d.f. $f_i(x_i)$ of X_i is obtained by summing and/or integrating $f_{\mathbf{X}}(x_1, \ldots, x_n)$ over all possible values of the other variables $x_1, \ldots, x_{i-1}, x_{i+1}, \ldots, x_n$.

- The marginal joint p.m.f., p.d.f., mixed p.m.f./p.d.f. $f_{i_1, \ldots, i_k}$ of $(X_{i_1}, \ldots, X_{i_k})$ is obtained by summing and/or integrating $f_{\mathbf{X}}(x_1, \ldots, x_n)$ over all possible values of the other variables besides $x_{i_1}, \ldots, x_{i_k}$, $k = 1, \ldots, n$.

Example 6.49 Suppose that three random variables X_1, X_2, and X_3 have a mixed joint distribution with the following p.m.f./p.d.f.:

$$f_{\mathbf{X}}(x_1, x_2, x_3) = \begin{cases} c(x_1 + 2x_2 + 3x_3), & \text{for } 0 \leq x_i \leq 1, i = 1, 2, \text{ and } x_3 = 0, 1, 2, \\ 0, & \text{otherwise.} \end{cases}$$

- To find c, we sum and integrate

$$\int_0^1 \int_0^1 \sum_{x_3=0}^{2} c(x_1 + 2x_2 + 3x_3) dx_1 dx_2 = \int_0^1 \int_0^1 (3cx_1 + 6cx_2 + 9c) dx_1 dx_2 = 13.5c.$$

  Thus, $c = 1/13.5$.

- To get the marginal joint p.d.f. of X_1 and X_2, we merely sum over the possible x_3 values. We did this already and got

$$f_{1,2}(x_1, x_2) = \frac{1}{13.5}(3x_1 + 6x_2 + 9), \text{ for } 0 < x_1, x_2 < 1.$$

- To get the marginal p.f. of X_3, we integrate over the other two variables

$$f_3(x_3) = \int_0^1 \int_0^1 \frac{1}{13.5}(x_1+2x_2+3x_3)dx_1dx_2 = \frac{1.5+3x_3}{13.5}, \text{ for } x_3 = 0,1,2.$$

■

■ **Example 6.50 — Multinomial distribution.** The multivariate version of a Binomial distribution is called a multinomial distribution. Consider n individuals, each of whom falls into exactly one of k categories, with probabilities $p_1,\ldots,p_k$ summing to 1. Let $\mathbf{X} = (X_1,\ldots,X_k)^T$ be the vector of counts for how many individuals fall into each category. We say that the random vector $\mathbf{X} = (X_1,\ldots,X_k)^\top$ has a *multinomial* distribution with parameters n and $\mathbf{p} = (p_1,\ldots,p_k)$. The p.m.f. of $\mathbf{X}$ is given by

$$f_{\mathbf{X}}(x_1,\ldots,x_k) = \binom{n}{x_1,\ldots,x_k} p_1^{x_1}\cdots p_k^{x_k}, \quad \sum_{j=1}^k p_j = 1, \quad \sum_{j=1}^k x_j = n.$$

- Note that X_i is the number of successes for category i, i.e. the number of individuals falling into category i. Thus, X_i has the binomial distribution with parameter n and p_i, $X_i \sim \text{Binomial}(p_i)$, $i = 1,\ldots,k$. Thus, $\text{E}(X_i) = np_i$, $\text{Var}(X_i) = np_i(1-p_i)$.

- Also, $Y = X_1 + \cdots X_\ell$ has the binomial distribution with parameters n and $p_1 + \cdots + p_\ell$, $\ell < k$ since it is the number of individuals falling into one of ℓ categories. In addition, since $X_i + X_j \sim \text{Binomial}(n, p_i + p_j)$, we have

$$\begin{aligned}\text{Var}(X_i+X_j) &= n(p_i+p_j)(1-p_i-p_j) = \text{Var}(X_i)+\text{Var}(X_j)+2\text{Cov}(X_i,X_j)\\ &= np_i(1-p_i)+np_j(1-p_j)+2\text{Cov}(X_i,X_j).\end{aligned}$$

  Solving for $\text{Cov}(X_i,X_j)$, we have $\text{Cov}(X_i,X_j) = -np_ip_j$.

■

■ **Example 6.51 — Trinomial.** Let $(X_1,X_2,X_3)^\top$ be a random vector with a multinomial distribution with parameters n and (p_1,p_2,p_3). Its joint p.m.f. is given by

$$f_{\mathbf{X}}(x_1,x_2,x_3) = \begin{cases} \frac{n!}{x_1!x_2!x_3!} p_1^{x_1} p_2^{x_2} p_3^{x_3}, & \text{if } x_1,x_2,x_3 = 0,1,\ldots,n \\ 0, & \text{otherwise.} \end{cases}$$

where $0 < p_1, p_2, p_3$, $p_1+p_2+p_3 = 1$ and $x_1+x_2+x_3 = n$.

- Notice the trinomial theorem, similar to the binomial theorem,

$$(a+b+c)^n = \sum_{x=0}^n \sum_{y=0}^n \binom{n}{x,y} a^x b^y c^{n-x-y}, \quad 0 \le x+y \le n.$$

- We know that X_1 has a binomial distribution with parameters n and p_1, and the marginal p.m.f. of X_1 is actually obtained as

$$f_{X_1}(x_1) = \sum_{x_2=0}^{n-x_1} \frac{n!}{x_1!x_2!(n-x_1-x_2)!} p_1^{x_1} p_2^{x_2} (1-p_1-p_2)^{n-x_1-x_2}$$

$$= \frac{n!\,p_1^{x_1}}{x_1!\,(n-x_1)!}\sum_{x_2=0}^{n-x_1}\frac{(n-x_1)!}{x_2!\,(n-x_1-x_2)!}\,p_2^{x_2}\,(1-p_1-p_2)^{n-x_1-x_2}$$
$$= \binom{n}{x_1} p_1^{x_1}\,(1-p_1-p_2+p_2)^{n-x_1} = \binom{n}{x_1} p_1^{x_1}\,(1-p_1)^{n-x_1}.$$

- We also compute the joint m.g.f. of X_1 and X_2 given by

$$M_{X_1,X_2}(t_1,t_2) = \mathrm{E}\left(e^{t_1X_1+t_2X_2}\right) = \sum_{x_1=0}^{n}\sum_{x_2=0}^{n} e^{t_1x_1+t_2x_2} f(x_1,x_2,x_3)$$
$$= \sum_{x_1=0}^{n}\sum_{x_2=0}^{n} e^{t_1x_1+t_2x_2}\frac{n!}{x_1!x_2!\,(n-x_1-x_2)!}\,p_1^{x_1}\,p_2^{x_2}\,(1-p_1-p_2)^{n-x_1-x_2}$$
$$= \sum_{x_1=0}^{n}\sum_{x_2=0}^{n}\frac{n!}{x_1!x_2!\,(n-x_1-x_2)!}\,\left(e^{t_1}p_1\right)^{x_1}\left(e^{t_2}p_2\right)^{x_2}(1-p_1-p_2)^{n-x_1-x_2}$$
$$= \left(1-p_1-p_2+p_1e^{t_1}+p_2e^{t_2}\right)^n,$$

where the last equality follows by the trinomial theorem.

■

Definition 6.5.2 — Conditional Distributions. Let $\mathbf{X}$ be divided into subvectors $\mathbf{Y}$ and $\mathbf{Z}$. That is, let $\mathbf{X} = (\mathbf{Y},\mathbf{Z})$. For example $\mathbf{Y}$ might be the first n_1 coordinates of $\mathbf{X}$ and $\mathbf{Z}$ might be the last n_2 coordinates, where $n = n_1+n_2$ is the number of coordinates of $\mathbf{X}$. Suppose that $\mathbf{X}$ has the joint p.m.f., p.d.f., or p.f./p.d.f. f. Let $f_2(\mathbf{z})$ stand for the marginal p.f., p.d.f., or p.f./p.d.f of $\mathbf{Z}$. The conditional p.m.f., p.d.f., or p.f./p.d.f. of $\mathbf{Y}$ given $\mathbf{Z}=\mathbf{z}$ is

$$f_{\mathbf{Y}|\mathbf{Z}}(\mathbf{y}|\mathbf{z}) = \frac{f_{\mathbf{Y},\mathbf{Z}}(\mathbf{y},\mathbf{z})}{f_{\mathbf{Z}}(\mathbf{z})},\quad \text{and}\quad f_{\mathbf{Y},\mathbf{Z}}(\mathbf{y},\mathbf{z}) = f_{\mathbf{Y}|\mathbf{Z}}(\mathbf{y}|\mathbf{z})f_{\mathbf{Z}}(\mathbf{z}).$$

■ **Example 6.52 — Trinomial (Revisited).** Let $(X_1,X_2,X_3)^\top$ be a random vector with a multinomial distribution with parameters n and (p_1,p_2,p_3). Its joint p.f. is given by

$$f(x_1,x_2,x_3) = \begin{cases} \frac{n!}{x_1!x_2!x_3!}\,p_1^{x_1}\,p_2^{x_2}\,p_3^{x_3}, & \text{if } x_1,x_2,x_3 = 0,1,\ldots,n \\ 0, & \text{otherwise.}\end{cases}$$

where $0 < p_1,\,p_2,p_3,\,p_1+p_2+p_3 = 1$ and $x_1+x_2+x_3 = n$.

- Since the marginal p.d.f of X_1 is $f(x_1) = \binom{n}{x_1}p_1^{x_1}(1-p_1)^{n-x_1}$, the conditional distribution of (X_2,X_3) given $X_1 = x_1$ is computed as

$$f_{X_2,X_3|X_1}(x_2,x_3|x_1) = \frac{\binom{n}{x_1,x_2}p_1^{x_1}p_2^{x_2}p_3^{x_3}}{\binom{n}{x_1}p_1^{x_1}(1-p_1)^{n-x_1}} = \binom{n-x_1}{x_2,x_3}\frac{p_2^{x_2}p_3^{x_3}}{(p_2+p_3)^{n-x_1}},$$

where $x_2,x_3 = 0,\ldots,n-x_1$ with $x_2+x_3 = n-x_1$, for $x_1 = 0,1,\ldots,n$. Hence $(X_2,X_3)^\top|X_1 = x_1 \sim$ Binomial $\left(n-x_1,\frac{p_2}{p_2+p_3}\right)$.

■

Definition 6.5.3 — Independent random Variables. We say that $X_1, \ldots, X_n$ are *(mutually) independent* if

$$P(X_1 \in A_1, X_2 \in A_2, \ldots, X_n \in A_n) = P(X_1 \in A_1)P(X_2 \in A_2)\ldots P(X_n \in A_n).$$

- Let F_{X_i} and f_{X_i} be the c.d.f. and p.d.f. of a continuous random variable X_i. Independence is equivalent to the statement that for all $x_1, \ldots, x_n$, the joint distribution function factors into the product of the marginal distribution functions:

$$F(x_1,\ldots,x_n) = F_{X_1}(x_1)\ldots F_{X_n}(x_n),\ f(x_1,\ldots,x_n) = f_{X_1}(x_1)\ldots f_{X_n}(x_n),\ \text{for all } x_1,\ldots,x_n.$$

Remark If $X_1, \ldots, X_n$ are independent and they all have the same distribution, we say that they are *i.i.d.* (independent and identically distributed.)

- A collection of random variables $X_1, \ldots, X_n$ is called a *random sample* from a distribution if they are i.i.d., in other words, mutually independent and they all have that particular distribution.

- $X_1, \ldots, X_n$ are conditionally independent given $\mathbf{Y}$

$$f_{\mathbf{X}|\mathbf{Y}}(x_1,\ldots,x_n|\mathbf{y}) = f_{X_1|\mathbf{Z}}(x_1|\mathbf{z})\cdots f_{X_n|\mathbf{Z}}(x_n|\mathbf{z}), \text{ for all possible values } \mathbf{y} \text{ of } \mathbf{Y}.$$

- A common usage of conditional independence is to model some observables $X_1, \ldots, X_n$ as conditionally i.i.d. given some other unknown of interest $Y = y$ with some conditional p.m.f. or p.d.f. $f_{X|Y}(x|y)$. Then the joint conditional p.m.f. or p.d.f. of $(X_1, \ldots, X_n)$ given $Y = y$ is

$$f_{\mathbf{X}|Y}(x_1,\ldots,x_n|y) = f_{X_1|Y}(x_1|y)\cdots f_{X_n|Y}(x_n|y).$$

In this case, we also refer to $X_1, \ldots, X_n$ as a random sample (given $Y = y$).

■ **Example 6.53 — Gamma-Poisson.** Let X_1 be the number of customers who arrive at a store during a specific time period, and let Y be the rate at which customers arrive in general. Let X_2 be the number of customers who arrive during a later time period of the same length. Assume that X_1 and X_2 are conditionally independent given $Y = y$ with conditional p.m.f. $f_{X_i|Y}(x_i|y) = y^{x_i}\exp(-y)/x_i!$, for $x_i = 0, 1, \ldots$, $i = 1, 2$, that is, $X_i|y \sim \text{Poisson}(y)$. Also, assume that Y has the p.d.f. $f_Y(y) = y\exp(-y)$, for $y > 0$, which means that $Y \sim \text{Gamma}(2, 1)$.

- Then, we see that $Y|X_1 = x_1 \sim \text{Gamma}(x_1 + 2, 2)$ since

$$f_{Y|X_1}(y|x_1) = \frac{f_{X_1|Y}(x_1|y)f_Y(y)}{f_{X_1}(x_1)} = \frac{2^{x_1+2}y^{x_1+1}}{(x_1+1)!}\exp(-2y), \quad \text{for } y > 0.$$

- Now, suppose that we want to predict X_2 conditional on $X_1 = x_1$. We need the conditional p.m.f. of X_2 given $X_1 = x_1$, namely

$$f_{X_2|X_1}(x_2|x_1) = \int f_{X_2|Y,X_1}(x_2|y,x_1)f_{Y|X_1}(y|x_1)dy.$$

Note that by conditional independence, $f_{X_2|Y,X_1}(x_2|y,x_1)$ is the same as the conditional distribution of X_2 given $Y = y$, which has p.m.f. $y^{x_2}\exp(-y)/x_2!$. Hence,

$$\begin{aligned} f_{X_2|X_1}(x_2|x_1) &= \int \frac{y^{x_2}}{x_2!}\exp(-y)\frac{2^{x_1+2}y^{x_1+1}}{(x_1+1)!}\exp(-2y)dy \\ &= \frac{2^{x_1+2}(x_2+x_1+1)!}{x_2!(x_1+1)!3^{x_2+x_1+2}}, \text{ for } x_2 = 0,1,\ldots. \end{aligned}$$

■

We have learned how to find the joint distribution of a transformation of two random variables. These ideas are fairly easily extended to multivariate random variables, and we have a generalization of multivariate transform.

Theorem 6.5.1 Let $\mathbf{X} = (X_1,\ldots,X_n)^\top$ have a continuous joint distribution with the joint p.d.f. is $f_{\mathbf{X}}(x_1,\ldots,x_n)$. Consider new random variables $Y_1,\ldots,Y_n$ as follows:

$$Y_1 = r_1(X_1,\ldots,X_n),\ Y_2 = r_2(X_1,\ldots,X_n),\ \cdots,\ Y_n = r_n(X_1,\ldots,X_n),$$

where we assume that the $r_1,\ldots,r_n$ define a one-to-one differentiable transformation. Let the inverse transform be given as follows:

$$x_1 = s_1(y_1,\ldots,y_n),\ x_2 = s_2(y_1,\ldots,y_n),\ \cdots,\ x_n = s_n(y_1,\ldots,y_n).$$

Then the joint p.d.f. g of $Y_1,\ldots,Y_n$ is given by

$$f_{\mathbf{Y}}(y_1,\ldots,y_n) = f_{\mathbf{X}}(s_1,\ldots,s_n)|J|,\ \ J = \det\begin{bmatrix} \frac{\partial s_1}{\partial y_1} & \cdots & \frac{\partial s_1}{\partial y_n} \\ \vdots & \ddots & \vdots \\ \frac{\partial s_n}{\partial y_1} & \cdots & \frac{\partial s_n}{\partial y_n} \end{bmatrix},$$

where J is the Jacobian and $|J|$ denotes the absolute value of J.

In particular, suppose that $\mathbf{Y} = (Y_1,\ldots,Y_n)^\top$ by $\mathbf{Y} = \mathbf{AX}$, where $\mathbf{A}$ is a nonsingular $n \times n$ matrix. Then, $\mathbf{Y}$ has a joint p.d.f. $f_{\mathbf{Y}}(\mathbf{y}) = \frac{1}{|\det\mathbf{A}|}f_{\mathbf{X}}(\mathbf{A}^{-1}\mathbf{y})$, where $\mathbf{A}^{-1}$ is the inverse of $\mathbf{A}$.

■ **Example 6.54 — Dirichlet distribution.** Let $X_1, X_2, \ldots, X_{k+1}$ be independent random variables, each having a gamma distribution with, $X_i \sim \text{Gamma}(\alpha_i, 1)$, $i = 1,\ldots,k+1$. The joint p.d.f. of $X_1,\ldots,X_{k+1}$ is written as

$$f_{\mathbf{X}}(x_1,x_2,\ldots,x_{k+1}) = \begin{cases} \prod_{i=1}^{k+1}\frac{1}{\Gamma(\alpha_i)}x_i^{\alpha_i-1}e^{-x_i}, & 0 < x_i < \infty, \\ 0, & \text{elsewhere.} \end{cases}$$

- Define $Y_i = \dfrac{X_i}{X_1+X_2+\cdots+X_{k+1}}$, $i = 1,2,\ldots,k$, and $Y_{k+1} = X_1+X_2+\cdots+X_{k+1}$ denote $k+1$ new random variables. Then, the inverse transformations are

$$x_1 = y_1y_{k+1},\ x_2 = y_2y_{k+1},\ \ldots,\ x_k = y_ky_{k+1},\ x_{k+1} = y_{k+1}(1-y_1-\cdots-y_k),$$

so that the Jacobian is

$$J = \det \begin{pmatrix} y_{k+1} & 0 & \cdots & 0 & y_1 \\ 0 & y_{k+1} & \cdots & 0 & y_2 \\ \vdots & \vdots & & \vdots & \vdots \\ 0 & 0 & \cdots & y_{k+1} & y_k \\ -y_{k+1} & -y_{k+1} & \cdots & -y_{k+1} & (1-y_1-\cdots-y_k) \end{pmatrix} = y_{k+1}^k.$$

- Hence the joint p.d.f. of $Y_1,\ldots,Y_k,Y_{k+1}$ is given by

$$f_{\mathbf{Y}}(y_1,\ldots,y_k,y_{k+1}) = \frac{y_1^{\alpha_1-1}\cdots y_k^{\alpha_k-1}(1-y_1-\cdots-y_k)^{\alpha_{k+1}-1}y_{k+1}^{\alpha_1+\cdots+\alpha_{k+1}-1}e^{-y_{k+1}}}{\Gamma(\alpha_1)\cdots\Gamma(\alpha_k)\Gamma(\alpha_{k+1})}.$$

Note that Y_{k+1} has a gamma distribution with parameters $\sum_{i=1}^{k+1}\alpha_i$ and 1 since it is a sum of independent gamma random variables with the identical rate parameter.

- Finally, thc joint p.d.f. of $Y_1,\ldots,Y_k$ is givcn by

$$\begin{aligned} f_{Y_1,\ldots,Y_k}(y_1,\ldots,y_k) &= \int_0^\infty f_{\mathbf{Y}}(y_1,\ldots,y_k,y_{k+1})dy_{k+1} \\ &= \frac{\Gamma(\alpha_1+\cdots+\alpha_{k+1})}{\Gamma(\alpha_1)\cdots\Gamma(\alpha_{k+1})}y_1^{\alpha_1-1}\cdots y_k^{\alpha_k-1}(1-y_1-\cdots-y_k)^{\alpha_{k+1}-1}, \end{aligned}$$

when $0<y_i<1$, $i=1,\ldots,k$, $y_1+\cdots+y_k<1$.

- Random variables $Y_1,\ldots,Y_k$ are said to have a *Dirichlet* distribution with $k+1$ parameters $\alpha_1,\ldots,\alpha_k,\alpha_{k+1}$, Dirichlet$(\alpha_1,\ldots,\alpha_k,\alpha_{k+1})$. When $k=1$, the Dirichlet distribution becomes a *beta* distribution with parameters α_1 and α_2. Moreover, Y_{k+1} is independent of $Y_1,Y_2,\ldots,Y_k$ since the joint p.d.f is the product of $f_{Y_1,\ldots,Y_k}(y_1,\ldots,y_k)$ and $f_{Y_{k+1}}(y_{k+1})$.

■

Remark The determinant of a square $(n\times n)$ matrix $\mathbf{A}$ is a special measure of the size of $\mathbf{A}$. Determinants are usually computed recursively. For $n=1$, $\det(\mathbf{A})=a_{11}$. For $n>1$, let $\mathbf{M}_{i,j}$ stand for the $(n-1)\times(n-1)$ matrix obtained by deleting from $\mathbf{A}$ the ith row and the jth column. Then, for each $i=1,\ldots,n$, $\det(\mathbf{A})=\sum_{j=1}^{n}a_{ij}(-1)^{i+j}\det(\mathbf{M}_{i,j})$.

Some elementary properties of determinants are :

- $\det(\mathbf{A})=\det(\mathbf{A}^\top)$ and $\det(\mathbf{A}^{-1})=1/\det(\mathbf{A})$.
- If you multiply a single row of a matrix by a scalar a, the determinant gets multiplied by a. If you multiply the whole matrix by a, the determinant gets multiplied by a^n.

- The determinant is unchanged if we replace a row by itself plus a multiple of some other row.

■ Example 6.55 — Bivariate beta (Dirichlet). A continuous bivariate random variable (X,Y) is said to have the bivariate beta distribution if its joint p.d.f. is of the form

$$f_{X,Y}(x,y) = \begin{cases} \frac{\Gamma(\alpha_1+\alpha_2+\alpha_3)}{\Gamma(\alpha_1)\Gamma(\alpha_2)\Gamma(\alpha_3)} x^{\alpha_1-1}y^{\alpha_2-1}(1-x-y)^{\alpha_3-1}, & \text{if } 0<x,y,\ x+y<1 \\ 0, & \text{otherwise,} \end{cases}$$

where $\alpha_1,\alpha_2,\alpha_3$ are positive parameters. The bivariate beta random variables X and Y with positive parameters α_1,α_2 and α_3 is actually the Dirichlet distribution with parameters α_1,α_2 and α_3 in Example 6.54.

- From the joint p.d.f. of X and Y, we get the marginal p.d.f. of X given by

$$\begin{aligned} f_X(x) &= \int_0^{1-x} f_{X,Y}(x,y)dy = \frac{\Gamma(\alpha_1+\alpha_2+\alpha_3)}{\Gamma(\alpha_1)\Gamma(\alpha_2)\Gamma(\alpha_3)} x^{\alpha_1-1}\int_0^{1-x} y^{\alpha_2-1}(1-x-y)^{\alpha_3-1}dy \\ &= \frac{\Gamma(\alpha)}{\Gamma(\alpha_1)\Gamma(\alpha_2)\Gamma(\alpha_3)} x^{\alpha_1-1}(1-x)^{\alpha_3-1}\int_0^{1-x} y^{\alpha_2-1}\left(1-\frac{y}{1-x}\right)^{\alpha_3-1} dy \end{aligned}$$

Now we substitute $u = 1-\frac{y}{1-x}$ in the above integral. Then we have

$$\begin{aligned} f_X(x) &= \frac{\Gamma(\alpha)}{\Gamma(\alpha_1)\Gamma(\alpha_2)\Gamma(\alpha_3)} x^{\alpha_1-1}(1-x)^{\alpha_2+\alpha_3-1}\int_0^1 (1-u)^{\alpha_2-1}u^{\alpha_3-1}du \\ &= \frac{\Gamma(\alpha)}{\Gamma(\alpha_1)\Gamma(\alpha_2)\Gamma(\alpha_3)} x^{\alpha_1-1}(1-x)^{\alpha_2+\alpha_3-1}B(\alpha_2,\alpha_3) \\ &= \frac{\Gamma(\alpha)}{\Gamma(\alpha_1)\Gamma(\alpha_2+\alpha_3)} x^{\alpha_1-1}(1-x)^{\alpha_2+\alpha_3-1}, \end{aligned}$$

which proves that the random variable $X \sim \text{Beta}(\alpha_1,\alpha_2+\alpha_3)$. Similarly, we can get the random variable $Y \sim \text{Beta}(\alpha_2,\alpha_1+\alpha_3)$.

■

■ Example 6.56 — Dirichlet and Uniform. Suppose that $(Y_1,\ldots,Y_n)^\top$ has a Dirichlet distribution, Dirichlet$(1,1,\ldots,1)$ with a joint p.d.f given by

$$f_{Y_1,\ldots,Y_n}(y_1,\ldots,y_n) = n!, \text{ for } 0<y_i<1,\ i=1,\ldots,n,\ y_1+\ldots+y_n<1.$$

Define $W_1 = Y_1$, $W_2 = Y_1+Y_2$, …, $W_n = \sum_{i=1}^n Y_i$. Find the joint p.d.f. of $(W_1,\ldots,W_n)^\top$ and the marginal p.d.f of W_i.

- Consider the inverse transformations, $Y_1 = W_1$, $Y_2 = W_2-W_1$, $Y_3 = W_3-W_2-W_1$, …, $Y_n = W_n - \sum_{i=1}^{n-1} W_i$. Then, the joint p.d.f. of $(W_1,\ldots,W_n)^\top$ is

$$f_{\mathbf{W}}(w_1,\ldots,w_n) = f_{\mathbf{Y}}\left(w_1, w_2-w_1,\ldots,w_n-\sum_{i=1}^{n-1} w_i\right)|J|,\ w_1<w_2<\ldots w_n,$$

where the Jacobian J is $J = \det \begin{pmatrix} 1 & 0 & \cdots & 0 & 0 \\ -1 & 1 & \cdots & 0 & 0 \\ \vdots & \vdots & & \vdots & \vdots \\ -1 & -1 & \cdots & 1 & 0 \\ -1 & -1 & \cdots & -1 & 1 \end{pmatrix} = 1.$

Hence, $f_{\mathbf{W}}(w_1,\ldots,w_n) = n!$, for $0 < w_1 < w_2 < \ldots w_n < 1$.

- Note that as shown in Example 6.54, $Y_i = \dfrac{X_i}{\sum_{i=1}^{n+1} X_i} \sim \text{Beta}(1,n)$, where $X_1,\ldots,X_n$ are i.i.d. with $X_1 \sim \text{Exp}(1)$ but that $Y_1,\ldots,Y_n$ are not independent. However, $W_k = \sum_{i=1}^{k} Y_i = \dfrac{\sum_{i=1}^{k} X_i}{\sum_{i=1}^{n+1} X_i} = \dfrac{\sum_{i=1}^{k} X_i}{\sum_{i=1}^{k} X_i + \sum_{i=k+1}^{n+1} X_i}$ has a beta distribution with parameters k and $n-k+1$, $W_k \sim \text{Beta}(k, n-k+1)$.

- Suppose we choose n random numbers independently from a Uniform$(0,1)$ distribution. Let X_k be the kth largest of the numbers and let N_t be the numbers of U_i's less than t. Then $\{X_k \le x\} = \{N_x \ge k\}$ since the kth largest is less than or equal to x implies that at least k of the U_i's less than or equal to k, and vice versa. Note that $N_x \sim \text{Binomial}(n,x)$ since $P(U_i \le x) = x$ and $U_1,\ldots,U_n$ are i.i.d. Thus,

$$F_{X_k}(x) = P(X_k \le x) = P(N_x \ge k) = \sum_{j=k}^{n} \binom{n}{j} x^j (1-x)^{n-j}.$$

Hence, we take the derivative of $F_{X_k}(x)$ to get the p.d.f. of X_k

$$\begin{aligned} f_{X_k}(x) &= \frac{d}{dx} F_{X_k}(x) = \frac{d}{dx} \sum_{j=k}^{n} \binom{n}{j} x^j (1-x)^{n-j} \\ &= \sum_{j=k}^{n} \binom{n}{j} j x^{j-1} (1-x)^{n-j} - \sum_{j=k}^{n} \binom{n}{j} (n-j) x^j (1-x)^{n-j-1} \\ &= n \left[\sum_{j=k}^{n} \binom{n-1}{j-1} x^{j-1} (1-x)^{n-j} - \sum_{j=k}^{n-1} \binom{n-1}{j} x^j (1-x)^{n-j-1} \right] \\ &= n \binom{n-1}{k-1} x^{k-1} (1-x)^{n-k} = \frac{\Gamma(n+1)}{\Gamma(k)\Gamma(n-k+1)} x^{k-1} (1-x)^{n-k}, \end{aligned}$$

which is a p.d.f. of Beta$(k, n-k+1)$ distribution. That is, for positive integers k,n with $n \ge k$, the kth largest of a set of n independent Uniform$(0,1)$ random variables has a Beta$(k, n-k+1)$ distribution, and it has the same distribution with W_k.

∎

Remark Let $X_1,\ldots,X_n$ be a random sample from with c.d.f $F_X(x)$ and p.d.f. $f_X(x)$. For each k between 1 and n, the kth order statistic, denoted $X_{(k)}$, is defined to be the kth largest of $X_1,\ldots,X_n$. The random variables $X_{(1)},\ldots,X_{(n)}$ are called the *order statistics* corresponding to $X_1,\ldots,X_n$.

- The c.d.f. of $X_{(k)}$ is $F_{X_{(k)}}(x_{(k)}) = \sum_{j=k}^{n} \binom{n}{j} (F_X(x_{(k)}))^j (1 - F_X(x_{(k)}))^{n-j}$, and its p.d.f. is $f_{X_{(k)}}(x_{(k)}) = \dfrac{n!}{(k-1)!(n-k)!} f_X(x_{(k)})(F_X(x_{(k)}))^{k-1}(1 - F_X(x_{(k)}))^{n-k}$, using the same reasoning as in Example 6.56, $P(X_{(k)} \le x_{(k)}) = P(N_{x_{(k)}} \ge k)$, where $N_{x_{(k)}} \sim \text{Binomial}(n, F_X(x_{(k)}))$.

- Alternatively, consider a quantity $P(x_{(k)} \le X_{(k)} \le x_{(k)} + \Delta x_{(k)})/\Delta_{x_{(k)}}$, where $\Delta x_{(k)} \to 0^+$. Then, since $\Delta x_{(k)} \to 0^+$, the event $x_{(k)} \le X_{(k)} \le x_{(k)} + \Delta x_{(k)}$ is equivalent to the event that $k-1$ of the X_i's are in $(-\infty, x_{(k)}]$, one in $(x_{(k)}, x_{(k)} + \Delta x_{(k)}]$ and the remaining $n-k$ in $(x_{(k)} + \Delta x_{(k)}, \infty)$ as shown in Figure 6.7.

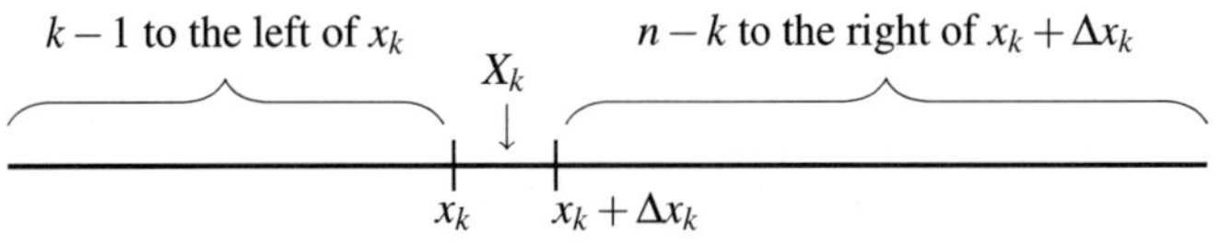

Figure 6.7: The event $\{x_k \le X_{(k)} \le x_k + \Delta x_k\}$

Thus, we have a multinomial distribution with three categories (intervals) as given in Figure 6.7, $(Y_1, Y_2, Y_3)^\top \sim \text{Multi}(n; p_1, p_2, p_3)$, where $Y_1 = k-1$, $Y_2 = 1$, and $Y_3 = n-k$ with $p_1 = F_X(x_{(k)})$, $p_2 = F_X(x_{(k)} + \Delta x_{(k)}) - F_X(x_{(k)})$, and $p_3 = 1 - F_X(x_{(k)} + \Delta x_{(k)})$, and the corresponding probability is computed as

$$P(x_{(k)} \le X_{(k)} \le x_{(k)} + \Delta_{x_{(k)}}) = \frac{n!}{(k-1)!1!(n-k)!} p_1^{k-1} p_2 p_3^{n-k}.$$

Hence, we have

$$\begin{aligned} f_{X_{(k)}}(x_{(k)}) &= \frac{d}{dx_{(k)}} F_{X_{(k)}}(x_{(k)}) = \lim_{\Delta_{x_{(k)}} \to 0^+} \frac{P(x_{(k)} < X_{(k)} \le x_{(k)} + \Delta_{x_{(k)}})}{\Delta_{x_{(k)}}} \\ &= \frac{n!}{(k-1)!(n-k)!} f_X(x_{(k)})(F_X(x_{(k)}))^{k-1}(1 - F_X(x_{(k)}))^{n-k}. \end{aligned}$$

■ **Example 6.57** Suppose that X_1, X_2, and X_3 are selected at random without replacement from N numbers $\{1, 2, \ldots, N-1, N\}$ where $N > 3$. Let $X_{(1)}$, $X_{(2)}$, and $X_{(3)}$ be the minimum, median, and maximum respectively among X_1, X_2, and X_3 such that $X_{(1)} \le X_{(2)} \le X_{(3)}$.

- Since all samples of X_1, X_2, and X_3 from N numbers are equally likely, and there are exactly $\binom{N}{3}$ of them, each event $\{X_{(1)} = x_{(i)},\ X_{(2)} = x_{(2)},\ \text{and}\ X_{(3)} = x_{(3)}\}$ has the same probability. Thus, the joint p.m.f. of $X_{(1)}$, $X_{(2)}$, and $X_{(3)}$ is given by

$$f_{X_{(1)},X_{(2)},X_{(3)}}(x_{(1)}, x_{(2)}, x_{(3)}) = P(\{X_{(1)} = x_{(i)},\ X_{(2)} = x_{(2)},\ \text{and}\ X_{(3)} = x_{(3)}\})$$

$$= \begin{cases} \frac{1}{\binom{N}{3}}, & \text{if } x_{(1)} < x_{(2)} < x_{(3)} \text{ and } x_{(1)}, x_{(2)}, x_{(3)} = 1, \ldots, N, \\ 0, & \text{otherwise.} \end{cases}$$

- Also, we can find the joint p.m.f. of $X_{(1)}$ and $X_{(2)}$ from the the joint p.m.f. of $X_{(1)}$, $X_{(2)}$, and $X_{(3)}$

$$f_{X_{(1)},X_{(2)}}(x_{(1)},x_{(2)}) = \sum_{x_{(3)}=x_{(2)}+1}^{N} f_{X_{(1)},X_{(2)},X_{(3)}}(x_{(1)},x_{(2)},x_{(3)})$$
$$= \begin{cases} \frac{(N-x_{(2)})}{\binom{N}{3}}, & \text{for } x_{(1)} < x_{(2)},\ x_{(1)},x_{(2)} = 1,\ldots,N-1, \\ 0, & \text{otherwise.} \end{cases}$$

- Further, we can get the marginal p.m.f. of $X_{(2)}$ that can take integer values from 2 through $N-1$

$$f_{X_{(2)}}(x_{(2)}) = \sum_{x_{(1)}=1}^{x_{(2)}-1} \sum_{x_{(3)}=x_{(2)}+1}^{N} f_{X_{(1)},X_{(2)},X_{(3)}}(x_{(1)},x_{(2)},x_{(3)})$$
$$= \begin{cases} \frac{(x_{(2)}-1)(N-x_{(2)})}{\binom{N}{3}}, & \text{for } x_{(2)} = 2,\ldots,N-1, \\ 0, & \text{otherwise.} \end{cases}$$

- Let $T = X_{(2)}/N$. Then, the p.m.f. of T is given by

$$f_T(t) = P(T=t) = P(X_{(2)} = Nt) = \frac{(Nt-1)(N-Nt)}{\binom{N}{3}}$$
$$= 3!\frac{N^2(N-3)!(t-1/N)(1-t)}{N!} \approx 3!t(1-t), \text{ as } N \to \infty,$$

which is the p.d.f. of $U_{(2)}$, the median of U_1, U_2 and U_3 when U_i's are i.i.d. uniform random variables on $(0,1)$ that follows from the remark discussed before.

■

Remark More generally, the joint p.m.f. of $(X_{(1)}, X_{(2)}, \ldots, X_{(n)})^\top$ is $\frac{1}{\binom{N}{n}}$, $X_{(1)} \le \cdots \le X_{(n)}$, where $X_1, \ldots, X_n$ are n numbers randomly chosen from $1, \ldots, N$ without replacement as in Example 6.57. Thus, for fixed n, the joint distribution of $(X_{(1)}/N, X_{(2)}/N, \ldots, X_{(n)}/N)^\top$ gets close to the joint distribution of the order statistics of n independent uniform random variables on $(0,1)$, as $N \to \infty$. In other words, by Stirling's formula given in Chapter 5, it follows that

$$\frac{1}{\binom{N}{n}} = n!\frac{(N-n)!}{N!} \approx n!\frac{\sqrt{2\pi}e^{-(N-n)}(N-n)^{N-n+1/2}}{\sqrt{2\pi}e^{-N}N^{N+1/2}}$$
$$= n!e^n\left(1-\frac{n}{N}\right)^{N-n+1/2}\left(\frac{1}{N}\right)^n \approx n!\Delta u_1 \cdots \Delta u_n$$
$$= f_{U_{(1)},\ldots,U_{(n)}}(u_{(1)},\ldots,u_{(n)})du_1 \cdots du_n,$$

where $U_{(1)} \le \cdots \le U_{(n)}$, order statistics from Uniform$(0,1)$, and du_i is interpreted as an infinitesimal length Δu_i near u_i, $i = 1, \ldots, n$. That is, $f_{U_{(1)},\ldots,U_{(n)}}(u_{(1)},\ldots,u_{(n)}) = n!$, which is also given as the joint distribution of W in Example 6.56.

Example 6.58 Let $X_1,\dots,X_n$ be a random sample from an exponential distribution with parameter λ. Define $Y_k=\sum_{j=1}^{k}X_j$ for $1\le k\le n$.

- It is obvious that the marginal distribution of Y_k is a gamma distribution with parameters k and λ. To obtain a joint p.d.f. of $(Y_1,\dots,Y_n)^\top$, consider inverse transformations, $X_1=Y_1$, $X_2=Y_2-Y_1$, $X_k=Y_k-Y_{k-1}$, $k=1,\dots,n$. Thus, the joint p.d.f of $\mathbf{Y}=(Y_1,\dots,Y_n)^\top$ is $f_{\mathbf{Y}}(y_1,\dots,y_n)=f_{\mathbf{X}}(y_1,y_2-y_1,\dots,y_n-y_{n-1})|J|$, where $f_{\mathbf{X}}=\prod_{i=1}^{n}\lambda e^{-\lambda x_i}=\lambda^n e^{-\lambda y_n}$ and $J=1$ as computed in Example 6.56. Hence, $f_{\mathbf{Y}}(y_1,\dots,y_n)=\lambda^n e^{-\lambda y_n}$, for $0<y_1<y_2<\cdots<y_n<\infty$, which also implies that $Y_1,\dots,Y_n$ are not independent.

- On the other hand, let $W_1,\dots,W_n$ be random variables with a joint p.d.f. $f_{\mathbf{W}}(w_1,\dots,w_n)=\lambda^n e^{-\lambda w_n}$ for $0<w_1<\cdots<w_n$. Define $X_1=W_1$ and $X_k=W_k-W_{k-1}$ for $2\le k\le n$. Then, the joint p.d.f. of $(X_1,\dots,X_n)^\top$ is given by

$$f_{\mathbf{X}}(x_1,\dots,x_n)=f_{\mathbf{W}}\left(x_1,x_2+x_1,\dots,\sum_{i=1}^{n}x_i\right)|J|=\lambda^n e^{-\lambda\sum_{i=1}^{n}x_i}.$$

■

Example 6.59 — Bivariate normal in Matrix. Consider the bivariate normal distribution $(X,Y)^\top\sim N_2(\mu_X,\mu_Y,\sigma_X^2,\sigma_Y^2,\rho)$ using matrix notation. To do so, we denote

$$\mathbf{X}=(X,Y)^\top,\ \mu=(\mu_X,\mu_Y)^\top,\ \text{and }\Sigma=\begin{pmatrix}\sigma_X^2 & \rho\sigma_X\sigma_Y\\ \rho\sigma_X\sigma_Y & \sigma_Y^2\end{pmatrix}.$$

- Then, the joint p.d.f of $\mathbf{X}$ is written as $f(\mathbf{x})=\dfrac{1}{2\pi}(\det\Sigma)^{-1/2}\exp\left[-\dfrac{1}{2}(\mathbf{x}-\mu)^\top\Sigma^{-1}(\mathbf{x}-\mu)\right]$.

- To confirm the representation of $f(\mathbf{x})$, we check the value of $(\mathbf{x}-\mu)^\top\Sigma^{-1}(\mathbf{x}-\mu)$ with the matrix operation

$$\begin{aligned}
&(\mathbf{x}-\mu)^\top\Sigma^{-1}(\mathbf{x}-\mu)\\
=\;&\frac{1}{\sigma_X^2\sigma_Y^2(1-\rho^2)}(x-\mu_X,y-\mu_Y)\begin{pmatrix}\sigma_Y^2 & -\rho\sigma_X\sigma_Y\\ -\rho\sigma_X\sigma_Y & \sigma_X^2\end{pmatrix}\begin{pmatrix}x-\mu_X\\ y-\mu_Y\end{pmatrix}\\
=\;&\frac{1}{\sigma_X^2\sigma_Y^2(1-\rho^2)}(\sigma_Y^2(x-\mu_X)^2-2\rho\sigma_X\sigma_Y(x-\mu_X)(y-\mu_Y)+\sigma_X^2(y-\mu_Y)^2)\\
=\;&\frac{1}{1-\rho^2}\left(\frac{(x-\mu_X)^2}{\sigma_X^2}-2\rho\frac{(x-\mu_X)(y-\mu_Y)}{\sigma_X\sigma_Y}+\frac{(y-\mu_Y)^2}{\sigma_Y^2}\right),
\end{aligned}$$

and $(\det\Sigma)^{-1/2}=(\sigma_X^2\sigma_Y^2-\rho^2\sigma_X^2\sigma_Y^2)^{-1/2}=\dfrac{1}{\sigma_X\sigma_Y\sqrt{1-\rho^2}}$. Hence, the matrix notation allows us to easily represent the bivariate normal p.d.f., and we can generalize it to the n-dimensional multivariate normal distribution.

■

Example 6.60 — Multivariate normal. An n-dimensional random vector $\mathbf{X} = (X_1, \ldots, X_n)^\top$ has a multivariate normal distribution with mean vector μ and covariance matrix Σ, denoted by $\mathbf{X} \sim N_n(\mu, \Sigma)$ with joint p.d.f. $f_{\mathbf{X}}(\mathbf{x}) = \dfrac{1}{(2\pi)^{n/2}(\det\Sigma)^{1/2}} \exp\left\{ -\dfrac{1}{2}(\mathbf{x}-\mu)^\top \Sigma^{-1}(\mathbf{x}-\mu) \right\}$.

- Let $Z_1, \ldots, Z_n$ be a random sample from the $N(0,1)$ distribution. Let $\mathbf{Z} = (Z_1, \ldots, Z_n)^\top$. Then, the joint p.d.f. of $\mathbf{Z}$ is given by

$$f_{\mathbf{Z}}(\mathbf{z}) = \prod_{i=1}^{n} \frac{1}{\sqrt{2\pi}} \exp\left(-\frac{1}{2}z_i^2\right) = \left(\frac{1}{\sqrt{2\pi}}\right)^n \exp\left(-\frac{1}{2}\mathbf{z}^\top\mathbf{z}\right).$$

- Let $\mathbf{X} = \mu + \mathbf{AZ}$, where μ is a fixed vector and $\mathbf{A}$ is a nonsingular fixed $n \times n$ matrix. Then, the joint p.d.f. of $\mathbf{X}$ is

$$\begin{aligned} f_{\mathbf{X}}(\mathbf{x}) &= f_{\mathbf{Z}}\left(\mathbf{A}^{-1}(\mathbf{x}-\mu)\right)|J| \\ &= \left(\frac{1}{\sqrt{2\pi}}\right)^n \exp\left(-\frac{1}{2}(\mathbf{x}-\mu)^\top \left(\mathbf{AA}^\top\right)^{-1}(\mathbf{x}-\mu)\right)|\det(\mathbf{A}^{-1})| \\ &= (2\pi)^{-n/2}(\det\Sigma)^{-1/2}\exp\left(-\frac{1}{2}(\mathbf{x}-\mu)^\top \left(\mathbf{AA}^\top\right)^{-1}(\mathbf{x}-\mu)\right), \end{aligned}$$

  where $\det\Sigma = \det\mathbf{AA}^\top = [\det\mathbf{A}]^2 > 0$. Thus, the distribution of X is the (n-dimensional) *multivariate normal* distribution, denoted as $\mathbf{X} \sim N_n(\mu, \Sigma)$.

- The vector $\mu = (\mu_1, \ldots, \mu_n)^\top$ is $\mathrm{E}(\mathbf{X})$ and Σ is called the *covariance matrix* since the covariance $\mathrm{Cov}(X_i, X_j) = \mathrm{E}[(X_i - \mu_i)(X_j - \mu_j)]$ is the (i, j)th entry of Σ as

$$\mathrm{E}[(\mathbf{X}-\mu)(\mathbf{X}-\mu)^\top] = \mathrm{E}[(\mathbf{AZ})(\mathbf{AZ})^\top] = \mathbf{A}\mathrm{E}[\mathbf{ZZ}^\top]\mathbf{A}^\top = \mathbf{AIA}^\top = \mathbf{AA}^\top = \Sigma.$$

- Note that $\mathbf{Z}$ has a multivariate normal distribution, $\mathbf{Z} \sim N_n(\mathbf{0}, \mathbf{I})$, where $\mathbf{0}$ is an n-dimensional zero vector and $\mathbf{I}$ is the $n \times n$ identity matrix.

- Suppose that $\mathbf{A}$ is an $n \times n$ orthogonal matrix in that $\mathbf{AA}^\top = \mathbf{A}^\top\mathbf{A} = \mathbf{I}$. Then, $\det(\mathbf{AA}^\top) = \det(\mathbf{A})^2 = \det(\mathbf{I}) = 1$. Also, $\mathbf{X} = \mathbf{AZ} \sim N_n(\mathbf{0}, \mathbf{I})$.

■

Remark Suppose that $\mathbf{B} = [b_{ij}]$ is an $n \times n$ symmetric matrix, $b_{ij} = b_{ji}$. Associated with $\mathbf{B}$ is the quadratic form $q(\mathbf{x})$ which maps each $\mathbf{x} = (x_1, \ldots, x_n)^\top \in \mathbb{R}^n$ into the real value

$$q(\mathbf{x}) = \mathbf{x}^\top\mathbf{B}\mathbf{x} = \sum_{i=1}^{n}\sum_{j=1}^{n} b_{ij}x_ix_j.$$

- Consider a function $f(\mathbf{x}) = Ce^{-q(\mathbf{x})/2}$. Then, $f(\mathbf{x}-\mu)$ is a multivariate normal p.d.f. with mean vector $\mu = (\mu_1, \ldots, \mu_n)^\top$ and a covariance matrix $\mathbf{B}^{-1}$, where the diagonal elements of $\mathbf{B}$ must be strictly positive, b_{ii}, $i = 1, \ldots, n$.

- Suppose $X_1, \ldots, X_n$ are independent random variables with each $X_i \sim N(0, \sigma_i^2)$ normal with mean zero and variance σ_i^2. Then $\mathbf{X} = (X_1, \ldots, X_n)^\top$ has p.d.f.

$$f_{\mathbf{X}}(x) = \frac{1}{(2\pi)^{n/2}\sigma_1 \ldots \sigma_n} \exp\left\{-\frac{1}{2}\left(\frac{x_1^2}{\sigma_1^2} + \cdots + \frac{x_n^2}{\sigma_n^2}\right)\right\},$$

which is the multivariate normal p.d.f.with mean vector $\mathbf{0}$ and the covariance matrix $\mathbf{B}^{-1}$, where $\mathbf{B} = \text{diag}(\sigma_1^{-2}, \ldots, \sigma_n^{-2})$ and $\det(\mathbf{B}^{-1}) = \prod_{i=1}^{n} \sigma_i^2$. That is, $X_1, \ldots, X_n$ are uncorrelated with $\text{Cov}(X_i, X_j) = 0, \forall\, i \neq j$.

- Suppose that a normal random vector $\mathbf{X} = (X_1, \ldots, X_n)^\top$ is uncorrelated with its mean vector μ and covariance matrix Σ. Then $\Sigma = \text{Cov}(X) = \text{diag}(\sigma_1^2, \ldots, \sigma_n^2)$, where $0 < \sigma_i^2 = \text{Var}(X_i)$, $i = 1, \ldots, n$. Then,

$$q(\mathbf{x}) = (\mathbf{x} - \mu)^\top \Sigma^{-1} (\mathbf{x} - \mu) = \sum_{i=1}^{n} \frac{1}{\sigma_i^2} (x_i - \mu_i)^2, \text{ and } \det(\Sigma) = \prod_{i=1}^{n} \sigma_i^2.$$

  Hence, it follows that the variables $X_1, \ldots, X_n$ are independent.

- Suppose that $\mathbf{X} = (X_1, \ldots, X_n)^\top$ has a multivariate normal distribution, $\mathbf{X} \sim N_n(\mu, \Sigma)$. Then, the respective marginal distributions of $X_1, \ldots, X_n$ are all normal distributions. Further, if $\mathbf{X} = \begin{pmatrix} \mathbf{X}_1 \\ \mathbf{X}_2 \end{pmatrix} \sim N_n\left(\begin{pmatrix} \mu_1 \\ \mu_2 \end{pmatrix}, \begin{pmatrix} \Sigma_{11} & \Sigma_{12} \\ \Sigma_{21} & \Sigma_{22} \end{pmatrix} \right)$, then $\mathbf{X}_k$ has the k-dimensional multivariate normal distribution, $\mathbf{X}_k \sim N_{n_k}(\mu_k, \Sigma_{kk})$ where n_k is the number of coordinates in $\mathbf{X}_k$ for $k = 1, 2$, $n_1 + n_2 = n$. In addition, the random vectors $\mathbf{X}_1$ and $\mathbf{X}_2$ are independent if and only if they are uncorrelated, that is, $\Sigma_{12} = \mathbf{0}$. In other words, uncorrelated multivariate normal random vectors are independent.

- Given $\mathbf{A}$ with $\mathbf{X} = \mathbf{AZ}$, the covariance matrix of $\mathbf{X}$ is $\Sigma = \mathbf{AA}^\top$. Conversely, given a covariance matrix Σ that is (semi) positive definite, we can find a lower triangular matrix $\mathbf{A}$ such that $\Sigma = \mathbf{LL}^\top$, called the *Cholesky decomposition.*

Example 6.61 — Cholesky decomposition of the bivariate covariance matrix. Consider the bivariate normal distribution $(X, Y)^\top \sim N_2(\mu_X, \mu_Y, \sigma_X^2, \sigma_Y^2, \rho)$ with the covariance matrix $\Sigma = \begin{pmatrix} \sigma_X^2 & \rho\sigma_X\sigma_Y \\ \rho\sigma_X\sigma_Y & \sigma_Y^2 \end{pmatrix}$. Find a lower triangular square-root matrix $\mathbf{L} = \begin{pmatrix} a & 0 \\ b & c \end{pmatrix}$ such that $\mathbf{LL}^\top = \Sigma$.

- We solve the equation $\mathbf{LL}^\top = \Sigma$ for a, b and c

$$\begin{pmatrix} a & 0 \\ b & c \end{pmatrix} \begin{pmatrix} a & b \\ 0 & c \end{pmatrix} = \begin{pmatrix} a^2 & ab \\ ab & b^2 + c^2 \end{pmatrix} = \begin{pmatrix} \sigma_X^2 & \rho\sigma_X\sigma_Y \\ \rho\sigma_X\sigma_Y & \sigma_Y^2 \end{pmatrix}.$$

- Then, we get $a = \sigma_X$, $b = \rho\sigma_Y$, and $c = \sqrt{\sigma_Y^2 - \rho^2\sigma_Y^2} = \sigma_Y\sqrt{1 - \rho^2}$. Thus,

$$\begin{pmatrix} X \\ Y \end{pmatrix} = \begin{pmatrix} \sigma_X & 0 \\ \rho\sigma_Y & \sigma_Y\sqrt{1 - \rho^2} \end{pmatrix} \begin{pmatrix} Z_1 \\ Z_2 \end{pmatrix},$$

  where $\mathbf{Z} = (Z_1, Z_2)^\top \sim N_2(\mathbf{0}, \mathbf{I})$. That is, $X_1 = \sigma_X Z_1$ and $X_2 = \sigma_Y[\rho Z_1 + \sqrt{1 - \rho^2} Z_2]$, which is also given in Theorem 6.4.1.

■

■ **Example 6.62 — Helmert transformation.** Let $Z_1,\ldots,Z_n$ be a random sample from the standard normal distribution. Let $\mathbf{A}$ be the $n\times n$ matrix defined by

$$\mathbf{A}=\begin{bmatrix} \frac{1}{\sqrt{n}} & \frac{1}{\sqrt{n}} & \frac{1}{\sqrt{n}} & \cdots & \frac{1}{\sqrt{n}} \\ \frac{1}{\sqrt{2}} & -\frac{1}{\sqrt{2}} & 0 & \cdots & 0 \\ \frac{1}{\sqrt{6}} & \frac{1}{\sqrt{6}} & -\frac{2}{\sqrt{6}} & \cdots & 0 \\ \vdots & \vdots & \vdots & & \vdots \\ \frac{1}{\sqrt{n(n-1)}} & \frac{1}{\sqrt{n(n-1)}} & \frac{1}{\sqrt{n(n-1)}} & \cdots & -\frac{(n-1)}{\sqrt{n(n-1)}} \end{bmatrix},$$

where the first row is given by $\mathbf{a}_1^\top=(1/\sqrt{n},\ldots,1/\sqrt{n})$, and the i-th row $(i\geq 2)$ by

$$\mathbf{a}_i^\top=\left(\frac{1}{\sqrt{i(i-1)}},\ldots,\frac{1}{\sqrt{i(i-1)}},-\frac{i-1}{\sqrt{i(i-1)}},0,\ldots,0\right),\ a_{ii}=-\frac{i-1}{\sqrt{i(i-1)}}.$$

Define $\mathbf{X}=\mathbf{AZ}$, which is known as the *Helmert transformation*, named after Friedrich Robert Helmert (1843-1917), a famous German geodesist who made important contributions to the theory of errors.

- By Helmert transformation, we have

$$X_1=\sqrt{n}\cdot\overline{Z}_n,\ X_2=\sqrt{\frac{1}{2}}(Z_2-\overline{Z}_1),\ X_3=\sqrt{\frac{2}{3}}(Z_3-\overline{Z}_2),\ldots,$$
$$X_n=\sqrt{\frac{n-1}{n}}(Z_n-\overline{Z}_{n-1}),\ \text{where } \overline{Z}_n=\tfrac{1}{n}\textstyle\sum_{i=1}^n Z_i.$$

- Consider the inner product of the i-th row $\mathbf{a}_i$ with itself, for $i=2,\ldots,n$,

$$\mathbf{a}_i^\top\mathbf{a}_i=\left(\sum_{j=1}^{i-1}\left(\frac{1}{\sqrt{i(i-1)}}\right)^2\right)+\left(\frac{i-1}{\sqrt{i(i-1)}}\right)^2=\frac{i-1}{i(i-1)}+\frac{(i-1)^2}{i(i-1)}=1,$$

  and $\mathbf{a}_1^\top\mathbf{a}_1=1$. Also, for $i>1$, and $1<k<i$,

$$\begin{aligned}\mathbf{a}_i^\top\mathbf{a}_1 &= \frac{1}{\sqrt{n}}\left(\left(\sum_{i=1}^{j-1}\frac{1}{\sqrt{j(j-1)}}\right)-\frac{j-1}{\sqrt{j(j-1)}}\right)=0\\ \mathbf{a}_i^\top\mathbf{a}_k &= \left(\sum_{j=1}^{k-1}\frac{1}{\sqrt{ik(i-1)(k-1)}}\right)-\frac{k-1}{\sqrt{ik(j-1)(k-1)}}=0,\end{aligned}$$

  which implies that $\mathbf{AA}^\top=\mathbf{A}^\top\mathbf{A}=\mathbf{I}_n$ and $\mathbf{A}$ is orthogonal.

- Hence, $X_1,\ldots,X_n$ are independent and identically distributed with $N_n(\mathbf{0},\mathbf{I})$ since $\sum_{i=1}^n x_i^2=\mathbf{x}^\top\mathbf{x}=(\mathbf{Az})^\top\mathbf{Az}=\mathbf{z}^\top\mathbf{z}=\sum_{i=1}^n z_i^2$, and

$$f_{\mathbf{X}}(\mathbf{x})=f_{\mathbf{Z}}(\mathbf{A}^{-1}\mathbf{x})|J|=\left(\frac{1}{\sqrt{2\pi}}\right)^n\exp\left(-\frac{1}{2}\mathbf{x}^\top\mathbf{x}\right).$$

■

Remark A set of vectors $\{\mathbf{x}_1, \mathbf{x}_2, \ldots, \mathbf{x}_n\}$ is said to be an orthogonal set if each pair of them is orthogonal and to be orthonormal if each $\mathbf{x}_i$ satisfies $\|\mathbf{x}_i\| = \langle \mathbf{x}_i, \mathbf{x}_i \rangle$. If $\mathscr{V}$ is a finite dimensional inner-product space, $\mathscr{V}$ always has an orthogonal basis, known as the *Gram-Schmidt orthogonalization*.

- Let $\mathbf{v}_1, \ldots, \mathbf{v}_n$ be linearly independent vectors in $\mathbf{R}^n$. We can find a collection of orthonormal vectors that span the same space.
- Let $\mathbf{x}_1 = \mathbf{v}_1/\|\mathbf{v}_1\|$. If $k = 1$, we are finished. If $k > 1$ proceed as follows. Start a recursion with $j = 2$. Let $\mathbf{x}_1, \ldots, \mathbf{x}_{j-1}$ be the orthonormal vectors that span the same space as $\mathbf{v}_1, \ldots, \mathbf{v}_{j-1}$. Let $\mathbf{x}_j = \mathbf{v}_j - \sum_{i=1}^{j-1} \mathbf{x}_i^\top \mathbf{x}_i \mathbf{v}_j$. By construction $\mathbf{x}_i^\top \mathbf{x}_j = 0$ for all $i < j$. Since the $\mathbf{v}_i$'s are linearly independent $\mathbf{x}_j \neq 0$.
- Let $\mathbf{x}_j = \mathbf{x}_j/\|\mathbf{x}_j\|$. If $j < k$, replace j by $j+1$ and go back to the start of the recursion. Since the $\mathbf{v}_i$'s and the $\mathbf{x}_i$'s can easily be recovered as linear combinations of each other, the two sets of vectors span the same spaces.
- For example, consider $\mathbf{a}_1^\top = \left(\frac{1}{\sqrt{n}}, \ldots, \frac{1}{\sqrt{n}}\right)$ with $\|\mathbf{a}\| = 1$. Then, we can construct an orthogonal matrix $\mathbf{A}$ with $\mathbf{a}_1^\top$ as the first row of $\mathbf{A}$ by the Gram-Schmidt orthogonalization.

Theorem 6.5.2 — Sampling distributions from a normal random sample. Let $\mathbf{Z} = (Z_1, \ldots, Z_n)^\top \sim N_n(\mathbf{0}, \mathbf{I})$, that is, $Z_1, \ldots, Z_n$ are a random sample from the standard normal distribution. Define $\overline{Z}_n = \frac{1}{n}\sum_{i=1}^n Z_i$ and $S_n^2 = \frac{1}{n-1}\sum_{i=1}^n (Z_i - \overline{Z}_n)^2$. Then,

① $\overline{Z}_n$ and S_n^2 are independent.

② $\overline{Z}_n$ has a normal distribution with mean 0 and variance $1/n$, and $(n-1)S_n^2$ has a chi-squared distribution with $n-1$ degrees of freedom.

③ Define $T_n = \sqrt{n}\frac{\overline{Z}_n}{\sqrt{S_n^2}}$. Then, T_n has a t-distribution with $n-1$ degrees of freedom, called the *Student*'s t-distribution.

Proof. Consider the Helmert transformation $\mathbf{X} = \mathbf{AZ}$ given in Example 6.62, where $\mathbf{A}$ is an orthogonal matrix and the first row of $\mathbf{A}$ is $\mathbf{a}_1^\top = \left(\frac{1}{\sqrt{n}}, \ldots, \frac{1}{\sqrt{n}}\right)$.

- Then, it follows that $X_1 = \sqrt{n}\,\overline{Z}_n$, $\sum_{i=1}^n X_i^2 = \sum_{i=1}^n Z_i^2$, and $\sum_{i=2}^n X_i^2 = \sum_{i=1}^n X_i^2 - X_1^2 = \sum_{i=1}^n Z_i^2 - n\overline{Z}_n^2 = \sum_{i=1}^n (Z_i - \overline{Z}_n)^2$. Since the joint p.d.f. of $\mathbf{X}$ is given by $f_{\mathbf{X}}(\mathbf{x}) = f_{\mathbf{Z}}(\mathbf{A}^{-1}\mathbf{x})|J| = \left(\frac{1}{\sqrt{2\pi}}\right)^n \exp\left(-\frac{1}{2}\mathbf{x}^\top\mathbf{x}\right)$, we know that $X_1, \ldots, X_n$ are independent and identically distributed with the standard normal distribution. Therefore, the two random variables X_1 and $\sum_{i=2}^n X_i^2$ are independent, and it follows that $\overline{Z}_n$ and

$S_n^2 = \dfrac{1}{n-1}\sum_{i=1}^{n}(Z_i - \overline{Z}_n)^2$ are independent, which proves ①.

- Note that $X_1, \ldots, X_n$ are i.i.d. with $N(0,1)$ and that $X_1^2, \ldots, X_n^2$ are i.i.d. with $\chi^2(1)$ distribution. Thus, $\overline{Z}_n = X_1/\sqrt{n} \sim N(0, 1/n)$ and $(n-1)S_n^2 = \sum_{i=1}^{n}(Z_i - \overline{Z}_n)^2$ has the χ^2 distribution with $n-1$ degrees of freedom, which proves ②.

- Alternatively, note that $Z_1, \ldots, Z_n$ are a random sample from the standard normal distribution. Since any linear combination of independent normal random variables has a normal distribution, $\mathrm{E}(\overline{Z}_n) = \mathrm{E}(Z_1) = 0$ and $\mathrm{Var}(\overline{Z}_n) = \frac{1}{n}$, it follows that $\overline{Z}_n \sim N(0, 1/n)$. In addition, consider the following transformations

$$X_1 = \overline{Z}_n = \frac{1}{n}(Z_1 + \cdots + Z_n),\ X_2 = Z_2 - \overline{Z}_n,\ \ldots, X_n = Z_n - \overline{Z}_n.$$

Then, the inverse transformations are $z_1 = x_1 - x_2 - \cdots - x_n$, $z_2 = x_1 + x_2$, $z_3 = x_1 + x_3$, $\ldots$, $z_n = x_1 + x_n$, and thus the Jacobian is computed as

$$J = \frac{\partial(z_1, \ldots, z_n)}{\partial(x_1, \ldots, x_n)} = \begin{vmatrix} 1 & -1 & -1 & \cdots & -1 \\ 1 & 1 & 0 & \cdots & 0 \\ 1 & 0 & 1 & \cdots & 0 \\ \vdots & & & & \\ 1 & 0 & 0 & \cdots & 1 \end{vmatrix} = \begin{vmatrix} n & 0 & 0 & \cdots & 0 \\ 1 & 1 & 0 & \cdots & 0 \\ 1 & 0 & 1 & \cdots & 0 \\ \vdots & & & & \\ 1 & 0 & 0 & \cdots & 1 \end{vmatrix} = n.$$

Now, notice that $\sum_{i=1}^{n} Z_i^2 = \sum_{i=1}^{n}(Z_i - \overline{Z}_n)^2 + (\sqrt{n}\,Z_n)^2$, $Z_1 - \overline{Z}_n = Z_1 - X_1 = -X_2 - X_3 - \cdots - X_n$ and that $Z_i - \overline{Z}_n = X_i$, $i = 2, \ldots, n$. Thus, we have

$$\sum_{i=1}^{n} Z_i^2 = \sum_{i=1}^{n}(Z_i - \overline{Z}_n)^2 + n\overline{Z}_n^2 = (-X_2 - X_3 - \cdots - X_n)^2 + \sum_{i=2}^{n} X_i^2 + nX_1^2.$$

Therefore, the joint p.d.f. of $\mathbf{X}$ is given by

$$f_{\mathbf{X}}(\mathbf{x}) = f_{\mathbf{Z}}(\mathbf{z})|J| = n\left(\frac{1}{\sqrt{2\pi}}\right)^n \exp\left\{\frac{1}{2}\left[-\left(\sum_{i=2}^{n} x_i\right)^2 - \sum_{i=2}^{n} x_i^2 - nx_1^2\right]\right\},$$

which implies that X_1 and $(X_2, \ldots, X_n)$ are independent. Hence, X_1 and a function of $(X_2, \ldots, X_n)$ are also independent, which proves that $\overline{Z}_n$ and S_n^2 are independent.

- On the other hand, we see that X_1 and $(X_2, \ldots, X_n)$ have a multivariate normal distribution and are uncorrelated, and thus conclude they are independent. Specifically, we evaluate each of covariances between X_1 and X_i,

$$\mathrm{Cov}(X_1, X_i) = \mathrm{Cov}(\overline{Z}_n, Z_i - \overline{Z}_n) = \mathrm{Var}(\overline{Z}_n) - \mathrm{Cov}(Z_i, \overline{Z}_n) = 0,\ i = 2, \ldots, n.$$

Thus, X_1 and $(X_2, \ldots, X_n)$ are independent since they are linear combinations of normal random variables. Indeed, we can rewrite this transformation with $\mathbf{X} = \mathbf{BZ}$, where $\mathbf{X} = \begin{pmatrix} \mathbf{X}_1 \\ \mathbf{X}_2 \end{pmatrix} = \begin{pmatrix} \frac{1}{n}\mathbf{1}^\top \\ \mathbf{I} - \frac{1}{n}\mathbf{1}\mathbf{1}^\top \end{pmatrix}\mathbf{Z}$, and $\mathbf{X}_1 = \overline{Z}_n$ and $\mathbf{X}_2 = (Z_2 - \overline{Z}_n, \ldots, Z_n -$

$\overline{Z}_n)^\top$. Thus, $\mathbf{X}$ has a multivariate normal distribution with mean $\mathbf{B0} = \mathbf{0}$ and covariance matrix

$$\Sigma = \mathbf{BIB}^\top = \begin{pmatrix} \frac{1}{n}\mathbf{1}^\top \\ \mathbf{I} - \frac{1}{n}\mathbf{1}\mathbf{1}^\top \end{pmatrix} \begin{pmatrix} \frac{1}{n}\mathbf{1} & \mathbf{I} - \frac{1}{n}\mathbf{1}\mathbf{1}^\top \end{pmatrix} = \begin{pmatrix} \frac{1}{n} & \mathbf{0}^\top \\ \mathbf{0} & \mathbf{I} - \frac{1}{n}\mathbf{1}\mathbf{1}^\top \end{pmatrix},$$

which implies that $\mathbf{X}_1$ and $\mathbf{X}_2$ are independent.

- Further, note that $\sum_{i=1}^n Z_i^2$ has a χ^2 distribution with n degrees of freedom and $n\overline{Z}_1^2$ has a χ^2 distribution with 1 degree of freedom. Since $(n-1)S_n^2 = \sum_{i=1}^n (Z_i - \overline{Z}_n)^2$ and $n\overline{Z}_n^2$ are independent, the m.g.f. of $(n-1)S_n^2$ is given by

$$M_{(n-1)S_n^2}(t) = \frac{M_{\sum_{i=1}^n Z_i^2}(t)}{M_{n\overline{Z}_n^2}(t)} = \frac{(1-2t)^{-n/2}}{(1-2t)^{-1/2}} = \left(\frac{1}{1-2t}\right)^{-(n-1)/2},$$

which is the m.g.f. of χ^2 distribution of $n-1$ degrees of freedom. Hence, by the uniqueness of the m.g.f., we conclude that $(n-1)S_n^2$ has a χ^2 distribution with $n-1$ degrees of freedom.

- For ③, consider $T_n = \sqrt{n}\frac{\overline{Z}_n}{\sqrt{S_n^2}}$. By ① and ②, we see that $\sqrt{n}\overline{Z}_n$ has the standard normal distribution, $(n-1)S_n^2$ has a χ^2 distribution with $n-1$ degrees of freedom, and $\overline{Z}_n$ and S^2 are independent. Hence, it follows from the definition of t-distribution that $T_n = \dfrac{\sqrt{n}\overline{Z}_n}{\sqrt{(n-1)S_n^2/(n-1)}}$ has a t-distribution with $n-1$ degrees of freedom.

■

Remark The t-distribution was discovered by William Sealy Gosset (1876-1936) while working for the Guinness Brewery in Dublin, who published his work under the pseudonym *Student*, "The probable error of a mean", Biometrika, pp.302-310, 1908 by Student. Therefore, this distribution is known as Student's t-distribution. The first rigorous proof was given by R.A. Fisher in 1923.

- The 1908 paper, "The probable error of a mean", as most of Gosset's statistical work, was the result of practical small-sample problems arising in the brewery. The main point of the 1908 paper was that the customary practice of calculating the standard error of the mean through the formula $\sqrt{s_n^2/n}$ and then assigning a normal distribution to $T_n = \sqrt{n}\overline{Z}_n / \sqrt{S_n^2}$ in ③ of Theorem 6.5.2 was not accurate when dealing with small sample sizes. This is because when n is small, there is sizable uncertainty introduced by having to estimate the population variance through the sample variance S_n^2, which is not being accounted for through the use of the normal distribution.

- A random variable X is said to have a t-distribution with ν degrees of freedom, $X \sim t(\nu)$, with p.d.f. as also given in Example 5.44

$$f_X(x) = \frac{\Gamma\left(\frac{\nu+1}{2}\right)}{\sqrt{\pi\nu}\,\Gamma\left(\frac{\nu}{2}\right)\left(1+\frac{x^2}{\nu}\right)^{\left(\frac{\nu+1}{2}\right)}}, \qquad -\infty < x < \infty,\ \nu > 0,$$

and we can use the scale mixture representation of X as $X|W = w \sim N(0, \nu/w)$ and $W \sim \chi^2(\nu)$ to compute the mean and variance

$$\begin{aligned} \mathrm{E}(X) &= \mathrm{E}(\mathrm{E}(X|W)) = 0, \ \nu = 2,3,\ldots, \\ \mathrm{Var}(X) &= \mathrm{E}(\mathrm{Var}(X|W)) + \mathrm{Var}(\mathrm{E}(X|W)) = \mathrm{E}(\nu/W) \\ &= \frac{\nu}{2^{\nu/2}\Gamma(\nu/2)} \int_0^\infty w^{-1} w^{\nu/2-1} e^{-w/2} dw = \frac{\nu}{\nu-2}, \ \nu = 3,4,\ldots. \end{aligned}$$

- The t-distribution is a generalization of the Cauchy distribution and the normal distribution. That is, if $\nu = 1$, then the p.d.f. becomes $f_X(x) = \frac{1}{\pi(1+x^2)}$, $-\infty < x < \infty$, which is the Cauchy distribution. Further, if $\nu \to \infty$, then $\lim_{\nu\to\infty} f_X(x) = \frac{1}{\sqrt{2\pi}} e^{-\frac{1}{2}x^2}$, $-\infty < x < \infty$, since $\lim_{\nu\to\infty}\left(1+\frac{x^2}{\nu}\right)^{\frac{\nu+1}{2}} = e^{-\frac{1}{2}x^2}$, which is the p.d.f. of the standard normal distribution. Here, we use the Stirling's formula $\Gamma(x+1) \approx \sqrt{2\pi} e^{-x} x^{x+\frac{1}{2}}$ to approximate the normalizing constant

$$\begin{aligned} \frac{\Gamma\left(\frac{\nu+1}{2}\right)}{\sqrt{\pi\nu}\,\Gamma\left(\frac{\nu}{2}\right)} &\approx \frac{\sqrt{2\pi} e^{-(\nu-1)/2} [(\nu-1)/2]^{\nu}}{\sqrt{\pi\nu}\sqrt{2\pi} e^{-(\nu-2)/2} [(\nu-2)/2]^{(\nu-1)/2}} \\ &= \frac{1}{\sqrt{\pi\nu}} \frac{1}{\sqrt{e}} \left(1+\frac{1}{\nu-2}\right)^{-(\nu-1)/2} \sqrt{\frac{\nu-1}{2}} \approx \frac{1}{\sqrt{2\pi}}, \quad \text{as } \nu \to \infty. \end{aligned}$$

In Section VI of Gosset's 1908 paper, Gosset investigated the small-sample behavior of T_n empirically with the dataset containing the height and left middle finger measurements of 3,000 criminals, from MacDonell(1902) published in Biometrika, also available as `crimtabl` from R (`http://127.0.0.1:21546/library/datasets/html/crimtab.html`). In Gosset's simulation, 4 measurements were taken each time as a sample at random without replacement, which is 750 samples of size 4. This is the earliest instance of the Monte Carlo methods, even before using computers and statistical packages, and we can perform similar simulations using R.

- Let $X_1,\ldots,X_n$ be a random sample from a $N(\mu,\sigma^2)$ distribution. The quantity $\sqrt{n}(\bar{X}_n-\mu)/S_n$ has the Student's t distribution with $n-1$ degrees of freedom.

```
nsim<-10^4 ;n<-4
x<-matrix(rnorm(n*nsim,1,2),nrow=nsim, ncol=n)
xbar<-rowMeans(x); sx<-apply(x,1,sd)
stx<-sqrt(n)*(xbar-1)/sx
hist(stx,probability=T, ylim=c(0,0.5),nclass=100,xlim=c(-6,6),
main="Simulating t(3) random variables")
lines(density(stx), lty=1, lwd=3, col=1)

x1 = seq(-4,4, 0.01); z1 = dnorm(x1, 0, 1)
lines(x1, z1, lty=5, lwd=3, col=2)
t1 = dt(x1, 3);lines(x1, t1, lty=3, lwd=3, col=4)
legend("topright", legend=c("studentized xbar", "N(0,1)",
"t(3)"), lty=c(1,5,3), lwd=c(3,3,3), col=c(1,2,4))
```

Alternatively, we can generate t-random variables using the scale mixture representation (or its construction by definition) and Box-Müller transformation with two independent uniform random variables. That is, the t-distribution with r degrees of freedom, $X \sim t(r)$, $r > 0$, can be generated by

$$X \stackrel{d}{=} \sqrt{r\left(U_1^{-2/r} - 1\right)} \cos(2\pi U_2) \sim t(r),$$

where U_1 and U_2 are independent and identically distributed as Uniform$(0,1)$.

- First, let $Y \sim \text{Exp}(1)$ be independent of U_2. Then, $Z = \sqrt{2Y}\cos(2\pi U_2)$ has a standard normal distribution by Box-Müller transformation in Theorem 5.3.4. Let W have a χ^2 distribution with r degrees of freedom, $W \sim \chi^2(r)$, or a gamma distribution with parameters $r/2$ and $1/2$, $W \sim \text{Gamma}(r/2, 1/2)$, which is independent of Z. Then, $\frac{Z}{\sqrt{W/r}} = \sqrt{r\frac{Y}{W/2}}\cos(2\pi U_2)$ has a t distribution with r degrees of freedom.

- Let $V = W/2$ that has a gamma distribution with parameters $\frac{r}{2}$ and 1, $V \sim \text{Gamma}(r/2, 1)$, which is also independent of Y. Thus,

$$\frac{Y}{W/2} = \frac{Y}{V} = \frac{1 - V/(Y+V)}{V/(Y+V)} = \frac{1-B}{B} = B^{-1} - 1,$$

where the random variable B has a beta distribution with parameters $r/2$ and 1, $B \sim \text{Beta}(r/2, 1)$.

- Since the c.d.f. of B, $F_B(x) = \int_0^x r/2t^{r/2-1}dt = x^{r/2}$, $0 < x < 1$, its inverse function $F_B^{-1}(x) = x^{2/r}$ and thus $U_1^{2/r}$ has a beta distribution with parameters $r/2$ and 1. Therefore, $U_1^{-2/r} - 1$ has the same distribution with $\frac{Y}{W/2}$. Hence, $\sqrt{r\left(U_1^{-2/r} - 1\right)}\cos(2\pi U_2)$ has t distribution with r degrees of freedom, that is, the same distribution with X.

```
nsim<-10^4; trv<-numeric(nsim); r<-3
for (i in 1:nsim){
 u<-runif(2)
 trv[i]<- sqrt(r*((u[1])^(-2/r)-1)) * cos(2*pi*u[2]) }

hist(trv,probability=T, ylim=c(0,0.5),nclass=100,xlim=c(-6,6),
main="Simulating t(3) random variables")
lines(density(trv), lty=1, lwd=3, col=1)
```

6.6 Exercises

Problem 6.1 Suppose that two random variables x and y have the following moments

$$\mathrm{E}(Y)=1,\ \mathrm{Var}(X)=1,\ \mathrm{E}(Y|X)=1+X, \text{ and } \mathrm{E}(\mathrm{Var}(Y|X))=2. \text{ Find } \mathrm{Cov}(X,Y).$$

Problem 6.2 Suppose that the marginal p.d.f. of Y is $f_Y(y)=\begin{cases} e^{-y}, & \text{if } y>0, \\ 0, & \text{otherwise,}\end{cases}$ and that the conditional p.m.f. of X given $Y=y$ is $f_{X|Y}(x|y)=\begin{cases} \frac{(2y)^x}{x!}e^{-2y}, & \text{if } x=0,1,\ldots, \\ 0, & \text{otherwise.}\end{cases}$ Find $\mathrm{E}(X)$ and $\mathrm{E}(Y|X)$.

Problem 6.3 Let X and Y have joint p.d.f. $f(x,y)=\frac{y}{4x}$ if $0<y<x<4$, and $f(x,y)=0$ otherwise.

① Find $P(Y>1|X=2)$.

② Compute the conditional expectation of Y^2 given $X=x$.

Problem 6.4 . Let X and Y be random variables for which the joint p.d.f. is as follows :

$$f(x,y)=\begin{cases} xe^{-y}, & \text{for } 0<x<y<\infty, \\ 0, & \text{otherwise.}\end{cases} \quad \text{Find } \mathrm{Var}[e^{\ Y}|X].$$

Problem 6.5 Let N be the geometric random variable with parameter p, $N\sim\mathrm{Geo}(p)$. Suppose that the conditional distribution of X given that $N=n$ is the gamma distribution with parameters n and λ. Find the conditional mean and variance of N given X, i.e. $\mathrm{E}(N|X)$ and $\mathrm{Var}(N|X)$.

Problem 6.6 Let X, Y, and Z be independent random variables that are uniformly distributed on (0,1), that is, $X,Y,Z \overset{i.i.d.}{\sim} \mathrm{Uniform}(0,1)$. Find the probability that the quadratic equation $f(t)=Xt^2+Yt+Z=0$ has at least one real solution.

Problem 6.7 Suppose that X and Y have the following joint p.d.f.

$$f_{X,Y}(x,y)=\begin{cases} 6(1-x-y), & \text{for } x>0,\ y>0,\ x+y<1. \\ 0, & \text{otherwise.}\end{cases}$$

① Are X and Y independent and identically distributed (i.i.d)?

② Find the conditional expectation and variance of X given $Y=y$, $\mathrm{E}(X|Y=y)$ and $\mathrm{Var}(X|Y=y)$.

Problem 6.8 Let $\mathrm{E}(Y|X=x)=2x$ and $\mathrm{Var}(Y|X=x)=4x^2$, and let X have a uniform distribution on the interval $(0,1)$. Find the variance of Y.

Problem 6.9 Let the joint p.d.f of X and Y be

$$f_{X,Y}(x,y) = \begin{cases} \frac{1}{y}e^{-y}, & \text{if } 0 < x < y < \infty, \\ 0, & \text{otherwise.} \end{cases}$$

Find $\mathrm{E}(X^k|Y)$ for $k = 1,2,\ldots$.

Problem 6.10 Let X and Y be random variables with correlation ρ. Show that

$$\mathrm{E}(\mathrm{Var}(Y|X)) \leq (1-\rho^2)\mathrm{Var}(Y).$$

Assume without loss of generality that $\mathrm{E}(X) = \mathrm{E}(Y) = 0$ and use the Cauchy-Schwarz inequality.

Problem 6.11 A nonnegative random variable W is said to follow an inverted beta distribution with parameters α and β, if its density is

$$f_W(w) = \frac{\Gamma(\alpha+\beta)}{\Gamma(\alpha)\Gamma(\beta)} \frac{w^{\alpha-1}}{(1+w)^{\alpha+\beta)}}, \quad w > 0, \ \alpha, \beta > 0$$

We will write $W \sim \mathrm{IBeta}(\alpha,\beta)$. Show that

① Let $Z \sim \mathrm{Beta}(\alpha,\beta)$. Then, $\dfrac{Z}{1-Z}$ has an inverted beta distribution with parameters α and β.

② Let $X \sim \mathrm{Gamma}(\alpha,\lambda)$ and $Y \sim \mathrm{Gamma}(\beta,\lambda)$. Suppose that X and Y are independent. Find the distribution of $\dfrac{X}{Y}$.

Problem 6.12 Let $(X,Y)^\top \sim \mathrm{Dirichlet}(\alpha_1,\alpha_2,\alpha_3)$, where α_1,α_2 and α_3 are positive constants.

① Compute the correlation coefficient of X and Y, $\rho = \dfrac{\mathrm{Cov}(X,Y)}{\sqrt{\mathrm{Var}(X)\mathrm{Var}(Y)}}$.

② Compute the conditional expectation of Y given X, $\mathrm{E}(Y|X)$.

Problem 6.13 Suppose that X_1 and X_2 have a bivariate normal distribution, $(X_1,X_2)^\top \sim N_2(\mu_1,\mu_2,\sigma_1^2,\sigma_2^2,\rho)$. Evaluate $\mathrm{E}(X_1^2X_2)$.

Problem 6.14 Let $X_1,\ldots,X_n$ be i.i.d. random variables having the normal distribution with mean μ and variance σ^2. Define $\overline{X}_n = \frac{1}{n}\sum_{i=1}^{n} X_i$. Find the conditional distribution of X_i given $\overline{X}_n = \bar{x}_n$.

Index

Symbols

A

B

C

D

저자 소개

최태련
현) 고려대학교 정경대학 통계학과 교수
E-mail: trchoi@korea.ac.kr

Essentials of Probability for Statistics - A primer -

저　　자　최태련
발 행 인　김지영
발 행 처　자유아카데미
주　　소　경기도 파주시 회동길 37-42
　　　　　파주출판도시
전　　화　031-955-1321
팩　　스　031-955-1322
홈페이지　www.freeaca.com
전자우편　main@freeaca.com (대표)
　　　　　editor@freeaca.com (편집)
　　　　　crm@freeaca.com (영업)
등　　록　제406-2003-017호, 1980. 7. 12
제1판1쇄　2020년 3월 16일 발행
제1판4쇄　2024년 12월 26일 발행
정　　가　24,000원

저자와의
협의하에
인지생략

ISBN 979-11-5808-258-1 93310

Summary of Common Distributions

Discrete Distributions

	Bernoulli(p)	Binomial(n,p)
p.m.f.	$p^x(1-p)^{1-x}$, $x=0,1$	$\binom{n}{x}p^x(1-p)^{n-x}$, $x=0,1,\dots,n$
mean	p	np
variance	$p(1-p)$	$np(1-p)$
m.g.f	pe^t+1-p	$(pe^t+1-p)^n$

	Geo(p)	Negbin(r,p)
p.m.f.	$p(1-p)^x$, $x=0,1,2,\dots$	$\binom{r+x-1}{x}p^r(1-p)^x$, $x=0,1,2,\dots$
mean	$\frac{1-p}{p}$	$\frac{r(1-p)}{p}$
variance	$\frac{1-p}{p^2}$	$\frac{r(1-p)}{p^2}$
m.g.f.	$\frac{p}{1-(1-p)e^t}$, $t<\log(1/[1-p])$	$\left(\frac{p}{1-(1-p)e^t}\right)^r$, $t<\log(1/[1-p])$

	Poisson(λ)	Hypergeo(n,A,B)
p.m.f.	$e^{-\lambda}\frac{\lambda^x}{x!}$, $x=0,1,\dots$	$\frac{\binom{A}{x}\binom{B}{n-x}}{\binom{A+B}{n}}$, $x=\max\{0,n-B\},\dots,\min\{n,A\}$
mean	λ	$\frac{nA}{A+B}$
variance	λ	$\frac{nAB}{(A+B)^2}\frac{A+B-n}{A+B-1}$
m.g.f.	$e^{\lambda(e^t-1)}$	

Continuous Distributions

	Uniform(a,b)	Beta(α,β)
p.d.f.	$\frac{1}{b-a}$, $a<x<b$	$\frac{\Gamma(\alpha+\beta)}{\Gamma(\alpha)\Gamma(\beta)}x^{\alpha-1}(1-x)^{\beta-1}$, $0<x<1$
mean	$\frac{a+b}{2}$	$\frac{\alpha}{\alpha+\beta}$
variance	$\frac{(b-a)^2}{12}$	$\frac{\alpha\beta}{(\alpha+\beta)^2(\alpha+\beta+1)}$
m.g.f	$\frac{e^{-at}-e^{-bt}}{t(b-a)}$	

	Exponential(λ)	Gamma(α,λ)
p.d.f	$\lambda e^{-\lambda x}$, $x>0$	$\frac{\lambda^\alpha}{\Gamma(\alpha)}x^{\alpha-1}e^{-\lambda x}$, $x>0$
mean	$\frac{1}{\lambda}$	$\frac{\alpha}{\lambda}$
variance	$\frac{1}{\lambda^2}$	$\frac{\alpha}{\lambda^2}$
m.g.f	$\frac{\lambda}{\lambda-t}$, $t<\lambda$	$\left(\frac{\lambda}{\lambda-t}\right)^\alpha$, $t<\lambda$

	Normal $N(\mu,\sigma^2)$	Chi-squared $\chi^2(m)$
p.d.f	$\frac{1}{\sqrt{2\pi\sigma^2}}e^{-\frac{(x-\mu)^2}{2\sigma^2}}$, $-\infty<x<\infty$	$\frac{1}{2^{m/2}\Gamma(m/2)}x^{(m/2)-1}e^{-x/2}$, for $x>0$
mean	μ	m
variance	σ^2	$2m$
m.g.f	$e^{\mu t+\frac{1}{2}\sigma^2t^2}$	$(1-2t)^{-m/2}$, $t<1/2$

Summary of Common Distributions

Discrete Distributions

	Bernoulli(p)	Binomial(n,p)
p.m.f.	$p^x(1-p)^{1-x}$, $x=0,1$	$\binom{n}{x}p^x(1-p)^{n-x}$, $x=0,1,\ldots,n$
mean	p	np
variance	$p(1-p)$	$np(1-p)$
m.g.f	pe^t+1-p	$(pe^t+1-p)^n$

	Geo(p)	Negbin(r,p)
p.m.f.	$p(1-p)^x$, $x=0,1,2,\ldots$	$\binom{r+x-1}{x}p^r(1-p)^x$, $x=0,1,2,\ldots$
mean	$\frac{1-p}{p}$	$\frac{r(1-p)}{p}$
variance	$\frac{1-p}{p^2}$	$\frac{r(1-p)}{p^2}$
m.g.f.	$\frac{p}{1-(1-p)e^t}$, $t<\log(1/[1-p])$	$\left(\frac{p}{1-(1-p)e^t}\right)^r$, $t<\log(1/[1-p])$

	Poisson(λ)	Hypergeo(n,A,B)
p.m.f.	$e^{-\lambda}\frac{\lambda^x}{x!}$, $x=0,1,\ldots$	$\frac{\binom{A}{x}\binom{B}{n-x}}{\binom{A+B}{n}}$, $x=\max\{0,n-B\},\ldots,\min\{n,A\}$
mean	λ	$\frac{nA}{A+B}$
variance	λ	$\frac{nAB}{(A+B)^2}\frac{A+B-n}{A+B-1}$
m.g.f.	$e^{\lambda(e^t-1)}$	

Continuous Distributions

	Uniform(a,b)	Beta(α,β)
p.d.f.	$\frac{1}{b-a}$, $a<x<b$	$\frac{\Gamma(\alpha+\beta)}{\Gamma(\alpha)\Gamma(\beta)}x^{\alpha-1}(1-x)^{\beta-1}$, $0<x<1$
mean	$\frac{a+b}{2}$	$\frac{\alpha}{\alpha+\beta}$
variance	$\frac{(b-a)^2}{12}$	$\frac{\alpha\beta}{(\alpha+\beta)^2(\alpha+\beta+1)}$
m.g.f	$\frac{e^{-at}-e^{-bt}}{t(b-a)}$	

	Exponential(λ)	Gamma(α,λ)
p.d.f	$\lambda e^{-\lambda x}$, $x>0$	$\frac{\lambda^\alpha}{\Gamma(\alpha)}x^{\alpha-1}e^{-\lambda x}$, $x>0$
mean	$\frac{1}{\lambda}$	$\frac{\alpha}{\lambda}$
variance	$\frac{1}{\lambda^2}$	$\frac{\alpha}{\lambda^2}$
m.g.f	$\frac{\lambda}{\lambda-t}$, $t<\lambda$	$\left(\frac{\lambda}{\lambda-t}\right)^\alpha$, $t<\lambda$

	Normal $N(\mu,\sigma^2)$	Chi-squared $\chi^2(m)$
p.d.f	$\frac{1}{\sqrt{2\pi\sigma^2}}e^{-\frac{(x-\mu)^2}{2\sigma^2}}$, $-\infty<x<\infty$	$\frac{1}{2^{m/2}\Gamma(m/2)}x^{(m/2)-1}e^{-x/2}$, for $x>0$
mean	μ	m
variance	σ^2	$2m$
m.g.f	$e^{\mu t+\frac{1}{2}\sigma^2t^2}$	$(1-2t)^{-m/2}$, $t<1/2$